# Alternative Strategies for Desert Development and Management

## Related Titles in the Environmental Sciences and Applications Series

**Biswas** CLIMATIC FLUCTUATIONS AND AGRICULTURAL AND WATER RESOURCES PLANNING
**Biswas** DESERTIFICATION STUDIES
**Hinckley** RENEWABLE RESOURCES IN OUR FUTURE
**Holy** EROSION AND ENVIRONMENT
**United Nations** DESERTIFICATION: Its Causes and Consequences

## Pergamon Titles of Related Interest

**Biswas/Biswas** ENVIRONMENT AND MAN
**Biswas** WATER AND ENVIRONMENT
**Samaha et al** WATER MANAGEMENT FOR ARID LANDS

## Related Journals*

ATMOSPHERIC ENVIRONMENT
CONSERVATION AND RECYCLING
THE ENVIRONMENTAL PROFESSIONAL
NUCLEAR & CHEMICAL WASTE MANAGEMENT
WATER SUPPLY AND MANAGEMENT

***Free specimen copy available upon request.**

# Alternative Strategies for Desert Development and Management

Proceedings of an International Conference held in Sacramento, CA, May 31-June 10, 1977

United Nations Institute for Training and Research

## Energy and Minerals volume 1

Volume 3 in the Environmental Sciences and Applications Series
Series Editors: Margaret R. Biswas
Asit K. Biswas

Pergamon Press New York Oxford Toronto Sydney Paris Frankfurt

Pergamon Press Offices:

| | |
|---|---|
| **U.S.A.** | Pergamon Press Inc., Maxwell House, Fairview Park, Elmsford, New York 10523, U.S.A. |
| **U.K.** | Pergamon Press Ltd., Headington Hill Hall, Oxford OX3 0BW, England |
| **CANADA** | Pergamon Press Canada Ltd., Suite 104, 150 Consumers Road, Willowdale, Ontario M2J 1P9, Canada |
| **AUSTRALIA** | Pergamon Press (Aust.) Pty. Ltd., P.O. Box 544, Potts Point, NSW 2011, Australia |
| **FRANCE** | Pergamon Press SARL, 24 rue des Ecoles, 75240 Paris, Cedex 05, France |
| **FEDERAL REPUBLIC OF GERMANY** | Pergamon Press GmbH, Hammerweg 6 6242 Kronberg/Taunus, Federal Republic of Germany |



**Library of Congress Cataloging in Publication Data**
Main entry under title:

Alternative strategies for desert development.

(Environmental sciences and applications ; v. 3)
Contents: v. 1. Energy and minerals -- v. 2. Agriculture -- v. 3. Water -- [etc.]
Proceedings of the Conference on Alternative Strategies for Desert Development and Management.
1. Arid regions--Congresses. 2. Arid regions agriculture--Congresses. 3. Deserts--Congresses. I. United Nations Institute for Training and Research. II. Conference on Alternative Strategies for Desert Development and Management (1977 : Sacramento, Calif.) III. Series.
GB611.A44 1982 333.73 81-23433
ISBN 0-08-022401-6 (set) AACR2
ISBN 0-08-022402-4 (Vol.1)
ISBN 0-08-022403-2 (Vol.2)
ISBN 0-08-022404-0 (Vol.3)
ISBN 0-08-022405-9 (Vol.4)



***Printed in the United States of America***

## CONTENTS



### Conference Papers

VOLUME I: ENERGY AND MINERALS

REMARKS OF

Edmund G. Brown, Jr.
Governor of the State of California
The Opening Session

For all of you who have traveled to Sacramento to work together and communicate your thoughts and to learn from those of others, it's significant that the United Nations should send this group to Sacramento and to California for it was here that the United Nations began right after the second World War. California itself understands and has confronted for many years the subject matter of your deliberations during this conference. Much of our State is a desert or, at least, semi-arid. Through many generations, efforts have been made to bring it into the service of human beings through agricultural development, water transmission and mineral development. It's a place where many things can be learned both from our successes and from our failures. Ever since gold was discovered -- not too far from where all of you sit -- waves of human migration have been coming to this State that are unparalleled in human history. Those waves of migration have continued through many decades after the gold rush had long since been completed.

Now we are here to discuss, not just a mineral, but issues that face all of us as citizens of the world; the past when we were just parts of one particular country, or one small ecosystem, is now transforming into a much larger participation into a world ecology and a world economy and a world political community. Certainly much of what you can do by way of investigation and learning and communication will bring us closer to the day when all of us can understand and sense our identity as citizens of the world as well as citizens of a nation or citizens of a locality or citizens of a particular family or ethnic or tribal group. That day I hope will come sooner rather than later. The desert is a very appropriate subject and a very appropriate symbol. It has been viewed by many as a formidable area, as a place to be avoided or, if traversed, as quickly as possible. Yet it holds many resources. It holds potential in the clean air, in the abundance of sun, in the potential for wind power, and in developments that could take place under the ground by way of geothermal -- geopressure. I notice in the papers that you have before you many investigations of those subjects.

Here in California in the private sector, in universities and through the government, both at the State and local level, efforts are now being made to develop and preserve and enhance our desert areas. In the southern San Joaquin Valley and in Imperial County you will find places where the deserts are blooming and where energy is being developed through geothermal power. What has been learned here certainly can be utilized in other parts of the country. If those who believe we are entering a cooling period are correct, then, of course, those areas that enjoy greater sun will become even more important in the decades to come.

As we look to our technology and our science, I think we also have to recognize the inherent limits that any human endeavor has. The vanity and the arrogance of power that have characterized so many human endeavors in both this era and those that have long passed should serve as a warning that as we develop a resource, or we embark upon a particular journey, we should do so with humility. Some would look at the effort to preserve the planet as some form of imperialism -- as some luxury that is only appropriate to the more advanced and developed countries. But I think that all of us, wherever we live, sooner or later have to come to the realization that the air and the water and the other resources of this planet are in fact limited. The system is closed. We can move the pieces around. We can readjust them. We can recycle them. But what we have is already in being and it is stretched, often unevenly, throughout the planet. Our effort must be to develop it, to share it, but also to preserve it. While growth and development are the watchwords, understanding what kind of growth, and quality as well as quantity, and long-term consequences as well as short-term gains have to be added into the equation.

Much is made of long-term planning, rational decision making, comprehensive undertakings and yet all of that, to some extent, implies that we can know far more than we really can. What may turn up in 1990 or the year 2000 is often very hard to predict. As we go I think we have to walk step by step with a great deal of humility and challenging our assumptions -- whether they be assumptions that are conventional wisdom in the western hemisphere, the southern hemisphere or any place on the planet. Often today's wisdom is tomorrow's folly. Unlike the past where nations and efforts were separated by distances of land and water, so much of what we do today is a part of people and places thousands of miles away.

That globalization, because of mobility and travel and communication will not diminish, it will accelerate. As we explore the deserts we will also explore the oceans. We will also explore the outer reaches of space because all of these things are there and human curiousity cannot be restrained. As we recognize the limits we must also delve into the possibilities. The hopes of the mind and the spirit that man has are unlimited. Potential is unimaginable. It is there if we are fortunate enough to find it and if, as we go, we are able to walk with a spirit of respect and gratitude for those who have come before and with a sense of stewardship for those yet to come.

# FOREWORD

Executive Director
UNITAR

United Nations Institute for Training and Research (UNITAR) and the State of California, U.S.A., co-sponsored a Conference on Alternative Strategies for Desert Development and Management, which took place at Sacramento, California, from May 31 to June 10, 1977.

The Conference, which was designed primarily to study development opportunities of arid and desert zones, brought together up-to-date information and the latest technology on many aspects of desert development. Included on the agenda were detailed discussions of possible obstacles to development, such as the lack of water, the depletion of fossil water, and relevant social problems. The Conference had before it approximately 125 papers and was attended by more than 200 participants representing most parts of the world involved in desert development. Geologists, hydrologists, biologists, industrial planners, government planners, civil engineers, economists, and social scientists were among those who attended. The interdisciplinary character of the meetings and the wide international participation (experts from 33 countries took part) ensured intensive international comparison of experiences, views, and outlooks.

The Conference, in the wide scope of its subject matter, attempted to provide a comprehensive view of what we know today of opportunities for desert development. It thus brought out a number of elements crucial to the choice of strategies, while stressing that a precise strategy must be designed and carried out according to each country's characteristics.

These proceedings of the Conference reflect the important information placed before the members and discussed by them. It confirmed the original concept that there are indeed in most deserts opportunities for development, and that desertification is not and need not be the fate of all arid lands. Indeed, modern arid zone development can, in many cases, overcome desertification wherever it is the result of human neglect or excessive use of the limited carrying capacity of the desert ecology. Water capital, modern technology, and an educated population can usually bring about a development of desert areas in harmony with long-term stability and with benefits to mankind as a whole.

UNITAR is grateful for the active support of the State of California, especially the Department of Water Resources and its staff, and for the contributions of the experts who participated in the preparatory meetings.

The Conference was organized as a contribution to the United Nations Conference on Desertification, organized under the United Nations Environment Programme, and was planned in consultation with UNEP officials. UNITAR is grateful to UNEP for this cooperation. It also wishes to thank the United Nations Development Programme (UNDP), the office of Technical Cooperation (OTC), and the Organization of Arab Petroleum Exporting Countries (OAPEC) for their generous assistance. It thanks, as well, the many experts from developing and developed countries for their participation and outstanding contributions.

Although UNITAR takes responsibility for the decision to publish these volumes, in keeping with its policy,the opinions expressed herein are those of the individual authors and do not necessarily reflect the views of the Institute, its Board of Trustees, the organizations or the governments with which the authors are identified.

# INTRODUCTION

Philippe de Seynes
Director, Project on the Future
United Nations Institute for Training and Research

If the doomsday prophets, so fashionable in the first half of the 70's, did not quite convince us that we could manage our social problems with a zero-growth policy, they did at least make us aware of some of the complexities of the earth's system which we had for too long overlooked or neglected.

They have relieved us of an unconditional and somewhat naive faith in the potentialities of technology. They have highlighted biological, chemical, and physical processes which interfere with the traditional economic calculus. After years of concentrating exclusively on the vagaries of demand, economists are now forced to recognize supply constraints. The maintenance of our small planet requires more care than we are ready to acknowledge.

The extreme case of ecological constraints is, of course, to be found in the desert and arid land areas. Our attention was focused on them as never before when recently droughts of unprecedented proportions covered a large number of countries along a very broad geographic arch. The fragility of ecological systems was brought home to millions of people on the television screens in countless pictures of sickness, death, desolation and tragedy.

One of the many reasons why we were so anxious to join with the Government of the State of California in the planning of this Conference is that this State has probably been more alert, and somewhat earlier, than many other places to the phenomenon of environmental degradation and it has gained unique experience in this respect, both in non-desert and desert areas. We greatly admire Governor Brown's dedication to the cause of environment, and it is a great inspiration for all of us that he should be here with us at the threshold of two weeks of hard work and fruitful exchanges. We also very much sympathize with him, as we do with all decision makers who have to make policy choices between two equally valid objectives, and must conciliate the aspiration towards development with the requirements of environmental protection. This is one of the great challenges of our time, and we have not as yet found reliable and rational approaches. There are

---

Presented 31 May 1977 in Sacramento, California.

cruel trade-offs between employment and income creation on the one hand, and the preservation of clean air and the beauties of nature on the other. Ever present is the risk of gross misallocation of resources. However, when we come to the desert, respect of the ecology is a matter of sheer survival. We can step beyond certain limits - not always, admittedly, easy to recognize - only at the risk of total destruction of life-supporting systems.

Yet, civilizations have endured and flourished in the deserts, some of them among the most remarkable which history has recorded. They were based on and intuitive and practical relationship between Man and Nature. They developed complex and fascinating forms of social organization. They extended their reach far beyond the territories of tribal communities or of existing empires. They may be viewed as forerunners of international trade, and they left behind cultural and artistic achievements which even today attract much interest. These well organized systems were destroyed during the colonial era, much before anything valid could be put in their place. Still, they were not completely lost, as they survived in certain ways of life and in the oral tradition from which much wisdom and ingenuity can be learned.

Modern ways, of course, have to be very different, and we can only hope that the more advanced technology will open up possibilities of organizing life in the deserts in a way which is worth living; we hope that desert development will not be left to the false magic of short-term profit maximization. Long-term planning cannot be avoided in the desert, and we must mobilize all the knowledge which we can derive from vastly diverse experiences and different ecological and historical conditions. May I add that this task is becoming urgent. Desert areas are the great reserve of unoccupied land today. And, land more and more must be viewed as an essential resource as the population of the globe continues to grow rapidly, and most probably will do so for a number of decades to come.

So, our joint purpose in organizing this Conference is simple: to acquire new knowledge for designing appropriate social, economic and technological strategies. If we are to reach this objective, it is very important that we hear about the mistakes and the failures as well as about the successes, and I hope this rule of candor will be adhered to by all.

We also hope that from this Conference new programs of cooperation for research and development will emerge, some of them national or local, but others on a regional and multi-national basis, combining the efforts of governmental agencies and private institutions. Research and Development is an important watchword today. It is becoming extremely expensive, and as we realize that not all technology implicit in our present state of scientific knowledge can be developed with existing human and financial resources, the process of selection becomes crucial. Desert areas and arid zones should receive high priority. There are many potentialities challenging our sense of innovation and cooperation, not just energy and mineral resources, but also new forms of agriculture, projects related to health, tourism, ecological preserves and, naturally, the ever-widening industry of knowledge and learning.

We are confident that we will all leave this Conference better prepared for these tasks.

# THE CHALLENGE OF THE DESERT

Dr. Joseph Barnea

The desert and arid areas of the world form about one-third of the land surface, perhaps more depending on the definition. These vast areas pose a challenge to our ingenuity at a time when we have a technology at our disposal unmatched by past experience, when we have began to realize the vast natural resource potential in many deserts, and when we have capital at our disposal and international communication and cooperation in providing manpower, technical know-how and managerial skills. The easy conclusion which might be drawn, that all the elements and all the tools in order to make the desert bloom are available, must be tempered by two basic facts, namely, the many failures of earlier desert civilizations and the recognition that water will remain the overriding and limiting factor for most deserts.

Deserts are strewn with remnants of past civilizations, from big dams, as (in Yemen, for example) to traces of ancient irrigation systems and roads, caravanserais serving ancient trade routes, and, more recently, to ghost towns and other settlements abandoned when either the minerals ran out or the water ran out or became too salty.

## THE ROLE OF WATER

Clearly, water is the key element, the key factor which will determine the extent of desert development, as it has determined it in the past. Water in itself is a complex commodity with no uniform price. It has many different sources, many different standards of quality, and the price depends on the end use of the water with the lowest price obtaining in traditional agriculture.

There are few reliable water sources in deserts because of the low rainfall and high evaporation rates; thus ground water and water brought in from the outside are the two main sources of water in arid and desert areas. Most of the ground water in arid or desert areas are water deposits with little or no recharge. Consequently, such water deposits have to be carefully mined, which is one of the crucial problems of present water use in some deserts. In some of the desert areas of the United States, water mining for agriculture is going on at a very high rate, a rate that in some areas indicates that the water will be mined out within one or two generations with consequences which could be serious and may represent a repetition of past failures of settlements in the desert.

The only large-scale, reliable source of water we have today in arid or desert areas is water brought in from outside areas through canals and in other ways diverted from water-rich areas. This has been done in the USSR, in India and in the United States, and its application in other parts of the world needs to be studied. But many desert areas have no water-rich areas nearby from which water could be carried to the desert, and thus, this large-scale reliable outside source of water is only a solution in certain favorable areas. It is necessary in desert and arid zones therefore to look to the utilization of all sources and all qualities of water. Specifically this includes brackish and saline water and, in the future, geopressure water and other sources of water wherever they can be found. It also leads to the conclusion that in desert areas reprocessing of water will have to be carried out on a much more extensive scale than in water-rich areas. Given the higher value of water in desert areas, the economic water reprocessing should be more favorable in deserts than in other areas.

The wide-spread application of technology, which is and will become the hallmark of modern desert development, will first of all have to be applied to water, its exploration, acquisition, use and reuse. This may involve in the foreseeable future rainmaking, as well as modification of flash floods and in the more distant future, changes in our weather system which should be designed to divert rain-clouds to desert areas. Today, the bulk of all rainfall falls over the oceans with no direct benefit to mankind. There are many other needed applications of modern technology. Among them is an effective process for retarding water evaporation and techniques for the fuller utilization of dew and fog. Drip irrigation is another area of modern water technology that can usefully be applied in desert areas; it can sharply reduce water needs, and this is possible also in protected agriculture. There is no room here to enumerate the many areas in which technology can, in the future, make a substantial contribution to the central problem of the desert, namely, the availability of water.

Given the fact that water is the overriding constraint to desert development, water considerations should and must form part and parcel of every project and every economic policy in desert areas. Some basic principles must be understood and accepted. These involve the recognition that the worker and his family, at a decent standard of living, is a fairly significant user and consumer of water. Thus, desert development should not be based on the concept of maximizing labor, and fortunately, this principle can be applied in the future, because almost all desert areas are still fairly thinly populated.

As machines need far less water than human labor, modern desert development should be machine intensive and not labor or animal intensive.

A second principle which needs to be considered is that for all production the value added or net income per ton should be calculated as well as water requirements per ton. If this is done, it will often lead to the conclusion that low value resources may not justify development in desert areas.

Water supplies systems in arid and desert areas need a high degree of safety including the provision of storage capacity either above or below ground, as a breakdown in water supplies in such areas may endanger life. Finally, in long-term water planning it might be useful for governments to retain unallocated supplies so that new needs can be met without depriving present users of their water supplies.

## OPPORTUNITIES IN THE DESERT

Within the framework of the water restraints necessary in arid and desert areas and with the tools of modern technology, arid and desert areas offer many advantages. They include: 1) the considerable natural resource potential of vast areas, many of which contain high value resources, such as hydro-carbon, copper, gold, zinc, lead, etc.; 2) a stable climate which allows industrial production practically uninterrupted by weather; 3) a healthy climate which can provide a basis for health industries; and 4) the transport potential which was significant in the past but still retains a potential in the seacoasts of arid and desert areas, as well as their attractiveness for airports due to the unique flying conditions of desert areas. Many of these areas have also historical assets such as religious centres as well as cultural and historic places which could become and are beginning to become centres of tourism. There is a considerable potential for industries, especially industries which are not water intensive, and if the educational standard in arid and desert areas is raised, deserts may become important centres for research and training.

## THE EDUCATIONAL STANDARD

The application of technology to the opportunities in arid and desert areas is a pre-condition for modern effective desert development. This is not an easy process but it is an important one. It is significant to note that in some desert countries, such as Kuwait and Saudi Arabia, constant steps are being taken to raise the educational level as fast as possible. This has already been done in the American deserts, in Israel and, to some extent, in the Latin American deserts. It is noteworthy that in all these cases research institutes have been established in desert areas, and this is a trend which needs full support and priority. It is also true that this is difficult to achieve in desert areas which do not have large income and where traditional forms of life and an alphabetism are still prevalent. It must be realized, however, that water-saving and sophisticated irrigation systems such as drip irrigation as well as modern industry and service industries require an educated population and thus a rapid advance in the educational standards of the desert population is a pre-condition to the large-scale application of modern technology in such areas. The social change involved in the transformation of a nomadic society to a high technology society is a difficult but necessary process, and the speed of this transformation will be one of the factors which will determine the rate of progress in such arid and desert areas.

## PROMISE OF TECHNOLOGY IN DESERT AREAS

The application of modern technology in desert areas is not restricted to water alone. It can transform practically every area of economic activity in desert areas. In agriculture the tools of breeding and genetic engineering will allow us to produce the type of crops best suited for the desert climate, high value by nature and low in water requirements. The future agriculture in desert areas will involve new types of industrial crops from the jojoba to many other crops on which research is going on now in several countries. The same principles will have to be applied to animal raising to the extent that it will prove economic in desert areas, and wildlife farming may become another branch of agriculture in the future. In the future, conventional farming using large quantities of water for growing wheat and similar crops would gradually disappear and modern desert agriculture will grow in the desert what best and most profitably can be grown in such areas.

Technology will be applied also to underground resources, especially to petroleum. Contrary to the wide-spread belief, it is unlikely that with the end of the primary production, oil production in desert areas will come to an end. Through enhanced recovery using thermal, chemical and microbiological methods, technology will allow secondary and tertiary recovery, and oil will continue to be produced in the foreseeable future. Similarly, for mineral deposits, modern technology will allow the utilization of even complex ores, and low-grade resources will also become minable. Tailing will also be reworked. Around desert institutes, research industries could develop, and along the seacoast industries can be established processing the resources of the sea and other local resources, provided the industrial processes are not water-intensive.

Hospitals will be established in deserts. Sanitoria, rest homes and the medical industry, based on local medical schools and research hospitals, can become an important industry. These are only a few examples of the many possibilities which the application of modern technology can create in desert and arid zones, provided governments understand the opportunities and the limitation of deserts.

## LONG-TERM PLANNING

In no other area is a need for long-term planning as important as in desert areas. The need arises first from the lack of reliable water supplies. Conventional agriculture, if based on mining ground water, has a limited life span and endangers the investments including the infrastructure in such agriculture and might create difficult social problems for the farming population and the country as a whole when the water begins to be depleted. The same problem may exist if industries are established using temporary resources cheaply available, as natural gas, which otherwise would be flared. The key question to ask in a development project in desert areas is - will the project still be around in thirty years? Will the water supply or other supplies still be available at reasonable cost after thirty years? If not, how do we overcome the dislocation which will be created when short-term resources come to an end or if the price of certain basic resources should sharply increase in the future due to a much higher cost than was expected. There is very little experience in long-term planning, but it is a need which we have to recognize and which we have to develop through careful planning in desert and arid zones.

We cannot afford today to develop a mineral deposit and, when it is mined out, to abandon the area, to throw out the workers and create a ghost town. The economic and social costs of unemployment and welfare payments, have caused governments to be interested in full employment, and the loss of the investments in infrastructure through the abandonment of an area is also something which should be avoided. This does not mean that we have to give up the search for minerals and their mining in desert areas. It means we have to develop systems under which the abandonment of areas is avoided, and this is not beyond our planning capacity.

## THE OUTLOOK FOR THE FUTURE

The development of the deserts in a sustained and careful manner is indeed a challenge as well as a task for all of mankind. Deserts are important centres of natural resources. They have climatic and many other advantages and their development calls for international cooperation. There is, first of all the need to cooperate in the exchange and distribution of desert experience and desert technology. There is a need to raise the educational standards of the desert population and, where possible, water-rich areas should share their water with desert areas. Therefore, we have to study more schemes for transporting water to desert areas. We have to cooperate on an international basis in the exploration and development of new water resources in desert areas, in better water technology and better water utilization.

We have also to realize that the arid and desert zones, with their long sunshine hours and sunny days and the availability in many arid and desert zones of suitable soil, could become a large-scale producer of food if the need should arise and the water would be available. If the cooling trend, which is now seriously discussed by many climatologists, should reduce the food production capacity of the northern and temperate zones of the world, then the desert areas could make up the short-fall in food production, providing water in sufficient quantities could be made available to such arid and desert zones. It is here in California that the most ambitious scheme of bringing water from Alaska through Canada to the deserts of the United States and Mexico was proposed many years ago. It is appropriate at this conference to look at such engineering dreams of the past as possible necessities in the future. The same applies to schemes designed to bring the surplus water of tropical Africa to the Sahara. In a world which has to share its food and has to cooperate not only for peace but also for development, these possible dreams for the future now deserve serious attention.

# AN OVERVIEW OF THE CONFERENCE

Joel Schechter

## INTRODUCTION

Effective strategies for desert development must be based on the entire complex of physical, socio-cultural and economic characteristics of the particular nation or region in which it is situated. Deserts have in reality one basic characteristic in common, namely, a large potential evapotranspiration rate relative to the amount of precipitation. Aside from that common factor there are as many differences among them as there are among the more climatically temperate nations or regions. One can characterize these on the basis of climate, as hot deserts or cold deserts; winter rain, summer rain or other seasonal rain deserts. There are coastal deserts and inland deserts, high deserts and low deserts and there are various gradations of aridity. Deserts may be a small part of a large country, a large part of a small country or, worse yet, a large part of a large country. They are situated in economically and culturally advanced nations and in newly developing nations. Matlock (1.2)* has further divided them into five categories depending upon their socio-economic status and the availability of exploitable resources. The importance of their arid areas, even to such technologically advanced nations as the U.S., Australia and Israel, differs completely in each of these nations. The Chinese perception of their desert as a strategic barrier (Chang 1.10) is still another outlook. In attempting to construct sets of alternate strategies for the development of these regions, care must be taken to define under which conditions and aspirations such strategic alternatives can be effective and what will be their impact on the environment. Since the basic needs of people for food and shelter are similar, many of these strategy constructs will have common elements but they too will require considerable mitigation to suit them to the particular case. Many of the technologies used will be similar or even identical but there are probably few strategies that can be universally applied.

For this reason it is not strange that this Conference brought out a wide spectrum of development programs and strategies each based upon the specific conditions, aspirations and perceptions of the individual nations involved.

---

*The number refers to the chapter of the paper in the forthcoming publication of the Conference proceedings. The full table of contents of the four-volume publication can be found at the back of this issue.

In general, the Conference was successful in bringing together a large international and interdisciplinary group of participants and in covering a wide spectrum of desert development strategies and technologies. Such seemingly disparate subjects as problems of road construction in deserts, plant and animal ecology, waste water re-use, industry and mining, desert agriculture, issues dealing with water and energy supply and exploitation, etc., were presented to the plenum and often the interrelationships among these were effectively brought out in the discussions. It would of course have been impossible to include all relevant subjects in the limited time allocated to the Conference, and with the already large number of subjects, not all were able to be treated as extensively as they deserve. However, many important issues were raised and while it was not expected that easy solutions and standardized programs would be developed, the discussions and the definition of the issues were an important contribution towards the evolution of strategies for development of arid and semi-arid areas.

## ISSUES FOR DEVELOPMENT

Desert civilizations have in the past been amongst the foremost contributors to the progress of mankind. Today, too, the arid and semi-arid regions of the world contribute significantly to the world's economy (Matlock 1.2). While a large part of the production of many agricultural crops such as cotton, sesame and groundnuts are produced in arid areas, the major contributions of these areas today are petroleum and minerals. Due to the large capital influxes to the petroleum exporting nations the usual constraint of lack of capital has virtually disappeared in several desert nations. Many of the less fortunate arid zone nations can also anticipate some form of aid for capital accumulation for development. The issue then arises as to the relative priorities which should be allocated to agricultural and industrial growth.

Shelton and Pratt (1.4) in their paper and in the ensuing discussion made a strong argument for the importance of food production and would place agricultural development as the major priority for national development. Alternate supplies of food are not always available, are usually more expensive and deplete available capital and foreign currency which could be used for development. Furthermore, it is claimed (Joekes, Singer and Swift, 1.3) that industrial development in these countries is dependent on an improved and expanded agricultural base. The authors claim that this is the case in the rich oil-producing nations as well as in the resource poor arid nations. Rapid industrialization without an adequate agricultural base can cause severe dislocations in the labor force with resultant widespread unemployment. This seems to be the case today in many oil rich nations where according to the authors there is a large measure of under-employment, a large rise in unproductive service activities and high rates of urbanization and emigration. In addition there is a need to build up internal markets for manufactured goods and the agrarian population is by far the major potential consuming sector by virtue of being a large proportion of the population. On the other hand, several participants pointed out (as have also Joekes, Singer and Swift) that these countries also import large numbers of skilled and semi-skilled workers.

It was pointed out (Barnea--IV) that in most desert areas water is the limiting factor for the development of the region and it will be important to consider the income effect of each unit of water consumed. This may lead to the conclusion that agriculture for the production of low value staple food which produces a low value added per unit of water used, may have the lowest development effect. Thus, in situations of limited water availability careful planning will be required in order to maximize national income on the basis of the water available.

No one can deny the advantages of self-sufficiency in staple foods and the need to reduce the consequences of recurring droughts. However, mining and industrial development also have their advantages. In effect they are among the few endeavors which can propel a nation into the twentieth century (if that is considered desirable--and it is considered so by many). They tend to promote a higher level of education and can accelerate economic progress and social change. Rates of capital accumulation and foreign earnings can more rapidly be increased. Unfortunately, there is no ready-made formula for determining the optimum proportionate effort between industrial and agricultural development. Historically, mining and/or agriculture have been the pioneers, but a more serious consideration of the potential industrial contribution to development of arid areas could possibly indicate the need for a stronger role for this type of activity.

Sea coasts comprising some 20,000 miles in arid and semi-arid areas pose many advantages (Brandanovic 1.5). The revolution in sea transport during recent decades has greatly lowered shipping costs thereby opening these areas to importation of new materials and export of local products. Saudi Arabia might, for example, cheaply import iron ore from as far away as Brazil, effectively use cheap natural gas in the direct reduction process and export the resulting iron at very competitive prices. Desert sea coasts also have various marine resources and can provide cooling water for power and industrial plants. In comparison landlocked desert nations are at a considerable disadvantage and will often suffer severe limitations on their industrial growth capacity.

## WATER RESOURCES, PLANNING AND MANAGEMENT

The largest number of papers presented at the Conference justifiably centered around the issue of water resources, planning and management. A vast diversity of water problems were presented by the various speakers. The papers presented by the U.S. participants concentrated almost exclusively upon the dwindling water supplies in the U.S. Southwestern region, while those from the Sahel and other developing arid and semi-arid regions were more concerned about the discovery and development of new water resources. However, an underlying theme concerning the need for rational planning and management prevails throughout this wide diversity.

The conference participants had an opportunity to visit and study the vast California water storage facilities which have allowed California, in spite of the two years of unprecedented drought, to maintain not only its urban and industrial activities but to maintain almost unimpaired its agricultural production. This was a vivid demonstration of how the effects of a drought can be largely mitigated in semi-arid and arid areas provided the capital is available and the effort is made to store water in large quantities.

## PLANNING METHODOLOGIES

Several planning methodologies were presented--mostly by U.S. contributors (2.1.6 to 2.1.10). Lack of planning which has characterized the exploitation of underground water resources in the U.S. has resulted in a crisis situation in several areas and promises catastrophic results if vigorous corrective measures are not implemented. The large number of U.S. papers on this subject is probably a symptom of this lack of planning and hopefully signals new directions for U.S. water conservation policies. Even so, as Patricia Paylore indicates in her paper (2.1.2), in some regions the possible occurrence of a major tragedy will be the only sufficiently acute incentive for the cessation of present water depleting practices. Newly developing nations which are presently in the process of developing their underground water resources would do well to take heed of the unsuccessful U.S. experience.

Many of the methodologies for water resources planning have several common ideologies. These call for long-range, regional, non-traditional, multi-disciplinary planning and utilization of the entire spectrum of modern sophisticated methodologies--Delphi, input-output analysis, computer modeling, etc. In addition, programs must meet the criteria of cost effectiveness and take into account non-hydrological as well as hydrological elements. They must consider appropriate institutional changes, flexibility, organizational development and a large number of economic, political and social constraints -- truly an enormous task for any planning agency anywhere.

## WATER RIGHTS

However, we have little choice but to plan. The consequences of unplanned water use have been too negative and the present situation cannot be allowed to continue. On the other hand, we do have some good examples in the encouraging results of several national programs for planned water use. The basis for any planning for water exploitation is contained within the system of "water rights" as practiced by any specific nation or region. These "rights" are usually defined by custom, tradition or law. Water resources, both surface and underground are 'communal' in nature in that the actions of any one utilizer affect the quantity and quality of water available to other users of the same stream or aquifer. Planning predicates a relinquishing of the individual's 'rights' to selfishly exploit what is essentially a communal water resource. It implies some mechanism for resource management for a more communal and long-range benefit. As P.A. Towner points out (2.1.1.), a system of water rights that can maximize effective management is a correct one. However, this must take into account the specific local situation and the long term goals of the society. If such a goal is the maximization of short range profits for a limited community, then the wider community as well as future generations could suffer irretrievable damage. As water demand grows and becomes acute, the number of economic and societal factors to be included within the sphere of planning will have to increase. So too, will the surrendering of traditional individual water rights have to become more widespread. It would be logical for water planners to foresee such a situation and prepare in advance the legal framework needed already in many situations for the effective water management which will be needed even more in the future.

Towner's paper brings to our attention the legal basis for underground water planning and management. It appears that only in Iran and Israel have water resources become the legal common property of the nation. It is well known that Israel's well-planned and coordinated water system could not have been achieved without this legal basis as embodied in the Israel Water Law. Other countries, such as Mexico and several States of the U.S. are moving in a similar direction. It is becoming abundantly clear that effective planning and management of a water resource will require severe limitation of traditional individual property rights to that water resource.

## WATER RESOURCES

Several optimistic surveys of potential water resources in arid and semi-arid areas were presented at the Conference. Especially impressive is the description of the water resource maps presented by the French Bureau de Recherches Geologiques et Minieres (Biscalde et al, 2.1.12) of large underground aquifers underlying the Sahel. They estimate natural renewable resources for eight Sahelian countries at 60 billion cubic meters per year at economically feasible costs. In addition another 770 to 1625 billion cubic meters of exploitable reserves are estimated to be available. These are considerable amounts of water. The renewable quantities alone could be sufficient to irrigate one crop per year on over 6 million hectares. Another large amount of water could be mined from the Nubian Sandstone in Egypt's Western Desert (Schneider, 2.1.10, Ezzat 2.4.1) over a period of several decades to several hundred years. The existence of additional economically available, replenishable or minable water resources in other areas is indicated in several of the papers presented (e.g. Hajrah 2.4.3, Golani 2.4.2, Thomas 2.1.11, etc.). The mining of an aquifer has to be planned with great caution and possible alternatives for future supply (e.g. water transport, desalination) after exhaustion of the aquifer carefully considered. Mining, for a period of say 100 years seems reasonable if pumping potential is built up slowly and the reaction of the aquifer is carefully monitored.

The intensive exploitation of underground water resources is an important alternative development strategy whose advantages will be discussed in a further section of this overview. The existence of such large water quantities and the possibilities for their safe and economic extraction and use should be further investigated and verified.

The Conference had the opportunity to discuss the large-scale transfer of water to desert areas such as the Karakum Canal in the Soviet Union, the Rajasthan Canal in India and the Colorado River system in the U.S. Whenever water from water-rich areas can be brought to the desert the potential of the desert area for development is increased. However, such schemes may have ecologically deleterious effects if not planned and managed properly. One paper (Bouchardeau 2.1.3) dealt with the better utilization of large rivers which cross deserts such as the Nile, the Senegal and Niger Rivers.

## NON-CONVENTIONAL WATER RESOURCES

The Conference also considered non-conventional water resources. Unfortunately, despite technological improvements we do not seem to be closer to an economically feasible method for ocean water desalination for agricultural use. Increases in the price of energy have reduced the potential value of desalination methods. There are some promising advances in the use of solar and wind energy for this purpose (Channabasappa 2.2.1) but it is doubtful if we shall see many instances of this technology used for agriculture in the near future. Desalination of ocean water and probably of brackish water as well will remain restricted to municipal and industrial uses or to those fortunate nations with abundant supplies of oil or gas. At best desalinated water may prove foreseeable for the limited highly sophisticated greenhouse culture of high value crops (Jensen 3.1.16).

The utilization of icebergs as a water source (Al-Faisal 2.2.9, Hult 2.2.10), a challenging and highly imaginative concept, does not as yet seem to be founded on a solid technological or potentially economic basis. There are major difficulties along each step of the process, from towing the iceberg, wrapping it with an insulating blanket, and slicing it where necessary into more maneuverable pieces. The major difficulty, as brought out in the discussion, is thought to be in melting the ice. Artificial means would prove too expensive and solar heat is too diffuse to cause rapid melting. There is a need for new ideas before this concept will become feasible.

A little known source of water which may be of great value under certain conditions is that which can be obtained accompanying the production of petroleum (Collins 2.2.11). Some of these waters are of sufficient quality to be used for agriculture or even for watering livestock. They are often suitable for processes used for enhanced recovery of oil. Furthermore, it appears that sea water or very brackish water could be suitable for many of the needs of petroleum production (Bye 4.2.14). Hence, already existing better quality water could become more available in these critical areas for other uses.

The potential use of municipal and industrial effluents as a water source was the subject of several papers and aroused considerable interest among conference participants. The simplest treatment appears to be primary processes (filtration and sedimentation) followed by infiltration and recharged into a suitable aquifer (Bouwer and Rice 2.2.4). This results in water whose quality is sufficiently safe for irrigation of forage and fiber crops. Alternatively waste waters could undergo secondary treatment (further filtration and sedimentation, biological oxidation, etc.) and even tertiary processes (desalination, activated carbon absorption) where necessary or desirable (DeCook 2.2.5, Brice 2.2.7). Depending upon the quality of the resultant water and its associated cost, it can then be used for unrestricted irrigation, artificial lakes, industrial cooling and where there are no feasible alternatives--as a municipal water supply.

The environmental hazards of waste water reuse may be severe and include: a build-up of salt content and toxic inorganic substances especially nitrates, increase of non-degradable organics, and the presence of various pathogens. Adequate technology seems to be available to overcome these problems, albeit sometimes at a considerable expense. For this reason processing must be carefully matched to end use and advanced management procedures used to ensure the safety of the population and of the environment.

As cities and industries grow in size and demand greater water supplies, the need for water recycling will become of greater importance. One could envisage cities, at some future date, with an almost closed system of water use and reuse. Already this is being done in one South African town with satisfactory results. Geraghty (2.2.6) proposes deep well injection of waste water into suitable aquifers where they could be stored for future use. The utilization of municipal effluents for the production of a high protein algal feed is in advanced experimental stages in Israel and the U.S. An integrated design for waste water reclamation with the coordinated production of various freshwater foods, animal feeds and fertilizer was presented by Serfling and Mendola (2.2.8). While much further experimentation will undoubtedly be necessary before establishing the economic feasibility of such an imaginative scheme it does already demonstrate the great potential value of these waste water resources.

## WATER CONSERVATION AND USE EFFICIENCY

Efficient use and conservation of a resource has often proved to constitute the most efficacious means for its economic utilization. This important topic probably received less emphasis at the Conference than it deserves. The amount of water evaporated from water reservoirs, streams, canals and other bodies of open water in arid areas are enormous, often amounting to far more than the proportion of water actually used. (Paylore 2.1.2, Ezzat 2.4.1) Evaporation control methods have been experimented with for many years now (Cluff 2.3.1) with only limited success. One would hope that the various agencies and governments involved in arid zone research and arid zone development projects would encourage more research on this problem or where feasible replacement of surface storage by underground storage.

The compartmented reservoir combined with evaporation barriers suggested by Cluff (2.3.2) would seem to be a feasible water-saving scheme for small reservoirs in areas with scarce water supplies. Other water-saving devices for sanitary purposes (e.g., Ortega and Lefebvre 2.3.5) again can be effective where water is exceedingly scarce. However, there is a great need for water-saving devices where the use of water conservation methods could extend the effective use of a limited supply.

Nor was the subject of efficient water use for irrigation sufficiently dealt with at the Conference. The greatly advanced methods of drip irrigiation were discussed (DeMalach et al 3.1.14, Menzel 3.1.15), and the exceptional water use efficiencies obtained were plausibly demonstrated. However, there are many techniques requiring less capital investment which could be used at the outset to establish more efficient irrigation practices in newly developed areas. More essential still would be a system of incentives, on the level of the individual farmer, designed to encourage the more efficient use of water. Such incentives could include water pricing, water quotas, subsidies, grants and loans for purchase of more efficient irrigation equipment.

One of the key factors in the rise and fall of the so-called "hydraulic civilization" (which developed in arid areas with large quantities of surface water originating in distant humid regions) was the improper utilization of irrigation technologies (Prah 2.1.4, Gibson 3.2.2). It would be wise to take into account the lessons of these former civilizations in our own water use practices, and this probably applies even more so to the advanced nations than to the newly developing one.

## AGRICULTURE

The development and expansion of agricultural production in arid and semi-arid zones is of course intimately related to the availability and use of water. This relationship was the major theme of most of the Conference papers relating to irrigated and rainfed cultivation, and range utilization. Various problems arise in determining overall strategies to overcome the difficult and complex obstacles to agricultural development. Should, for example, priority of resource use be given to rainfed agriculture and/or to developing new irrigated areas? Should new irrigation projects be constructed or should the existing ones be improved? What priority to rangelands or woodlands? What level of technology is appropriate, However, behind all these problems loom the social and cultural barriers of archaic systems of land tenure, rigid water rights, lack of development capital and of technological skills needed to absorb new technologies. Unfortunately, but perhaps understandably few strategies were presented to attempt to deal with these latter problems.

Several methodologies were presented for strategy development, many of which have been referred to in the preceding pages. These are known and accepted methodologies in advanced societies and should be applied with care until more experience is gained in less developed areas. However, the disciplined thinking that they require can only increase the effectiveness of planning procedures. Several papers called for technologies which would lead to combined use, multiple purpose and sustained yields. Considering the wide disparities among the nations with arid regions it is apparent that there will be no single overall methodology or technology level appropriate to all of these regions and there will possibly be great differences in approach within the region itself.

Several of the technologies presented are worth stressing. For desert nations with the ability to make large capital investments a major priority would be that of affecting a significant increase in the efficiency of agricultural water use. This would include drip irrigation systems (DeMalach, et al, 3.1.14, Menzel 3.1.15) and under exceptional circumstances controlled environment agriculture (Jensen 3.1.16, Pasternak et al, 4.6). For most arid zone nations the installation of drainage systems still remains at the top of the priority list. The resultant salinity and waterlogging caused by poor drainage is causing a yield reduction of 30-50% in those areas which have not already gone completely out of production. Drainage and better irrigation technology combined with some increase in overall level of agro-technology could probably result in doubling agricultural production in the presently irrigated areas of the arid and semi-arid regions.

As mentioned in the last section, one of the important facts brought out at the Conference was the large amount of underground water of economic cost found in many regions under the desert surface. While great caution has to be taken in verifying these estimates (especially since most of these resources are non-renewable fossil waters accumulated over the last 30,000 years) these resources can undoubtedly have tremendous impact on those regions where they exist. Ground water development has several advantages over use of surface resources. Large capital investments in dams and canals are not necessary as the wells can be drilled where needed and at a rate commensurate with capital availability. The lack of extensive water surface reduces considerably the development of water borne diseases so prominent whenever large dams have created new water bodies. These diseases have in many cases debilitated local populations to an alarming extent (Nicols and Lambrecht, 2.1.13). The same large surface area results in an enormous loss of water due to evaporation. Dams suffer from sedimentation from upstream and may cause drainage and ground water problems downstream. Surface water development, if not properly planned and managed often may cause more problems than it solves.

Surface water use is often encumbered by a fabric of established water rights and customs, accompanied by legal structures of unequitable land policy resulting in tenant farming and its associated adverse effects. The development of new ground water resources might be a means of opening up new lands to settlement of nomads or landless tenants, based upon irrigated parcels and integrated with rain-fed farming and/or cattle raising.

Even though there are large endangered fresh water resources available, often a disadvantage of these resources (aside from the dangers of over-exploitation) is the possibility that they contain large amounts of soluble salts. However, great strides have been made in the U.S., Tunisia, and Israel in the efficient utilization of saline water resources (Robinson 3.1.13, DeMalach et al 3.1.14). Above all is the paramount need for drainage which is even more acute where the applied irrigation water is saline (Hermsmeier 3.1.7). Research on problems of direct use of saline water has shown encouraging results during the past years adding confidence to the possibility for long range use of this natural resource by elimination of adverse effects on soil and plants.

In many semi-arid regions rain-fed agricultural development will warrant highest priorities. In most developing countries in the semi-arid zone this is the mainstay of the economy -- producing a large proportion of the food and giving employment to the vast majority of the population. Hence any incremental increase in productivity in this activity would have a far-reaching effect in the nation's economy. Conflicts in this sector relate to producing food for subsistence economy as alternatively growing more commercial crops (e.g., cotton, peanuts, etc.) for a cash economy. While most papers considered food self-sufficiency as being of great importance (Shelton and Pratt 1.4) it is difficult to see how essential agricultural inputs needed to raise productivity will be financed without some amount of cash earnings. A second problem consists of the proportion and integration of rain-fed with irrigated agriculture and the possibility of combining these with range and cattle raising into a coordinated production system (Arar and Huss 3.1.3, Young and Evans 3.1.5). There is a great need for better crop rotation, the introduction of valuable nitrogen-fixing legumes and for new grain varieties giving optimum accumulated yield over a period of several years. (A variety developed only for its "drought resistance" could give less accumulated yield over the same period.) Good grain storage facilities are vital to this type of strategy.

Range management will probably always remain one of the main activities for utilizing arid and semi-arid land resources. In many areas it is the only way of using these resources. However, more attention will have to be paid to proper grazing controls and management as well as integration with other aspects of the economy. (Young and Evans, 3.1.5, Uddin 3.1.4, Arar and Huss 3.1.3.) Gringof (3.1.6) presented an interesting methodology for predicting range capacity as a function of meteorological parameters. The results obtained in the arid south of the U.S.S.R. seem to indicate that these methods could be of value in other areas as well.

We owe a debt to Cayla and LeFloch (3.1.11) for their description of water harvesting methods as experimented with in North Africa. They warn us that such a project is "costly, delicate and hazardous" and should be based upon years of hydraulic study and analysis of the operation area. The high rate of failure encountered in Tunisia was due to silting, erosion, and the high labor requirements for operation and repair. Better results were obtained with more sturdy constructions but at a far greater cost. Careful economic analysis should be carried out before considering such schemes.

The use of desert plants as a source of industrial raw materials was another development strategy discussed at the Conference. This field of endeavor has been suggested many years previously, but possibly it is an activity whose time has finally come. It appears that several enterprises utilizing desert plants for raw materials are being undertaken in the U.S., Israel and Mexico. These include jojoba wax production and rubber manufacture from guayule latex. (Johnson 3.1.8, Schechter 3.1.9, Yermanos 3.1.10.) Among the possibly potentially valuable species are a large number which could fulfill a multi-purpose role. They could be used for sand dune stabilization, firewood production (e.g., Acacia senegal) and as a pasture reserve in addition to their value as an industrial material. More important still, they may provide the raw material and the stimulus for the development of industrial enterprises within the nation resulting in greater income and the development of greatly needed industrial skills.

An innovative proposal which aroused considerable controversy at the Conference was that of wildlife production in place of the conventional cattle herds (Hopcraft 3.1.23). Proponents of this scheme view cattle as being responsible for the widespread desertification we are witnessing and claim that wildlife raising for meat, hides, etc. would be more in harmony with "nature's" intent. While a case could be made for the need for further experimentation and even for field trials, it is difficult to believe that simply by changing the instrument of man's exploitation of his environment he will at one stroke eliminate mismanagement and all those abuses for which man alone and not the animal is responsible.

In general, where water is available agriculture in arid and semi-arid regions has many comparative advantages over more humid areas. Water can be applied when needed by the crop and in proper quantities. Sunlight is abundant and winters mild, offering opportunities for multiple yields. There is little or no interference to farm operations due to untimely rainfall. The majority of arid zone inhabitants are experienced in various farming techniques and while it is difficult to "unlearn" old ways the human resources are already potentially available. Intensive agricultural development accompanied by suitable social and cultural changes could be a major strategy for arid zone development.

## INDUSTRY AND MINING

Industrial development in arid and semi-arid areas has historically been associated with mining activities. Deserts seem to have accumulated large deposits of minerals and salts which have been preserved by their natural aridity and whose locations are relatively easily discovered (Smith 4.2.1, 4.2.4, Girandon 4.2.3, Ericksen 4.2.5, Hiebert 4.2.6, Sheldon 4.2.7). Unfortunately, most of these resources have been mined and exported as raw materials or at best as semi-finished products. These mineral resources could be used today as basic materials for more sophisticated manufactured products and become a far more significant source of foreign currency income for the arid region. However, as both Ellet (4.2.12) and Pelah (4.2.13) have pointed out, an extensive transportation, power and water infrastructure is required. To this could be added the need for skilled labor and management with a strong inclination for entrepreneural activities.

Mining is based upon non-renewable resources with finite life terms. With the exhaustion of the resource material entire communities have disappeared or have suffered disastrous economic repercussions. To avoid such an occurrence it has been proposed (Rose et al, 4.2.8) to plan in advance the creation of secondary industries and activities which will decrease the rate of resource exhaustion and provide other economic activities when depleted. In the discussion two other possible mining development strategies were suggested: (a) regional surveys, identifying a number of deposits but developing deposits in stages so that miners can be shifted from a depleted mine to a new mine; (b) restricted development of a big mine so that the life of the mine is extended to 50 to 100 years. It has also been suggested that mining communities in the U.S. should plan for future conversion to tourist and retirement centers (Carman 4.2.10). These suggestions are certainly suitable to those nations bordering the sea or those having easy accessibility for transport into and out of the mining area. Alternatively, such planning is possible in those nations fortunate enough to have large petroleum deposits and the resulting capital to finance such secondary development. However, it is difficult to perceive the possibility for a plan of capital and technology to inland isolated desert areas for secondary development despite the existence of valuable raw materials at such a location. The description of the difficulties encountered in proving and planning the development of a mineral resource in a remote area in Pakistan (Aslam and Bilgrami 4.2.9) indicates the serious barriers to such development. Governments and international agencies will have to consider programs to award extremely attractive incentives to attract capital and especially to obtain industrial technology for these remote areas. In addition, major programs for infrastructure development will be necessary as an essential base for attracting industry.

However, deserts do have many advantages for industrial development. Land is readily available at low prices and the mild winters could possibly ease construction costs. Industries may be able to easily and safely dispose of their waste products by simple evaporation of industrial effluents which in other areas could contaminate surface or ground water (Pelah 4.2.12). Low air humidities are often useful and solar energy could be utilized for drying processes resulting in considerable reduction of energy costs.

The shortage of water is not as limiting for industry as it is for agriculture. Income per unit of water is much higher for industrial products and water is proportionately a lower cost factor in product price. Industry can afford to pay more for water so that deeper or more distant water resources can be used. There is a great deal of active research on minimizing water consumption in industry and there already exists a wide spectrum of "dry" technologies especially for mining of minerals and metals in desert areas (Bhappu 4.2.2). The recycling of industrial waste water seems to be at a practical stage (Fisher 4.2.13) and water of low quality could be used often. A good example of this is the use of waste water and possibly sea-water by the petroleum industry for many of its needs and also for enhanced recovery of petroleum (Bye 4.2.14, Collins 2.2.11). Agricultural products and wastes and agro-industrial crops could be a suitable basis for initiating industrial development.

The human factor remains, as in other fields of endeavor, the most critical element. In many areas there is a lack of sufficiently trained personnel. Many desert areas are hampered bv lower educational levels or do not have suitable institutions to utilize that portion of the population which has been educated or adequately trained. Obviously this is a cardinal direction for future aid to these nations.

## ENERGY

The Conference heard reviews from several experts on the progress being made in the area of wind and solar energy (Meinel and Meinel 4.1.1, Chatel 4.1.2, Hughes 4.1.3). As the discussion brought out, progress has been disappointingly slow and it appears that neither wind nor solar energy will be economically available for wide use in the near future. On the other hand a considerable amount of enthusiasm was displayed for the possibility of tapping underground geothermal resources. These resources have several distinct advantages, the first being their wide distribution (Barnea and McNitt 4.1.4). The technology needed for the exploitation of geothermal energy is readily available and while there are sometimes specific problems of corrosion and disposal of possibly concentrated brines, solutions are also on hand. In fact, in many cases the water produced could be utilized for agriculture or for industrial purposes as well as for energy production.

Especially impressive are the reported geopressure areas (Jones 4.1.5) where hot water with dissolved methane are to be found. Such areas are more ubiquitous than previously suspected and also have a wide distribution. Many of these areas are in arid and semi-arid zones and one would hope that more effort would be made to verify their locations and their potentials for economic utilization.

Geothermal energy is not a new source of energy and is already being utilized in places such as Iceland, Italy, U.S.A., New Zealand, Japan, Mexico, El Salvador, and more recently large-scale exploration and development is being carried out in California's Imperial Valley (Edmunds 4.1.7) which is a desert area. An interesting example is the use of a low temperature geothermal source in Israel where greenhouses and even cultivated fields were heated for the production of high price winter vegetables for export (Pasternak et al, 4.1.6). Considerable optimism was expressed for the future role of geothermal energy.

The Conference also heard an interesting paper on the possibility of soil shale and tar sand development in arid areas (Donnell 4.1.8), and that there are excellent chances of further oil and gas discoveries in deserts (King 4.1.9). However, this Conference did not address itself directly to the critical problem of a substitute fuel to replace firewood. This most important problem was mentioned (Schechter 3.1.9) in reference to possible multiple uses of desert plant species for field as well as for utilizable raw materials, but no immediately usable strategies were proposed for dealing with this problem. This is a subject which should receive greater attention in the future.

## ECOLOGY AND DEVELOPMENT

In all of the Conference sessions on the various aspects of desert development and management, considerable concern was expressed for the safety of the environment. Desertification has already affected a considerable portion of the fringes of the deserts and threatens even further incursions into still productive lands. The general approach expressed or implicit in the discussions was that the development accompanied by proper management could constitute a tool to combat desertification. This has not always proved to be the case up to the present time. In fact, it has been the expansion of man's activities that has been the direct cause of environmental deterioration. Several participants expressed the need for rebuilding of the very herds that in the past decimated desert ranges on an almost unprecedented scale. It seems that despite the large number of conservation technologies that we have developed we still do not know how to effectively get them into operation in those areas where they are so desperately needed.

Still there is some hope. The Sacramento Conference on Alternative Strategies for Desert Development and Management and the forthcoming Nairobi Conference on Combating Desertification are signs of a growing concern and that may be the first stage leading to real progress in the amelioration and development of the world's arid and semi-arid areas.

# SOLAR ENERGY OPTIONS FOR ARID LANDS

Aden B. Meinel and Marjorie P. Meinel
Optical Sciences Center
University of Arizona
Tucson, Arizona 85721

## ABSTRACT

Arid lands are by nature ideal locations for use of solar energy because of its availability. A variety of active and passive modes of using solar energy are reviewed. Uses vary from domestic use for heating and cooling to industrial and agricultural uses, power production and export of hydrogen. Time scales for these different uses vary considerably. Economic barriers are considered limiting factors for all solar uses other than certain passive modes.

## INTRODUCTION

Desert lands are by nature ideal locations for maximum annual insolation. Some lands are both clear and arid; others are deficient in rainfall but frequently have thin clouds veiling the sun. Both types, however, need energy to sustain satisfactory levels of civilization. The solar asset has been neglected as a resource in the years when more compact forms of energy were available at a reasonable price. The rapid rise in cost of traditional energy forms forecast for the next few decades means that a new assessment of solar energy is timely. In spite of this renewed interest solar energy use is an old topic, receiving much attention at the turn of the century, before petrofuels became abundant. In this report we will review current modes of utilization, and refer to the reader to the history elsewhere, for example our textbook (1976).

Arid lands are not homogeneous in matters of climatic details, wealth, technical sophistication, even insolation. The only common thing seems to be the lack of sufficient natural rainfall to sustain a population without the injection of additional water. Some have rivers to provide this extra water and some use non-renewable ground water. These differences make it hard to address them as a homogenous group in describing solar energy options. In this look at the solar options we will not attempt to categorize the different topics into those appropriate to one arid region or another, since such a differentiation will obviously require a more extensive analysis than is possible in a short paper at this meeting.

## COOKING

Solar cookers have been popular items for inventors but ill adapted to customs followed by women of the world. The time of sunlight is not matched to cooking needs, and storage to shift heat availability sufficiently is impracticably expensive for other than in the patios of affluent societies where solar cooking might hold sway but even there only as a conversation piece.

In most societies the evening meal is cooked. Some morning heating is required even for the breakfast meal. Cooking while the sun is high in the sky is generally avoided, and unfortunately that is when solar energy is most available.

The main option for using solar cooking is where the meal can be prepared in the morning, slowly cooked during the main part of the day, and consumed after sunset from heat stored in the meal itself. Chile beans is one such meal, but it is heard to imagine the bulk of any attractive diet being regularly cooked by long processing.

Storing heat in a large mass of adobe or stone is an ancient way of providing enough thermal inertia to provide a high enough temperature, close to boiling, for an evening meal. To reach higher temperatures, concentrating mirrors will be needed; however, their price and upkeep of reflectivity make them of doubtful utility in many societies. In Arizona researchers have used concentrating parabolic cylinder collectors to heat vegetable oil hot enough to cook meat overnight from energy stored at 150°C.

In our opinion the eventual use of solar energy for cooking can be done more effectively through the use of electric hot plate cookers and ovens, since the energy can be delivered and incorporated into life patterns that exist or are currently being introduced.

## DOMESTIC HOT WATER

Heating of domestic hot water is an application of growing importance in the temperate zone. Simple collectors can be made of indigenous materials to yield enough hot water for household needs. Commercial collectors in the United States are not as inexpensive as once hoped. Current costs of collectors, storage tank (84 gal.), controls and values range between 160 and 350 $\$/m^2$, shipping and installation being extra. This cost is equivalent to an electric hot water heater at a price of 6 to 12 cents per kwh. Lower costs are expected, but as of this date actual costs are rising rather than falling.

The basic problms of hot water collectors, aside from price, but closely tied to price, is durability. Corrosion by water, degradation of plastic parts and windows by sunlight and heat, sealing, dust accumulation, loss of blackness of the absorber, breakage of glass, all are problems, some of which are not adequately solved in the case of many collectors tried to date.

## FLAT PLATE COLLECTORS

Flat plate solar collectors are easy to build with indigenous materials in almost every country. They are capable of producing hot water at 60 to 80°C. Their level of sophistication is low. These types of collectors have been used for heating domestic water in many countries. The California home I grew up in had a solar hot water heater, homemade. It was deficient, as evidenced by the fact that when natural gas lines came into the community everyone switched to the fuel that gave them reliable hot water, in any quantity, at lower cost and with no collector maintenance.

The fact that collectors need maintenance but seldom get it can be seen in countries where they still are used. The glass can be broken or totally missing, and the optical black absorbing surface rusted red. The interesting thing is that they can still yield some hot water, not as much as when new, but still enough to avoid the $1000 needed to replace them. The reason they operate even when degraded is clear if you examine the situation. A new solar hot water heater is designed to heat its tank of hot water quickly during a clear day, so its efficiency is high and utilization factor low. When the collector degrades the efficiency is lowered, but this is partially compensated by an increase in the utilization factors.

The real economic factor that affects solar energy utilization is not the absolute cost in dollars, but the fraction of income or productivity represented by the solar devices. Thus the dollar cost may vary widely in different countries, but the leveling factor is the average income in that country.

In the United States we are experiencing a persistent gap between hopeful projection of solar hot water heater costs and real costs. A recent analysis of actual advertized costs for solar hot water heater systems, all components supplied at the sellers location, showed the cost ranged between 19 and 38 $\$/ft^2$ before shipping and local installation costs were added. This is far above the early claims that hot water systems would cost less than 10 $\$/ft^2$. If one does the labor of installing them himself, then the final costs will be in the range of 20 to 40 $\$/ft^2$. If a building contractor or plumber makes the installation the cost could double. When one then considers the value of the collector, based on the costs of competing energy, solar hot water heaters are only marginally competitive with expensive electricity, depending on the lifetime assumed for the collector and peripheral tax benefits. At the present time some buyers of solar hot water heaters have been disappointed with early failure of parts of the system, such as from corrosion. Reliable units are beginning to be found, but since solar hot water heaters are so easy to build in a small shop many units offered the public do not meet the necessary reliability requirements. These growing pains are not unexpected, but they could easily damp the current enthusiasm of many people about solar energy and adversely affect the long-range future of the field.

Since the cost of solar energy depends to a large extent on the cost of the collectors and their installation each country considering the use of solar energy should evaluate the cost in terms of the materials and labor markets in their country. The final work on economics depends upon such an evaluation.

## SPACE HEATING

Space heating using solar energy is not an attractive application in desert lands. The cold part of winter is generally brief and consequently one does not get good utilization of the capital investment in the collector system. Passive use of solar energy through good architectural design is a more promising application. Windows can let in winter sun and exclude summer sun. One also can add a greenhouse on to a home to derive winter heat and protect frost-sensitive plants at the same time.

Liquid and air solar collectors can be used. Early efforts in the U.S. have emphasized liquid (water) collectors for space heating. Large tanks of stored hot water are needed to provide heat for night and a few cloudy days. Water collectors can present freezing problems even in desert lands because of the large swings in temperature between day and night and additional radiation cooling of the collector because of clear and dry night skies. Draining the collectors, use of antifreeze, or use of organic or silicone liquids adds expense to already marginal economics.

Hot air solar collectors with rock storage appear to be a more cost effective way of using direct solar heat for space heating. They can be more easily made of indigenous materials than water collector systems.

## COOLING

### Gas Absorption Space Conditioning

Cooling is generally much more important than heating in an arid climate. Preservation of food is foremost, but amelioriation of heat for people is close behind if summer stagnation of activity is to be avoided. Solar cooling is being tried in the United States primarily through gas absorption refrigeration systems, but primarily for buildings and not for food storage. House cooling systems that have been built thus far are bulky, expensive and pose maintenance problems for continued operation at high performance. They would not be applicable to areas where capital is scarce and expensive and where trained refrigeration engineers are not readily available at acceptable cost to the user. It is possible that solar gas absorption refrigeration systems will eventually become reliable and will moderate somewhat from the present high cost, but such refrigeration systems for solar energy are not a substitute for electrical compressor refrigeration in the present decade.

### Food Coolers

The use of gas absorption refrigeration on small scale for food coolers would appear closer to use, through adaptation of the old flame-operated refrigerator. These flame systems, however, require higher temperatures than can be obtained from simple flat-plate solar collectors, but the use of non-tracking concentrating collectors and heat pipe or vegetable oil heat transfer fluids might make possible small food coolers at reasonable cost. Very little work is being done on this application today.

### Evaporation Cooling

Most arid climates have low humidities for a significant part of the hot season. This situation makes evaporation cooling practical. If electrical power is available a fan can pull air through a saturated pad, cooling outside air from on the order of 35-40°C to a reasonably comfortable temperature without raising the final humidity to over 50-70%. We use this form extensively in Tucson, more so lately as persons formerly using compressor refrigeration switch to evaporation cooling for economic reasons. The amount of electricity used is but a small fraction of that required for compressor refrigeration. This means that it might become feasible to operate an evaporative cooler using solar electricity. Generating small amounts of solar electricity is not feasible using heat engines, because of efficiency, capital cost and maintenance problems. Solar cells combined with simple optical concentrators may be practical and reasonably cost-effective for this purpose in the next decade. Intensive development of the concentrated cell approach is being done under guidance of the Sandia Laboratories of ERDA, and the goal of power at $2000 per peak kilowatt appears likely within the next five years. An evaporative cooler, operating on a 50% duty cycle over 24 hours, for a 100 square meter dwelling would require a cell array generating about ½ kw during the day hours to drive a ½ hp motor on the fan and a 1/100 hp water pump to keep the pads wet. This still means an investment of on the order of $1000 in solar cells, high for most people in most arid lands. It will still provide more cooling power per dollar of investment than gas absorption refrigeration. If the cost of solar cells drops below the target for the 1985 period of $1000 $kw_p$, then the use of solar evaporative cooling in arid regions will find a wider market.

The basic problem with evaporative cooling for arid lands is that most arid zones have summer rainfall and associated periods of higher humidity, high enough to cause evaporative coolers to lose much of their efficiency or comfort. This seems unavoidable, being something people will have to put up with. Some current work at the Univeristy of Arizona on evaporation systems that avoid direct injection of wet air into the dwelling appears promising, but it is at additional expense over the basically simple evaporation unit.

## PASSIVE SOLAR HEATING & COOLING

The expense and reliability of active solar energy use, i.e., solar collectors, has led many researchers and especially architects to look at how best to use solar energy through good architectural design and construction materials and methods. Allowing sunshine into a house during winter and excluding it during summer is possible in the arid regions, that is, at 20 to 40° latitude. Roof overhang alone does a reasonably good job of controlling solar heating. Thermal inertia of houses of adobe and related materials is sufficient to damp out daily temperature swings and delay the time of maximum heating into the afternoon.

A number of homes have been constructed in which the passive use of solar heating is increased by enclosing a massive wall with windows so that more of the heat is held within the house in winter.

Passive cooling, like heating, is of ancient origins. The Mediterranean village with narrow streets and large enclosed courtyards naturally traps night-cooled air, making them pleasant until afternoon. Augmenting night cooling by use of special surfaces shows much scientific promise, but in practical situations natural dirt tends to render these surfaces ineffective, or no better than ordinary surfaces. Passive cooling by night effects has a maximum energy rate of less than 100 watts/$m^2$, about 1/10th as much as daytime solar input to a black surface. Passive cooling is also lowered to near zero during days and nights of high humidity, affecting this mode of cooling much like evaporative cooling.

A compromise between effective daytime heating and nighttime cooling was tried at the University of Arizona in the 1960's. This pioneer house used a bare roof (no cover glass surfaces) painted dark green to avoid the esthetic impact of black. It produced hot water in the daytime that was stored in the top of a large water tank. It produced cold water at night that was stored at the bottom of the same tank. The house then drew water from the well for a few years, but needed a heat pump to boost the hot and cold zones during the humid part of summer. It was thus the first solar-boosted heat pump, a serious contender once more for solar house heating. After a few years the house developed problems of the same type being rediscovered today. The copper collectors began to leak at pinholes that were impossible to easily identify and fix. A second problem was encountered since the heat exchanges were also the inner ceiling of the rooms. When these surfaces became colder than the dew point in the humid part of summer a gentle rain developed that distressed the occupants. Clearly these problems could be alleviated in a second generation of this type of house, but to date none has been built.

Passive control of solar energy is an old art, being revived as energy conservation receives close attention. The use of a simple coating of lime, familiar in many cities along the southern shore of the Mediterranean, does much to reduce heat input to building walls. The diffuse sunlight scattered by the many walls also filters pleasant light into rooms through small windows. More sophisticated coatings than lime are possible, such as titantium dioxide pigmented paints, but for cost it is hard to beat a water wash of lime.

## SWIMMING POOL HEATING

Swimming pools are useful for recreation and health in some arid countries while in most the thought of using energy to heat a swimming pool seems the height of irresponsible use of energy. The replacement of fossil fuel heaters by solar heaters has been popular, but with indifferent results. We have tried this solution with our pool and found how hard it is to heat a swimming pool in an arid region. The collector we built was slightly larger than the "traditional" half the pool area, but it heated it less than 3°C. The reason for disappointing performance is that the low humidity and clear nights in the desert remove heat energy very efficiently. We later found that if one covers the pool the result is much better. The simplest pool heater is merely a layer of polyethylene floating on the pool surface. Ours was made black since water circulation was vigorous enough to mix the hot layer adjacent with the cover with cooler deep water. A clear cover works well with an un-pumped pool since the sunlight is absorbed in the water or at a dark bottom, thereby naturally mixing the heat through the volume. Clear plastic, though, allows vigorous growth of algae.

The above comments about swimming pool heating can be applied to heating water ponds for other purposes, as during winter for farming operations. The water can be heated to about 10°C above the daily maximum temperature with a single cover floating on the water. The cost of such covers is low, only a few cents per square foot of plastic.

## WATER PUMPING

Solar water pumping appears to be an ideal application of solar energy since no storage of energy is required. One can pump when sunshine is available and wait when it is cloudy. There are a number of solar water pumps in operation in parts of the world, mainly the low-temperature Sofretes type. These units are barely 1-2% efficient and extremely expensive per cubic meter of water pumped even from shallow wells. Intermittant pumps operated from solar cells have been tried. Their economics, while close to the low temperature types, are also not favorable except in locations of dire need and no alternative source of pumping power. The present costs are in the range of $10,000 to $20,000 per kilowatt of generating capacity.

The most promising new option on the horizon for water pumping is the concentrating photovoltaic system, where costs in the range of 1,000 to 2,000 $/kw are projected for 1980. An example is the Motorola/ASR 8X module using new ribbon silicon solar cells. Construction of a test pump powered by 100 kw of these cells is projected for 1980, dependent on funding and verification of technical questions. These systems would best be operated on an intermittant basis, thus avoiding the need for matching real-time cell output to pumping horsepower demands, as has been demonstrated by recent CNRS experiments.

Wind is a form of solar energy that is adapted to water pumping. It has a long history of successful and economic application. Where wind is available it is to be preferred.

## BIOCONVERSION

Growing crops or using crop waste for energy is a form of solar energy. It has limited potential for arid lands unless abundant water is available for growing the crop. Most crops grow with good efficiency under desert skies when there is enough water to provide the plant respiration needed to keep the leaves cool, and enough fertilizer to provide crop nutrients. Desert soils are usually deficient in organic matter, so fertilizers are essential. The need for adding organic material for maintaining good mechanical properties of some desert soils means that to divert the plant stalks to fuel production would deny the soil of needed conditioning. The universal growing shortages of desert underground water supplies, as in Arizona, threatens future crops for commercial markets regardless of the possible additional need to grow energy.

Desert and arid regions generally do have some natural species that grow under natural rainfall or with a minimum of added water. These species should be re-examined to see if some can be useful for energy uses even if no food values are derived from them. We have examined salsola (tumbleweeds) as such a possible crop in view of its high water efficiency, being one of the best for converting natural rainfall into solid mass. The selection of natural plants and their improvement would appear to be an overlooked field for generating some energy from arid lands. Production per hectare may be small, 5 to 10 metric tons per year, but since there is potentially much un-usable land in arid regions the conversion of some land to indigenous energy crops may be environmentally acceptable.

## CROP PROCESSING

Solar energy is being studied as a means of processing crops in some parts of the world. In arid lands the natural drying of crops is generally not a problem, unless a rainy season encroaches into a drying season for some crop. In the case of drying grains, recent work in the United States midwest indicates that solar drying of corn, wheat and sorgum shortens the drying period about two days in twenty. The cost of the collectors required to do this, even simple polyethylene tubes, however, does not seem cost-effective.

## SOLAR PONDS

The use of solar ponds in arid regions at first sounds rather impractical until one remembers that quite a few arid lands have saline playas or permanent saline lakes. These natural lakes frequently show large thermal gradients between their top and bottom. Bottom temperatures can approach 20 to 40°C above the noon ambient temperature because the salinity gradient is large enough to prevent natural convection from stirring the waters, and the waters are clear enough that sunlight penetrates a meter or so.

The enhancement of thermal gradients is being done in order to heat large volumes of saline water to as high as 80°C, this water to be used for such things as space heating of buildings or greenhouses in winter or domestic hot water. Research has even indicated that saline ponds might be cost effective for electrical power production with low temperature organic fluid turbines. Such systems have very low efficiencies, between 1 and 4%, but the cost of forming a saline pond out of a pre-existing saline sink can be low since no liner would be required to keep the saline solution from invading underground fresh water levels.

Every site of a natural saline sink is a potential solar pond site of considerable importance. The area of these sinks is also generally large enough that the energy output can measure into the equivalent of tens of megawatts each.

## ELECTRICAL POWER

### Thermal Conversion

Electrical power is the ultimate usage of solar energy. This, along with hydrogen production, is the final goal of the high technology approaches to solar energy conversion. Both electricity and gas could fit readily into advanced society energy usage. Hydrogen could even be an exportable form of solar energy that could power the world as a product of the arid lands. The only problem between now and this goal is less a matter of technological than of economic barriers. The main problem is that the efficiency of conversion into either power of hydrogen is apt to be in the 10-30% range. Direct use of solar heat, on the other hand can yield efficiencies in the 50-80% range. To use solar electricity for heating, for example, therefore is very inefficient and prohibitively expensive unless solar power can be produced very cheaply.

Early demonstrations of solar power conversion are being done in several countries. In the United States the first 10 $MW_e$ system is under advanced design and a site has been selected in the arid zone of California. It is designed to answer technical questions of operation. The first solar cell demonstration power project is also under contract, relatively small 10kw systems to be installed in another arid zone site in New Mexico. It too is to answer technical questions. Under the current policies the shift of emphasis to economic goals for solar power will not occur until the later years of the 1980's or the 1990 decade. Clearly the first demonstration systems will be far from the economic goals. Intensive research should be encouraged so that the second generation systems will benefit from a baseline of information that was not at hand when decisions were made concerning the first demonstration systems.

The 5-$MW_t$ Solar Thermal Test Facility (STTF) large-scale thermal applications of solar energy is nearing completion at Albuquerque, New Mexico. Much of the research yield of the next decade will be tested with this impressive facility before entering use in actual power producing systems.

At the present time the most efficient power systems appear to be high-temperature thermal systems using Brayton or Rankine cycle engines. These systems are very sophisticated in design, construction, operation, and interfacing with existing power grids. They will be suitable only for operation by utilities in appropriately developed countries. Smaller organic engine thermal systems, such as have already been tested for water pumping and total energy system uses require less skilled organizations for their operation, but still far above the level of skill of small communities of individuals.

### Photovoltaic Conversion

The most easy to use system appears to be the solar cell power system. Batteries and power inverters appear to be the maximum level of sophistication involved. Here the large cost of solar cells bars their immediate use for a wide variety of arid regions needs, ranging from driving cooling fans to cooking the evening meal and preserving food in the cooler. The potential for major reduction in cost of solar cells is large, and a corresponding large effort is being directed to this task.

Solar cells have one desirable property not possessed by thermal power systems. Their response to the intensity of sunlight is immediate and linear. This means that one could directly attach a solar cell to an electrolysis cell to produce hydrogen, every instant that the sun shines. This accumulated hydrogen can later be collected and utilized, so a very simple (in principal) unit for producing energy could be devised that would need only occasional tending. One might consider the solar cell to be the essential ingredient in an "analog plant" for arid zone farming. The cell would be filled with water and the hydrogen stored in a titantium sponge, the sponge to be stripped of its hydrogen on a routine basis.

## TOTAL ENERGY SYSTEMS

Total energy systems may prove to be one of the most attractive uses of solar energy in the long run. Solar energy suffers from being capital intensive. The fuel is free, but the total cost of the system must be provided before any energy is delivered to the user. In years of high interest rates on money, capital investment is costly. Solar energy would benefit from low interest rates. The benefit derived by the user depends on how much he can use the system each year - the utilization factor. Solar space heating in an arid region is not attractive since the system can be used relatively few weeks of the year, being idle most of the year. Heating doemstic hot water is a better use of solar energy since some hot water is needed all year. The best use, by far, of a solar energy system is termed the "total energy system". The goal of a total energy system is to use all collectable sunlight every day of the year.

A hypothetical total energy system is one in which electrical power is produced first. The waste, or excess, heat is then used for industrial or domestic uses, such as gas-absorption cooling and hot water heating. Such a total energy system looks attractive for a community "mini-utility" or as part of an industrial complex. It looks best in the industrial context where industrial process heat is needed each day. For community uses one faces the problem that while electrical energy is needed each day one rarely uses heating or cooling during the change between summer and winter, so system utilization would be poor during the spring and fall.

The first solar total energy system has been in operation for more than a year at the Sandia Laboratories, Albuquerque, New Mexico. It provides electrical power and heating and cooling for part of the load of a laboratory building. While it has proven technically successful, its economics suffer from the credibility gap that affects almost all solar applications today, and from the problem of attaining given economic goals in the next decade for even total energy systems.

The application of the total energy system to small communities, especially those in the LDC's, may be one that deserves closer attention than it has received to date. We discussed the village energy center as a concept several years ago but received no encouragement to work further in this direction. The emphasis then was still on individual applications of solar energy. We still think that in most arid land countries the individual unit approach is not viable because of the capital costs which the individual cannot readily absorb, and also because of societal incompatibility with customs.

The village energy center, applicable to communities in the cities of industrialized arid lands as well as communities in the LDC arid lands, meets the demands rather well. Raising the capital to install the system is easier than for individuals, unless the unit is a gift from some governmental entity, home or foreign. Training professional operators is feasible, rather than expecting each family unit to learn the care required. Distribution of the benefits - power, heat or cold - is easier. Power grids are available in most communities, and are rapidly spreading into new territories. Delivering heat or cold is harder, but the using industrial groups could be those in the vicinity of the solar installation.

Small electrical devices to use the solar power, such as hot plates, ovens, radiant heaters, fans, refrigerators and coolers can all operate from the distributed power.

Maintenance, repair and backup energy are all easier when there is a multiplicity of community systems, connected to constitute a local power grid. While one of the ultimate goals of solar energy was to free the individual from the evils of corporate power and give utopian freedom we think that history will favor cooperative attack on our future energy problems, in the arid lands as well as the world at large. You should consider solar energy as having great potential, but be cautious about prematurely assuming that the Age of Solar Energy is here. The promise is great but so are the barriers that are yet to be surmounted.

## CONCLUSIONS

Generalizations such as are necessary in a review of this type are not adequate when one wishes to examine feasibility of any particular option and a specific location. When such a need arises then the exact input data must be assessed along with all of the practical matters that will affect the success of the project and its cost. There is a growing body of expert consultants available both from universities, non-profit corporations and commercial companies that can provide this type of service in most countries.

We have deliberately avoided encumbering this summary report with names of the many contributors to the developments discussed. They can be found in the literature, especially the abstract volumes prepared by ERDA to summarize the literature.

There are now several books available that present solar application in various levels of technical detail, ranging from popular ones on simple solar applications to formal textbooks, such as Daniels (1964), Meinel & Meinel (1976), Duffie & Beckman (1974), and Kreith & Kreider (1976).

## REFERENCES


Daniels, F. (1964) Direct Use of the Sun's Energy, Yale U. Press, New Haven, Conn. Available in paperback, Ballentine Books.

Duffie, J.A. and Beckman, W.A. (1974) Solar Energy Thermal Processes, Wiley, New York. 386p. $18.95.

Kreith, F. and Kreider, J.F. (1976) Solar Heating and Cooling: Engineering, Practical Design and Economics. McGraw-Hill, New York. 324p. $22.50.

Meinel, A.B. and Meinel, M.P. (1976) Applied Solar Energy: An Introduction, Addison-Wesley, Reading, Mass. 651p. $17.95.

# WIND AND SOLAR ENERGY POTENTIAL FOR DESERT DEVELOPMENT

Bertrand H. Chatel, Chief Technology Applications
Office for Science and Technology
United Nations, New York, N.Y.

## ABSTRACT

The paper reviews the potentials for wind utilization in underground water pumping and generation of electricity for deserts development. Solar energy utilization is also reviewed from the point of view of water pumping (by using flat-plate collectors and concentration), as well as recent progress in solar electricity (photovoltaic and thermodynamic processes) of particular relevance to desert development.

## INTRODUCTION

In April 1975, a mission was organized by the United Nations Development Programme (UNDP) to study the possible role of solar and wind energy to solve the problems of the Sahelian drought-stricken countries: Senegal, Gambia, Mauritania, Upper Volta, Mali and Niger.

The mission was composed of representatives of UNDP, UNESCO, the Economic Commission for Africa (ECA), the United Nations Industrial Development Organization (UNIDO), the United Nations Office for Technical Co-operation (OTC) and the United Nations Office for Science and Technology.

The mission found that a number of applications of solar and wind energy would be particularly useful, notably for pumping underground water and generating small outputs of electric power for villages and rural areas.

The paradox of the present situation of desert development is the lack of energy practically available for desert dwellers although sunshine abounds and wind energy remains mainly untapped.

The traditional approach of western engineers has been to build large centralized power plants in cities and pumping stations run by diesel engines with imported fuel. There is a need for small power sources disseminated on large desert areas where a scattered population lives at low densities. The Governments have expressed their interest and expectations for such approaches. For instance, in Niger, the Government emphasized the water requirements of families breeding a dozen cattle or cultivating an area of 1 ha, grouped in villages of some 10 families. Solar and wind energy appear to be good candidates to fulfil these local needs, due to their capability for decentralization and small scale utilization.

## WIND POTENTIAL

The use of the wind has been known for centuries, for propulsion at sea and for water-pumping or milling along the sea-coast (Holland, Denmark, Greek islands). In the United States many farms were equipped with windmills during the drive to the West. Six million windmills were built between 1880 and 1940 to pump water and generate electricity (15) until the Rural Electrification Act (1956) brought public utility power to the agricultural outposts and shut down many home power sources.

In desert development, the wind could be particularly useful for pumping underground water and generating electricity in the areas where measurements show that the wind has sufficient strength and stability.

### Wind Measurements

A map of wind measurements (figure 1) shows that, in many desert areas, the wind could provide a significant contribution to the energy needs. For instance, Mongolia, which has important wind potentials mostly untapped at present is considering several programmes of development. In Mauritania, wind measurements show that a number of areas are favourable for wind mills, particularly along the coast where speeds frequently attain 7 m/sec. The United Nations Office for Technical Co-operation has undertaken a project of wind development in the Cape Verde Islands.

In Upper Volta, the World Meteorological Organization (WMO) has undertaken wind measurements for one year. It is necessary to pursue wind measurements during a five-year period to obtain reliable information to plan wind activities in desert areas on a national basis. The WMO may assist Governments with wind measurements for national and international purposes.

### Water Pumping

In Senegal, windmills have been installed by the Ecole Polytechnique of Thies (E.P.T.) to test their performances. In Dakar, the Institut Universitaire de Technologie (I.U.T.) is developing windmills made of local raw materials. In Mauritania, windmills will produce electrical supply for pumps with a manometric height of 45 m. The propellers with a diameter of 7 m, placed on masts 15 m high, will operate for winds between 4 and 7 m/sec. Beyond this speed, the propeller stops automatically to avoid excessive mechanical strains. The propellers are mounted on sealed bearings and protected against sand storms by neoprene. In Upper Volta, the Engineering School of Ouagadougou (EIER) (1) is developing experimental windmills to pump underground water. The use of local materials such as bamboo and canvas is emphasized so that they can be easily produced in the form of kits to be assembled locally.

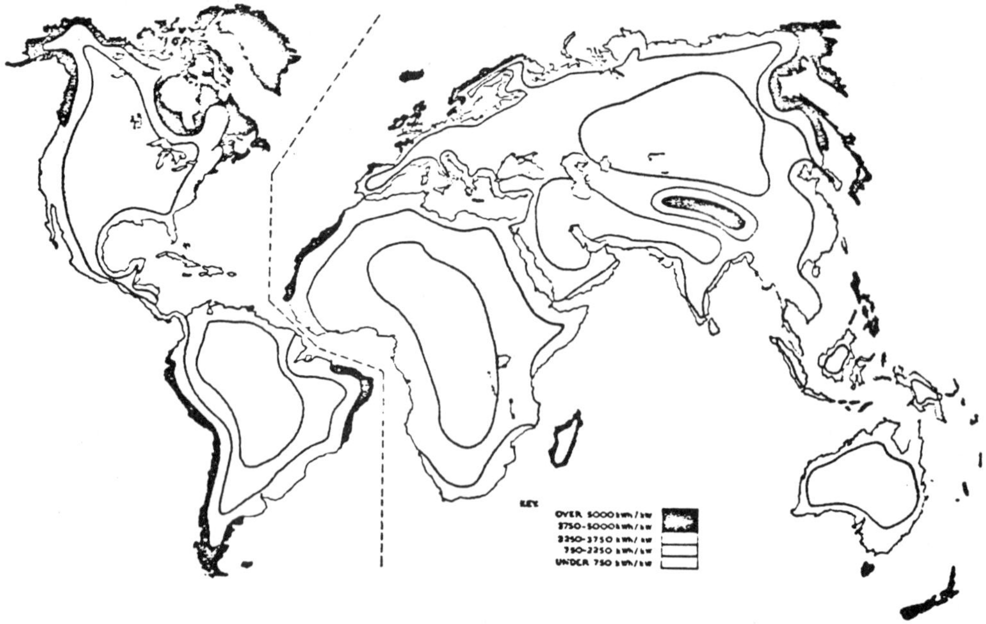


***Fig. 1.*—Availability of wind energy**
*Annual specific output of windmills rated at 25mile/h*

Figure I: Wind data for wind driven man
by Arthur H. Stodhart, Electric Research Association,
Surrey, England

Electrical Generation

The technology of electrical generation by windmills is well-known since the beginning of this century. In the United States, wind generators were used extensively in farms during the period 1925-1955 (2). Recently, the U.S. National Aeronautics and Space Agency (NASA) developed an experimental 100 KW wind electric power plant with horizontal axis, and aerodynamically designed rigid propeller-like rotors. A larger prototype of 1.5 Megawatts is now under development by the US Energy Research and Development Administration (ERDA). Another design, of interest for cost reduction in smaller sizes, is the Darrieus vertical axis design (15). This represents new progress in wind technology with a wind turbine of 18 m diameter, generating 60 KW, now under test in Albuquerque, New Mexico (5, 14).

Windmills produce direct current (D.C.), which necessitates the use of batteries for storage of electricity when wind is lacking. A converter DC/AC has now been developed to turn the D.C. into alternating current which can be introduced into the power grid, thus avoiding the need for batteries and providing economies when the user is connected to an electrical distribution network. In deserts, the use of sealed batteries would be required in isolated remote areas; in addition, the converter would provide the possibility of using standard A.C. electric tools, pumps and equipments, instead of being limited to D.C. devices, which are more difficult to maintain in a sandy environment.

### Conclusions

In desert areas, the problems of maintenance of windmills should not be underestimated. It is frequent to see windmills not in operation due to a mechanical breakage and lack of repairs. This problem should be overcome by the organization of maintenance teams duly trained in this basically simple technology. In the manufacturing of windmills, emphasis should be placed on the use of local raw materials and the production of kits to be assembled in villages with appropriate training.

## SOLAR POTENTIAL

### Solar Measurements

A map of solar intensities in the world (figure II) shows that deserts occupy a privileged situation on the globe as regards the availability of solar energy for development. It is significant to note that this abundant source is, at present, mostly untapped for practical applications. The World Meteorological Organization (WMO) is undertaking a world-wide programme of solar energy measurements.

### Water Pumping by Solar Energy

Pumping underground water by solar energy can be done directly by using a thermodynamic process, either with flat plate collectors or by using concentration.

(a) Flat plate collectors

Water, heated in flat plate collectors at a temperature of 70°C, exchanges its calories in a heat exchanger where a liquid (butane or freon) vaporizes at this temperature. The steam thus produced activates a steam engine which, in turn, moves the pump.

This process is used in a number of Sahelian countries where several prototypes have been in operation for a decade. The characteristics of the pumps are given in Figure III. In Mexico, pumps of 25 KW and 50 KW are now being installed to supply the needs of a rural village for irrigation and local requirements of the inhabitants. The pumps can deliver 1300 $m^3$ per day from a depth of 20 m. The characteristics of this equipment are given in Figure IV.

(b) Concentration

(i) The U.S. Energy Research and Development Administration (ERDA) is developing a solar pumping plant of 25 HP which will be used for the New Mexico Solar Irrigation Project in the Estancia Valley (New Mexico), 40 miles east of Albuquerque (4), where the average rainfall is 11 inches. The equipment (5) is composed of a solar collection field, an energy storage tank, a solar engine, an irrigation pump and a water storage pond. The sun-tracking collectors raise the temperature of the heat exchange fluid of the turbine (freon) (13). After exhausting from the turbine, the freon is condensed by the pumped water passing through the condenser. The solar engine, a Rankine cycle turbine, uses the freon at a temperature of 325°F and a pressure of 220 psig, and drives a pump which can deliver 880 gals per minute, from a well 75 feet deep. When the system is not operating the irrigation pump, it can deliver electric power: 200 Kilowatts hours/day for other uses, or deliver a heat of 10 million BTU/day.

Figure II: **SAHELIAN REGION INSOLATION**

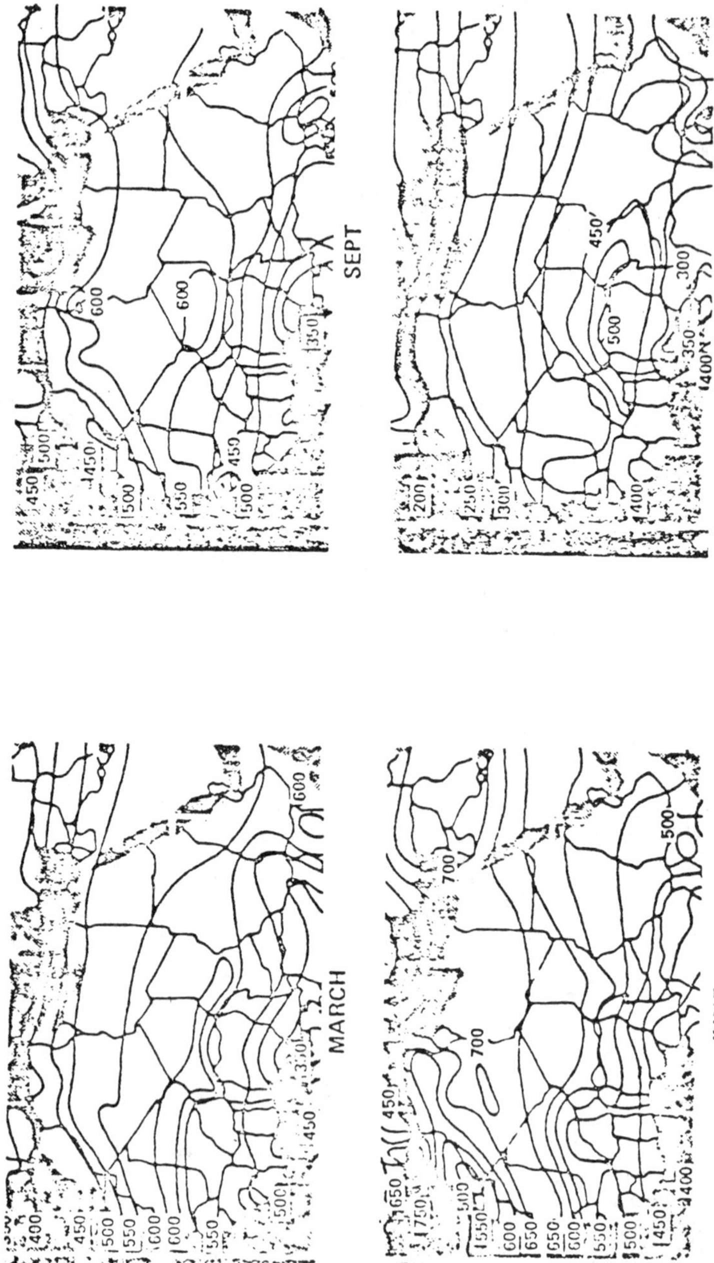


ANNUAL MEAN DAILY SOLAR RADIATION (GM. CAL./CM$^2$ DAY)

Projects Already Completed

| Place and Date | Collector Surface | Output | Manometric Height | Time of Operation |
|---|---|---|---|---|
| Dakar, Senegal (1966) | 12 $m^2$ | 1,200 l/h | 15 m | 5 - 6 h |
| Dakar, Senegal (1968) | 88 $m^2$ | 6 $m^3$/h | 25 m | 5 - 6 h |
| Ouagadougou, U.V. (1971) | 30 $m^2$ | 2 $m^3$/h | 20 m | 5 - 6 h |
| Niamey, Niger (1973) | 60 $m^2$ | 6-7 $m^3$/h | 12 m | 5 - 6 h |
| Chinguetti, Mauritania (1973) | 72 $m^2$ | 8-10$m^3$/h | 23 m | 5 - 7 h |

Projects Currently Under Construction or Under Study

| | | | | |
|---|---|---|---|---|
| Upper Volta | 50 $m^2$ | 5 $m^3$/h | 30 m | 6 h |
| Senegal | 85 $m^2$ | 8-10$m^3$/h | 20 m | 5 - 6 h |
| Algeria | 350 $m^2$ | 21$m^2$/h | 30 m | 5 - 6 h |
| Chad | 70-200$m^2$ | 5-10$m^2$/h | 15-60 m | 5 - 7 h |

Figure III: Solar Pumps Using Thermodynamic Processes, Flat Plate Collectors (3)

(ii) In Phoenix (Arizona), an irrigation project has been undertaken by the Arizona Cotton Growers Association in co-operation with the University of Arizona (4, 6). An array of 100 solar parabolic collectors, made of aluminium (4 x 10 ft) rotates to track the sun; the collectors serve as concentrators to heat pipes where the temperature of the water is raised to 400°F. This water provides calories to a heat exchanger. The steam produced activates a steam engine which delivers power to the pump, through a standard right angle drive. The pump is designed to produce 75 HP, and could be scaled up to 400 HP. This solar carrousel rotates 180 degrees each day so as to track the sun in two modes: East-West and North-South.

(iii) The Battelle Memorial Institute (4, 7) has undertaken the construction of a solar pump of 50 HP, to be operational in 1977 in a farm in the Southwest of Phoenix (Arizona).

(iv) Other ERDA projects include the development of a deep-well solar pumping system of 200 HP.

(v) In Niamey (Niger), Professor Moumouni is developing an engine for water pumping by a thermodynamic process, via concentrating collectors (17).

| Suction head (meters) | 5 | 10 | 20 | 30 |
|---|---|---|---|---|
| Delivery (cubic meters per day) | 5,200 | 2,600 | 1,300 | 900 |
| Intensive Crop Farming (acres) | | | | |
| Irrigation potential in continuous crop growing in cumulative areas: | | | | |
| - market garden produce | 740 | 370 | 185 | 125 |
| - corn, sorghum | 420 | 210 | 100 | 75 |
| - rice (total irrigation plus make up) | 495 | 250 | 125 | 85 |

25 KW Solar Power Unit
25 KW turbine - Flat Collectors 1,500 $m^2$ (0.37 acres)

| Suction Head (meters) | 5 | 10 | 20 | 30 |
|---|---|---|---|---|
| Delivery (cubic meters per day) | 11,000 | 5,200 | 2,600 | 1,800 |
| Intensive Crop Farming (acres) | | | | |
| Irrigation potential in continuous crop growing in cumulative areas: | | | | |
| - market garden produce | 1,480 | 740 | 370 | 250 |
| - corn, sorghum | 840 | 420 | 200 | 150 |
| - rice (total irrigation plus make up) | 1,000 | 500 | 250 | 170 |

50 KW Solar Power Unit
50 KW turbine - Flat collectors 3,000 $m^2$ (0.75 acres)

Figure IV: Solar Power Units Using Thermodynamic Processes Flat Plate Collectors (3)

(c) Electric pumps

Another solution for solar water pumping is to use standard electric pumps supplied by solar electric generation, as described below.

### Solar Electricity

Solar electricity can be produced by the photovoltaic or by thermodynamic processes.

(a) Photovoltaic process

The solar cells which directly convert the photons of solar energy into D.C. current, have been used in space technology for twenty years where they have acquired a high degree of reliability. But their high cost was prohibitive for most earth applications, although they decreased from $500/watt initially, to $100/watt. Further reductions in costs to $20 and $15/peak watt have made it possible to use them for practical terrestrial applications such as providing small scale power supplies to remote and isolated telecommunications units.

The present objective of the U.S. Energy Research and Development Administration (ERDA) is to reduce the costs of solar cells from $20/watt to a few dollars per watt, and even $0.5/watt by 1985. When this objective is attained, through mass production, solar cells will represent a competitive source of energy, particularly well adapted to the needs of desert development. They could power electric pumps to supply water in arid areas. They could provide a local source, which will not necessitate large central power stations and costly electric distribution grids in deserts where the population is scarce and widely scattered. In addition, they do not consume fuel and necessitate little maintenance.

(i) Educational Television (ETV)

A study made by the World Bank Office for Science and Technology (8) shows that solar cells, in many cases, can be already competitive to provide power for operating educational television (ETV) receivers and auio-visual equipment in rural schools and community centres, or for refrigerators in rural health dispensaries, or small electric appliances in remote tourist hotels.

The study shows that solar cell systems could be competitive in costs per kilowatt-hour for operating ETV receivers when compared to other commercial alternatives, and that, sometimes, they can also compete with gasoline generator sets, depending on the price of gasoline and the difficulty of generator maintenance.

(ii) Pumping Water

Many studies have been made of the use of solar cells to power electric pumps and draw underground water. For instance, an electric pump with a power of 400 Watts could draw 2,700 litres/hour from a well 20 metres deep. With 4 hours of operation per day of sunshine, one could obtain a flow of 10,000 litres per day and a power of 1.5 Kilowatt-hours. The solar cell array required to power this electric pump, located in a country at 40° latitude, receiving 700 kilojoules per year, would have a surface of 5 $m^2$ (9).

An experimental pumping station, composed of a standard electric pump powered by solar cells, has been installed in a farm in the south of Corsica. It provides data on the present state of the art and on the economics of a practical operation (10).

Electric pumps, supplied by solar cells, can range from 10 $m^3$/day for a depth of well of 20 metres, for example, to 30 $m^3$/day for a depth of 40 m (11).

(iii) <u>Conclusions</u>

(a) Recent developments in the United States (12) show that the price of photovoltaic cells could be accelerated by introducing mass production methods in the manufacturing of silicon solar cells, so that the price could reach $4/w in 1980 and $2/w in 1981. This is due to large procurements by the Federal Energy Administration (FEA), anticipated to reach 150 megawatts peak (MWp) of installed photovoltaic arrays during a period of 5 years. These prices would bring the solar cells into close competition with the traditional electric power systems within a few years, and would therefore place solar cells among the key candidates for electric supply particularly in desert development.

(b) In the meanwhile, it is proposed to experiment with pilot solar power systems operating in remote areas or villages located in deserts. An array of solar cells, equipped with battery storage and voltage regulator (figure V) would provide power to packages of equipment designed to meet some of the needs of the desert dwellers (13). For instance:

- Package I would include: a refrigerator, some lighting, an immersion heater, and a food grinder (Fig. VI).
- Package II would include the same equipments as above and in addition, a water pump, communication equipment (television or radio), and power tools.

Figure V:

SOLAR POWER SYSTEMS

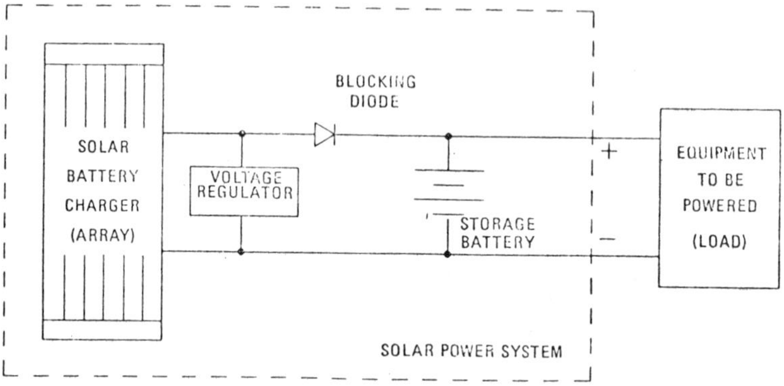


Spectrolab Solar Power Systems Include:

- ☐ Solar Power System Engineering (free-of-charge)
- ☐ Solar Battery Charger (array)
- ☐ Voltage Regulator
- ☐ Anti-Discharge "Blocking Diode
- ☐ Storage Battery

Package I (13) (see figure VII)

| | | |
|---|---|---|
| (a) | 12 volts DC supply: | - Refrigerator 6 cu ft. 6 Amp<br>- lighting: 2 tubes of 15 watts |
| (b) | 115 volts DC: | - Immersion heater: 600 watts |
| (c) | 120 v A.C. | - Food grinder - 480 watts |

Package II (13) (see figure VIII)

Same as package I plus:

| | | |
|---|---|---|
| (a) | 12 volts DC supply: | Communication equipment: TV receiver (20 w), or wide band radio receiver (10 w), or radio telephone or tape recorder |
| (b) | 36 volts DC supply: | Water pump (1/3 HP - 1800 RPM) (water tank of 18000 gallons) (see figure IV and Table 5). |
| (c) | 120 volts AC | Power tool 600 watts; saw, drill or tool grinder |

- - - -

The electric supply of these packages requires the following solar power systems (figure I):

| | | |
|---|---|---|
| (a) | DC supply: | solar cells arrays of 12 v, 36 v with voltage regulator, batteries and blocking diodes |
| (b) | AC supply: | 120 volts solar cells arrays (with voltage regulators and batteries) supply an inverter 120 v DC/AC - 500 VA (see figures I, II, III, IV, V and VI, attached). |

Figure VI.

The pilot would include, in addition to the supply of hardware, the services of an engineer for the training of the local personnel, the initial installation and the follow-up required.

(b) Thermodynamic processes

(i) Principle

The generation of electricity by thermodynamic processes is the subject of intensive research and development activities in several developed countries such as the United States, the USSR, Europe, Japan, Australia, and in some developing countries such as Senegal, Niger. The heat gathered from the sun in the solar collectors is transferred, in a heat exchanger, to steam which activates a turbine and an alternator. The solar collectors may be of the "flat plate" type, with a temperature limited to 70°C, or "flat plate with trapping", to increase the operating temperature. Other types of collectors, aiming at increasing the temperature by concentration up to 200°C, are: parabolic troughs, cylinders, paraboloid dishes, or flat mirrors tracking the sun and concentrating the reflected rays on a central point located in the top of a tower.

(ii) Central tower solar power plant

At present, central tower solar power plants are being developed by the U.S. ERDA for a power of 10 megawatts (5). It is anticipated that the prototype would be operational in 1980, and that solar power plants of 100 MW might be available commercially by 1983.

For the Sahelian region, for example, a proposal has been submitted for a central solar power plant of 5 megawatt (16). It would include 1320 heliostats composed of flat plate mirrors of 6 m diameter in a field of 370 m x 450 m (figures XI, XII, XIII).

(iii) The need for smaller units

For deserts it would be useful to develop solar electric generators utilizing the same thermodynamic principle but applied to smaller sizes, such as : 10 KW, 20 KW, 50 KW and 100 KW.

Several prototypes are now being experimented with in Europe, in the United States and in Niamey (Niger) (17).

## CONCLUSIONS

### 1. Wind Energy

It is suggested to develop small size windmills designed for operation in the deserts, with little or minimum maintenance requirements, and for a dual purpose:

(a) generating electricity: 1 KW, 5 KW, 25 KW, 50 KW
(b) pumping underground water: 1 to 5 HP.

Wind measurements would have to be undertaken systematically, during a five-year period, by Governments, with possible assistance by the World Meteorological Organization (WMO).

### 2. Solar Energy

Projects for desert development should include the identification of the solar equipment needed to pump underground water or produce electricity. The characteristics of underground water reserves underneath the deserts, particularly under the Sahara, are now becoming better known. A comprehensive survey of the solar equipment needed in the desert areas should be undertaken accordingly. It is possible already to classify the needs in three categories:

(a) for families and isolated farms, it would be useful to develop solar pumps and electric generators of 0.5 and 1 KW;
(b) for villages, small scale solar pumps and solar electric generators of 5 and 10 KW could be used for small scale irrigation projects, for family agriculture, cattle breeding and for the needs of the inhabitants;
(c) for larger agricultural projects, solar pumps of 25 HP, 50 HP and 75 HP would be particularly useful for irrigation.

ELECTRICAL SCHEMATIC

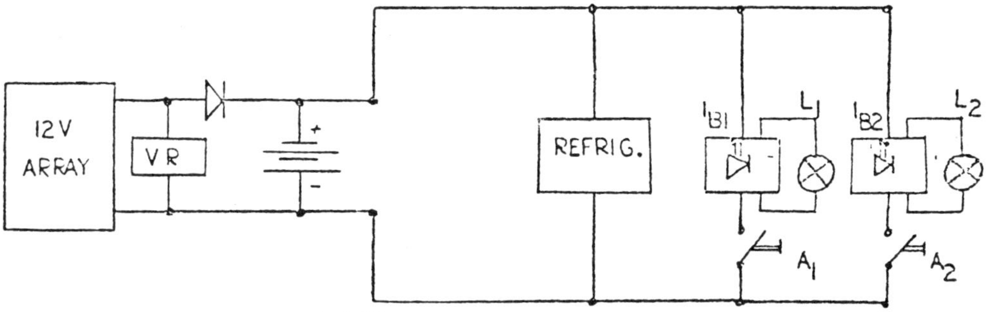


$A_{1,2}$ = LAMP SWITCH

$I_{B1,B2}$ = LAMP INVERTER BALLASTS, GE 6G 5026

$L_1, L_2$ = LAMPS T15T12

VR = VOLTAGE REGULATOR

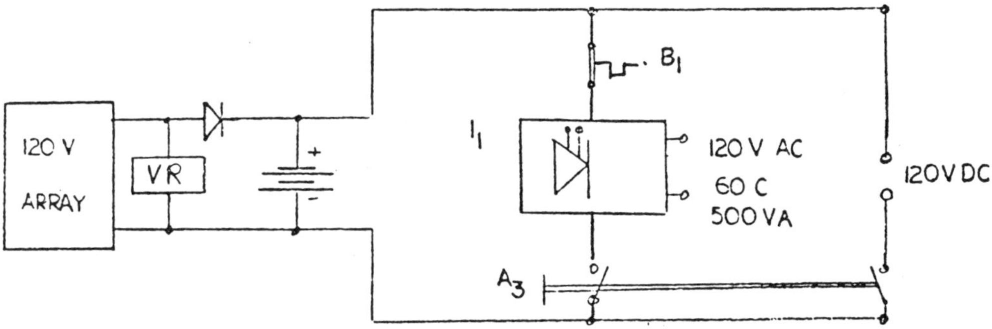


$A_3$ = SWITCH Choice of 120V AC, 120 VDC

$B_1$ = OVERLOAD PROTECTION

$I_1$ – INTERTER DC-AC ACDCo #A56-1

Figure VII

ELECTRICAL SCHEMATIC

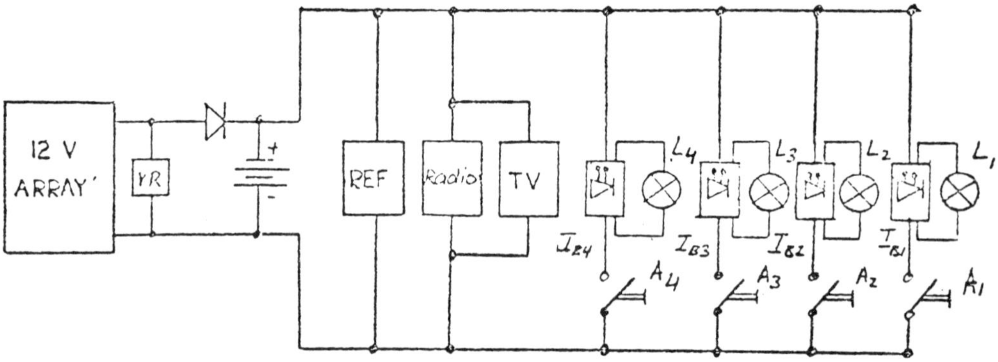


$A_{1,2,3,4}$ = LAMP SWITCHES

$I_{B_{1,2,3,4}}$ = LAMP INVERTER BALLASTS GE 6G 5026,7

$L_{1,2,3,4}$ = LAMPS F15T 12 F30T12

VR = VOLTAGE REGULATOR

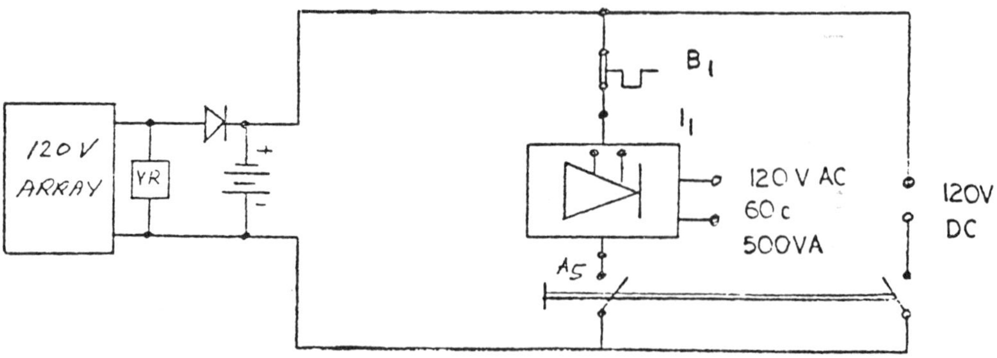


$A_5$ = SWITCH Choice of 120 AC, 120 DC

$B_1$ = OVERLOAD PROTECTION

$I_1$ = INVERTER DC-AC

ACDCo #A56-1

Figure VIII

PUMPING SYSTEM SCHEMATIC AND CAPACITY VERSUS DEPTH

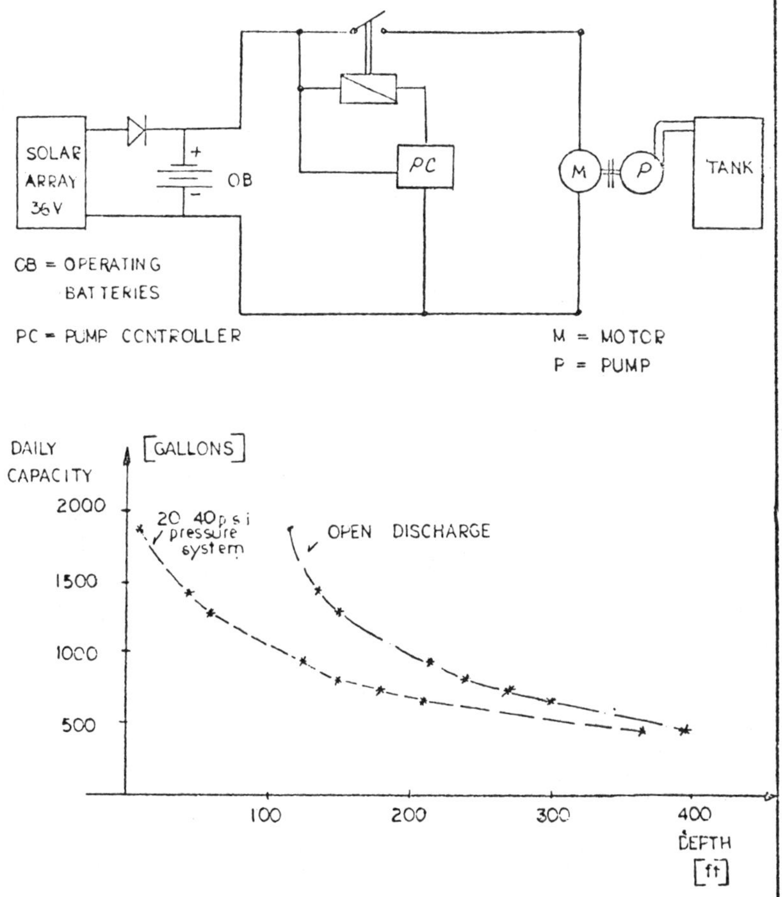


Figure IX

Figure X

Pumping System

***SPECTROLAB SOLAR POWER SUPPLY DESIGN ANALYSIS*** 4/21/76

U N
PUMPING SYSTEM
SENEGAL
14.72 DEGREES N
15.0

***SOLAR RADIATION DATA***

| JAN | FEB | MAR | APR | MAY | JUN | JUL | AUG | SEP | OCT | NOV | DEC |
|---|---|---|---|---|---|---|---|---|---|---|---|
| 395 | 461 | 539 | 567 | 556 | 539 | 456 | 382 | 415 | 417 | 401 | 364 |
| 450 | 503 | 556 | 553 | 520 | 494 | 423 | 367 | 417 | 443 | 450 | 420 |

***SYSTEM VOLTAGE AND CURRENT DEMAND***
FLOAT VOLTAGE: 40.5 MINIMUM BATT TEMP 15 DEG F.

| DE | AMPS | MINUTES PER HOUR | HOURS PER DAY | DAYS PER WEEK | WEEKS PER YEAR | AMP-HRS PER DAY |
|---|---|---|---|---|---|---|
| 1 | 8.33 | 60.000 | 4.00 | 7 | 52.00 | 33.332 |
| | | | | AVERAGE AMP-HRS PER DAY | | 33.332 |

***DESIGN ANALYSIS***

| | TANK GAL | STD BATT (A-H) | STD MODULE VOLTS- | CURRENT | MODULES SERIES | MODULES PARALLEL | TOTAL MODULES |
|---|---|---|---|---|---|---|---|
| 8.50 | 18000. | 85. | 12.0 | .50 | 3 | 17 | 51 |

| MONTH | START STORAGE GAL | MONTHLY LOAD GAL | WATER PUMPED GAL | STORAGE LOSS GAL | END STORAGE GAL | AMP-HOURS GENERATED (A-H) |
|---|---|---|---|---|---|---|
| SEP | 18000. | 56880. | 57333. | 453. | 18000. | 929. |
| OCT | 18000. | 56880. | 60853. | 3973. | 18000. | 986. |
| NOV | 18000. | 56880. | 61792. | 4912. | 18000. | 1001. |
| DEC | 18000. | 56880. | 57682. | 802. | 18000. | .935. |
| JAN | 18000. | 56880. | 61870. | 4990. | 18000. | 1003. |
| FEB | 18000. | 56880. | 69050. | 12170. | 18000. | 1119. |
| MAR | 18000. | 56880. | 76438. | 19558. | 18000. | 1239. |
| APR | 18000. | 56880. | 75951. | 19071. | 18000. | 1231. |
| MAY | 18000. | 56880. | 71409. | 14529. | 18000. | 1157. |
| JUN | 18000. | 56880. | 67812. | 10932. | 18000. | 1099. |
| JUL | 18000. | 56880. | 58077. | 1197. | 18000. | 941. |
| AUG | 18000. | 56880. | 50486. | 0. | 11606. | 818. |
| SEP | 11606. | 56880. | 57333. | 0. | 12059. | 929. |
| OCT | 12059. | 56880. | 60853. | 0. | 16032. | 986. |
| NOV | 16032. | 56880. | 61792. | 2944. | 18000. | 1001. |
| DEC | 18000. | 56880. | 57682. | 802. | 18000. | 935. |
| JAN | 18000. | 56880. | 61870. | 4990. | 18000. | 1003. |
| FEB | 18000. | 56880. | 69050. | 12170. | 18000. | 1119. |
| MAR | 18000. | 56880. | 76438. | 19558. | 18000. | 1239. |
| APR | 18000. | 56880. | 75951. | 19071. | 18000. | 1231. |
| MAY | 18000. | 56880. | 71409. | 14529. | 18000. | 1157. |
| JUN | 18000. | 56880. | 67812. | 10932. | 18000. | 1099. |
| JUL | 18000. | 56880. | 58077. | 1197. | 18000. | 941. |
| AUG | 18000. | 56880. | 50486. | 0. | 11606. | 818. |

Figure XI: **SOLAR THERMAL ELECTRIC SYSTEM**

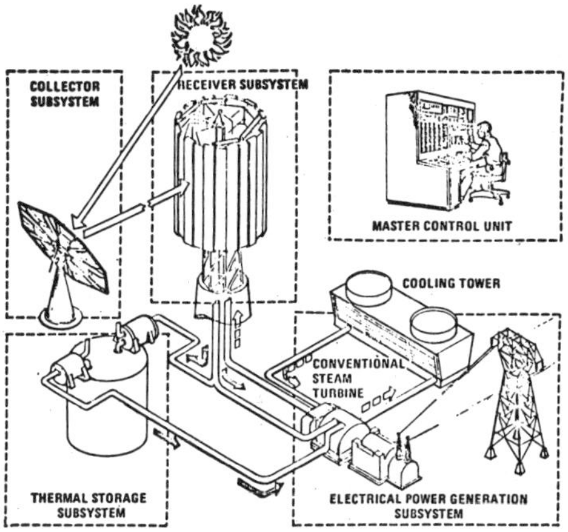


Figure XII: **PLANT SYSTEM SCHEMATIC**
NORMAL OPERATION

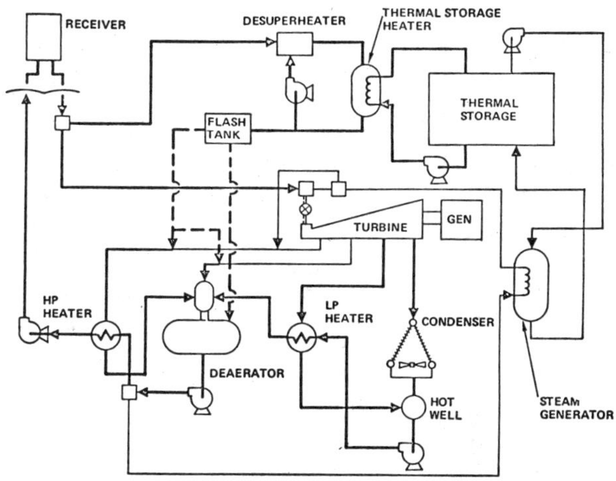

| | |
|---|---|
| FIELD DIMENSIONS | 370 m (N-S) x 450 m (E-W) |
| NUMBER OF HELIOSTATS | 1320 |
| HELIOSTAT SIZE | 6.1 m |
| TOWER HEIGHT (TOP OF RECEIVER) | 87.5 m |
| RECEIVER TEMPERATURE (PEAK) | 485°C |
| THERMAL STORAGE TEMPERATURE | 302°C |
| TURBINE INLET TEMPERATURE | 482°C |
| TURBINE THROTTLE PRESSURE | 5.84 $MN/m^2$ |
| NET PLANT EFFICIENCY | |
| DIRECT | 27.8 |
| FROM STORAGE | 24.4 |

Figure XIII: System Characteristics

Solar energy may also be of significant importance for desert development in the fields of water desalination, food conservation, (crop drying or refrigeration) and for the cooling and air-conditioning of buildings.

### 3. The United Nations Conference on Desertification

The Secretariat of the United Nations Conference on Desertification has prepared a draft "Plan of Action to Combat Desertification" (UNEP/GC/95/Add.1) which will be submitted for consideration by Governments at the forthcoming Conference (Nairobi, 29 August - 9 September 1977). This Plan includes the following recommendations in the chapter on "Alternative Sources of Energy":

"The use of alternative or unconventional energy sources in drylands, usually so favoured with sunlight and wind, should be vigorously investigated as a means of preserving organic materials, of reducing the tedious human labour so often involved with fuel collection, and of providing the peoples of the drylands with simple, inexpensive and convenient devices to serve their daily lives.

It is recommended that research be vigourously pursued on the use of alternative or unconventional energy sources in the drylands that will yield simple, inexpensive and useful devices to serve the needs of dryland peoples.

This recommendation implies national action to:

(a) Establish manufacturing facilities for the local production of whatever simple and efficient devices are produced by investigations into the use of alternative energy sources in the drylands.

(b) Establish distribution facilities to ensure that such devices reach the people who can use them, at a subsidized price when necessary. Distribution should be accompanied by instruction in the use of the devices which could be delivered by such agencies as extension services and fuel distributors.

This recommendation also implies international action to:

(a) Carry out research on the use of alternative energy sources, especially wind and solar energy, in the drylands. Such research should be keyed to the production of simple, inexpensive and efficient devices for the use of dryland peoples (1978-85).

(b) Carry out trials on the use of devices employing alternative energy sources and see that devices which prove efficient are brought to the attention of Governments, specially of those facing desertification problems (1978-85)."

REFERENCES

(1) Ecole Inter-Etats d'Ingenieurs de l'Equipement Rural, B.P. 7023, Ouagadougou (Upper Volta).
(2) "A Jacobs generator - Gone with the wind and back", New York Times, 21 April 1977; Jacobs wind electric company, Minneapolis, Minnesota.
(3) SOFRETES, Societe Francaise d'Etudes Thermiques et d'Energie Solaire, P.O. Box 163, Montargis 45203, France.
(4) "Solar Energy Digest", Vol. 7, No. 6, December 1976, P.O. Box 17776, San Diego, California 92117. "Solar irrigation pumps multiplying in the Southwest.
(5) SANDIA Laboratories, Box 5800, Albuquerque, New Mexico 87115.
(6) SUNPOWER Systems Corporation, 2123 S. Priest Road, Suite 216, Tempe, Arizona 85282.
(7) Battelle Memorial Institute, 505 King Avenue, Columbus, Ohio 43201.
(8) "Solar Cells seen as a power source", article page R-6 in "News of the World Bank Group", March-April 1976.
(9) "Solar Cells", R.T.C., Radio Technique, 140 Avenue Ledru-Rollin, Paris, France.
(10) Pompes Guinard, B.P.I. 92402, Courbevoie, France.
(11) "Pomp units with photopills generators", Brian, Tours, France.
(12) Federal Energy Administration, Task Force on Solar Energy Commercialization, March 11, 1977 "Costs and benefits of the Federal Photovoltaic Utilization Program".
(13) Spectrolab Inc., 12500 Gladstone Avenue, Sylmar, California 91342.
(14) ERDA News, 18 April 1977, page 6.
(15) "Non-conventional sources of energy for developing countries", a report prepared for the UN Office for Science and Technology, dated 31 August 1975, by Jerry Grey, Consultant, 359 West 21st Street, New York, N.Y. 10011.
(16) McDonnell Douglas Astronautics Company, 5301 Bolsa Avenue, Huntington Beach, California 92647.
(17) ONERSOL: Office du Niger pour l'Energie Solaire, Niamey, Niger, Professor Moumouni.

# WIND ENERGY IN DESERT AREAS

Wm. L. Hughes
Clark A. Dunn Professor of Engineering
Director, Engineering Energy Laboratory
Oklahoma State University
Stillwater, Oklahoma 74074 U.S.A.

## INTRODUCTION

The only alternate energy sources that may be of consequence in most desert areas are most likely to be direct solar energy, and wind energy. It is thus important to make a realistic preliminary evaluation of these sources so that their use, if economic, can be part of future planning processes.

At the time of this writing, the author has valid data on only two places in the world which have significant desert areas. They are the Southwestern United States and that part of North Africa that lies within the Egyptian borders.

The Oklahoma State University has been working with the Egyptian government (actually the Ministry of Electricity) for some years in the study of wind energy utilization, and we have a continuing program with them now. Discussions have barely begun with Tunisia tooking towards the possibility of setting up a similar program there. Inquiries have been received from other similar areas.

Before beginning the discussion of desert areas, it is important to make a few things rather explicit. Many groups, agencies, and governments are approaching the use of alternate energy sources with the idea that they are going to provide cheap sources of energy in the future. It must be understood that alternate sources (wind, solar, geothermal, tidal, etc.) are not going to be cheap, they are going to be expensive. In some cases they are going to be very expensive by today's standards. What is worse, they are going to be highly front-end capital intensive. Thus anyone who wishes to take advantage of wind and solar energy on a large scale must have considerable borrowing power with his sources of risk financing. He must also have considerable capability for maintaining mechanical systems which may not be complex by some standards, but which will require continuous maintenance by people of some skill.

Those developing countries which will get the most from alternate sources will be the ones which have at least some small amount of heavy industry now, and which have reasonably developed higher educational systems and international trade systems. Two such countries are Egypt and India.

Those developing countries which will get the next best benefits from alternate sources will be those which can develop products such as agricultural produce, minerals, or other raw materials such as timber. Such economic endeavors can provide leverage on the developed countries and can thus be the mechanism of obtaining the capital goods required to exploit alternate energy sources.

Those developing countries least likely to get reasonable benefits from alternate sources will be those that have neither of the preceding characteristics. It is precisely this third type of L.D.C. which has received the most attention in much literature which has presented alternate energy sources as a sort of "do it yourself with your two bare hands" approach. Of course many things can be done by scavenging old truck transmissions and the like. The author in no way wishes to discourage this approach as long as it is remembered that the "truck transmission" or "old car alternator" or what have you, represents an input of capital (at some time) to a factory of very sophisticated technology. We must not delude ourselves into thinking that an L.D.C. can lift itself out of an energy problem by its bootstraps with such methods.

The real use of alternate energy sources in a particular developing nation will come about when that country maximizes three sources.

1. It must maximize the use that it makes of its own facilities and labor.
2. It must seek to raise, dig, or perhaps hand-make products which can be marketed internationally.
3. It must take advantage of (1) above and the capital gained by (2) above to buy those capital goods which, in turn, make (1) and (2) above most effective.*

The adherance to the above principles is most critical in the L.D.C.'s with large desert areas because the resources of (2) tend to be meager and they must not be used ineffectively.

The whole world, not just the L.D.C.'s, is turning to alternate sources. This is not because the so called alternate sources are more economic, but becuase there is no long-term alternative for either the developed or developing nations. In particular most L.D.C.'s will for the most part, not be able to take advantage of such things as nuclear power, or coal gasification and liquifacation, unless they buy the "finished energy" from some of their more fortunate neighbors.

## ANALYSIS OF DESERT AREAS

In desert areas particularly, it is important to evaluate wind resources both in quantity (annually) and in quality (meaning continuity or lack thereof). Only in this way can the wind machines which are ultimately used be applied such that the rather high cost energy produced will be of maximum benefit to the user.

Fortunately, such an evaluation has been made of the Northern Sahara Desert, between the city of Alexandria and roughly the border of Libya. Similar evalu-

*This discussion in no way refers to one type of government or another. L.D.C.'s must adhere to the fundamentals whether or not they are dealing with socialist or capitalist developed countries and whether or not they prefer one kind of economy or another.

ation has been made of the Red Sea coastal area, across the Red Sea from the Sinai Desert. This evaluation was made as part of a joint project between the Engineering Energy Laboratory of the Oklahoma State University and the Ministry of Electricity of the Arab Republic of Egypt. Some of the results of that study are as follows:

We will be concerned primarily with eight major parameters in wind measurements.

1. Stations at which wind is measured.
2. The years over which wind is measured.
3. The monthly average winds at selected locations.
4. The percentage of the hours of the year which reach or exceed specified velocities in a selected number of places.
5. The mean wind speeds at all hours of the day.
6. The percentage frequency of winds within specified velocity ranges.
7. Average annual duration.
8. "Possible" usable electrical energy in kilowatts/meter$^2$.

A "simplified" map of Egypt is shown in Figure 1. The locations of a number of key stations are shown on this map. A list of stations at which wind measurements are taken and the year at which measurements were started is given in Table I.

We are interested primarily in three general areas:

1. North Coastal areas.
2. Red Sea Costal areas.
3. Desert interior areas.

On the North Coast, we will be concerned primarily with Sallum, Mersa Matruh, and Alexandria. On the Red Sea coast, the best data is from Hurghada. Interior measurements of primary interest are from Cairo and Minya, and Aswan along the Nile River, and at Dakla in the interior western desert.

It can be surmised from Table II (which gives monthly and annual averages) that Sallum, Sidi Barrani, Mersa Matrun, and Dabba, all on the North Coast and all west of Alexandria are promising sites. Alexandria does not look as promising. It is, of course, close to the delta, although not in it strictly speaking. Hurghada on the Red Sea looks most promising, but the interior locations, on both a month-by-month and annual average basis, are not encouraging. Experience indicates that generally, the favorable coastal winds penetrate less than ten or twenty kilometers into the desert.

Figure 2 gives the average monthly wind speeds in several localities. Notice that the coastal cities (with the exception of Alexandria for some reason) look reasonably good but the interior cities such as Cairo, and especially Minya, are not as favorable.

It must be remembered that these are generally airport measurements. Airports are not chosen to be in optimum wind locations so a significant variation is to be expected. Nevertheless, Table I implies with a high certainty level that in Egypt at least, the coastal areas will be much more fruitful sources of wind energy than in any interior area we know about. Further detailed studies will thus be concerned primarily with those areas.

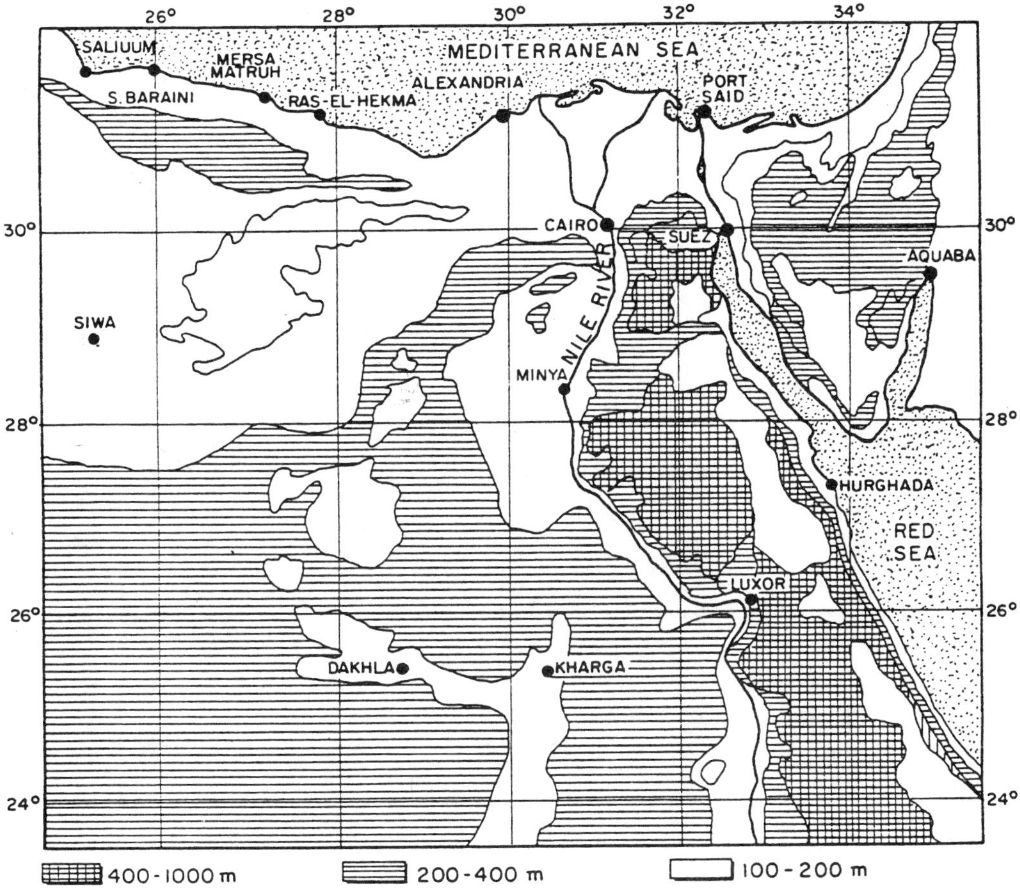


Figure 1: Map showing Orography of the Egyptian Region

Re-examination of Figure 2 shows that the summer dip in wind energy we experience in the United States occurs also in Egypt (with the exception of Hurghada on the Red Sea).

An examination of diurnal wind variations throughout the year (for Alexandria, Hurghada, Sallum, and Mersa Matruh) is extremely interesting. These diurnal variations in January, April, July, and October are shown in Figures 3, 4, 5, and 6 respectively.

With the single exception of Hurghada, the daily peaks come between 09:00 and 15:00 hours throughout the year. Hurghada invariably comes three to six hours earlier. One would conclude that the Red Sea Meteorological system has somewhat different governing forces than that of the rest of Egypt. It will be interesting, and important, to examine the Sinai desert in the future to see which forces govern its wind regimen.

Table I: Meteorological Stations in Egypt Recording Surface Winds

| Station | Lat. °N | Long. °E | Wind Vane at a Height (Meters) Above Ground | S. Wind Records (Year) |
|---|---|---|---|---|
| Salum | 31/32 | 25/11 | 14/0 | 1965 |
| Sidi-Barrani | 31/38 | 25/58 | 14.0 | 1955 |
| Mersa-Matruh | 31/20 | 27/13 | 15.0 | 1952 |
| Dabea | 31/16 | 28/28 | 13.0 | 1964 |
| Alexandria | 31/12 | 29/57 | 22.0 | 1952 |
| Rosetta | 31/24 | 30/24 | 14.0 | 1964 |
| Damietta | 31/25 | 31/49 | 15.0 | 1966 |
| Port Said | 31/17 | 32/14 | 17.0 | 1952 |
| Damanhur | 31/02 | 30/28 | 14.0 | 1965 |
| Tahrir | 30/39 | 30/42 | 15.0 | 1963 |
| Tanta | 30/49 | 30/56 | 14.0 | 1961 |
| Cairo | 30/08 | 31/34 | 18.0 | 1952 |
| Abbasyla | 30/05 | 31/37 | 17.0 | 1958 |
| Helwan | 29/52 | 31/20 | 20.0 | 1961 |
| Fayoum | 29/18 | 30/51 | 14.0 | 1962 |
| Minya | 28/05 | 30/44 | 10.0 | 1952 |
| Asyout | 27/11 | 31/06 | 20.0 | 1962 |
| Luxor | 25/40 | 32/42 | 15.0 | 1952 |
| Aswan | 23/58 | 32/47 | 15.0 | 1963 |
| Siwa | 29/12 | 25/29 | 15.0 | 1955 |
| Dakhla | 25/59 | 29/00 | 15.0 | 1958 |
| Kharga | 25/27 | 30/32 | 14.0 | 1966 |
| Hurghada | 27/17 | 33/46 | 15.0 | 1952 |
| Quseir | 26/08 | 34/18 | 15.0 | 1964 |

Finally, it is interesting to examine a proportional power duration curve for a few locations. This is given in Figure 7. The clear superiority of the coastal areas over the interior is illustrated, although again Hurghada appears as almost an anomaly.

The primary interest of this paper is clearly concentrated on the Middle East, but it is interesting to make some quick comparisons with the situation in the U.S.

An often seen, and often criticized, map of wind energy density is seen in Figure 8. This map gives a rough estimate of annual wind energy densities over the Continental United States. It can be clearly seen that the high energy density areas are the western high plains and the coastal areas, neither of which can be classified as truly desert areas. Our truly desert areas (from Southeastern California to New Mexico) are seen as relatively low wind energy areas on an annual basis.

The author does not have enough data to assure the following statement is a good generalization everywhere, but he will make it if only to create discussions. It seems to him that good wind areas for energy extraction will be some seacoasts, the windward sides of mountain ranges, and the high elevation plains. The fact that an area is (or is not) a desert will have little to do with whether wind energy will have high or low utilization value.

| | Jan. | Feb. | Mar. | Apr. | May | June | July | Aug. | Sept. | Oct. | Nov. | Dec. | Annual Average |
|---|---|---|---|---|---|---|---|---|---|---|---|---|---|
| Salum | 12.0 | 9.5 | 10.5 | 8.4 | 7.1 | 8.0 | 9.8 | 8.5 | 7.0 | 7.2 | 8.7 | 11.4 | 9.0 |
| Sidi-Barrani | 11.0 | 10.8 | 11.5 | 10.6 | 9.1 | 8.6 | 9.9 | 8.5 | 7.2 | 7.8 | 8.8 | 11.0 | 9.6 |
| Mersa-Matruh | 12.2 | 11.9 | 12.3 | 11.0 | 9.6 | 9.9 | 10.2 | 9.0 | 8.5 | 8.4 | 9.2 | 11.8 | 11.2 |
| Dabea | 11.0 | 11.2 | 12.4 | 11.3 | 9.9 | 11.6 | 12.1 | 10.6 | 9.0 | 7.8 | 8.4 | 10.6 | 10.5 |
| Alexandria | 8.7 | 8.5 | 9.1 | 8.4 | 7.8 | 7.9 | 8.6 | 7.8 | 7.0 | 6.2 | 6.5 | 8.0 | 7.9 |
| Rosetta | 5.8 | 5.3 | 6.4 | 5.6 | 4.9 | 5.3 | 5.9 | 4.6 | 4.1 | 3.9 | 4.2 | 5.0 | 5.1 |
| Damietta | 6.2 | 6.1 | 8.4 | 7.5 | 6.8 | 6.4 | 5.6 | 4.9 | 4.1 | 5.0 | 4.6 | 5.4 | 5.9 |
| Port Said | 9.5 | 10.3 | 11.6 | 10.7 | 9.6 | 9.0 | 8.6 | 7.6 | 7.5 | 8.0 | 8.5 | 8.4 | 9.1 |
| Damanhour | 4.7 | 4.1 | 5.4 | 5.1 | 5.0 | 4.7 | 4.0 | 3.7 | 3.5 | 3.7 | 4.0 | 4.5 | 4.4 |
| Tahrir | 6.0 | 6.0 | 7.1 | 6.3 | 5.9 | 5.5 | 5.0 | 4.3 | 4.0 | 3.8 | 4.1 | 5.0 | 5.3 |
| Tanta | 3.8 | 4.1 | 3.0 | 4.8 | 4.6 | 4.4 | 3.2 | 3.2 | 3.2 | 3.2 | 3.0 | 3.6 | 3.8 |
| Cairo | 8.0 | 8.1 | 8.7 | 8.7 | 8.8 | 8.1 | 6.7 | 6.4 | 6.6 | 7.1 | 6.5 | 7.6 | 7.6 |
| Abbasiya | 4.0 | 4.3 | 4.9 | 4.9 | 4.5 | 4.1 | 3.5 | 3.5 | 4.0 | 3.9 | 3.5 | 3.7 | 4.1 |
| Helwan | 5.6 | 7.1 | 8.6 | 9.5 | 9.3 | 9.8 | 8.5 | 8.0 | 8.5 | 8.5 | 7.9 | 6.2 | 8.1 |
| Fayoum | 3.5 | 4.0 | 5.0 | 4.7 | 5.4 | 5.5 | 5.1 | 4.9 | 5.1 | 4.3 | 4.4 | 3.6 | 4.6 |
| Minya | 4.8 | 5.8 | 6.5 | 7.6 | 8.4 | 9.0 | 7.7 | 6.4 | 7.2 | 6.2 | 5.6 | 4.4 | 6.6 |
| Asyout | 6.6 | 7.1 | 8.2 | 8.0 | 7.9 | 7.5 | 7.0 | 6.4 | 8.2 | 7.2 | 6.6 | 6.0 | 7.2 |
| Luxar | 3.5 | 4.0 | 4.8 | 5.1 | 4.6 | 4.3 | 4.0 | 3.7 | 3.3 | 3.0 | 3.3 | 3.5 | 3.9 |
| Aswan | 7.5 | 8.0 | 8.8 | 9.2 | 8.6 | 9.0 | 8.1 | 8.0 | 8.3 | 8.1 | 8.0 | 7.5 | 8.3 |
| Siwa | 5.8 | 6.5 | 7.8 | 7.8 | 7.0 | 6.5 | 6.0 | 5.4 | 4.6 | 4.2 | 4.0 | 5.2 | 5.9 |
| Dakhla | 3.6 | 4.4 | 4.9 | 5.0 | 5.9 | 6.3 | 5.4 | 4.8 | 5.4 | 4.8 | 4.1 | 3.2 | 4.8 |
| Rharga | 6.2 | 7.7 | 8.4 | 8.6 | 9.6 | 11.5 | 9.1 | 8.3 | 9.4 | 9.2 | 7.4 | 6.0 | 8.4 |
| Hurghada | 11.4 | 12.5 | 12.3 | 12.7 | 13.6 | 14.5 | 13.3 | 13.2 | 13.6 | 11.3 | 9.9 | 10.5 | 12.4 |
| Quseir | 9.4 | 9.2 | 9.4 | 8.8 | 9.6 | 9.8 | 7.7 | 7.6 | 9.0 | 8.7 | 9.4 | 9.4 | 9.0 |

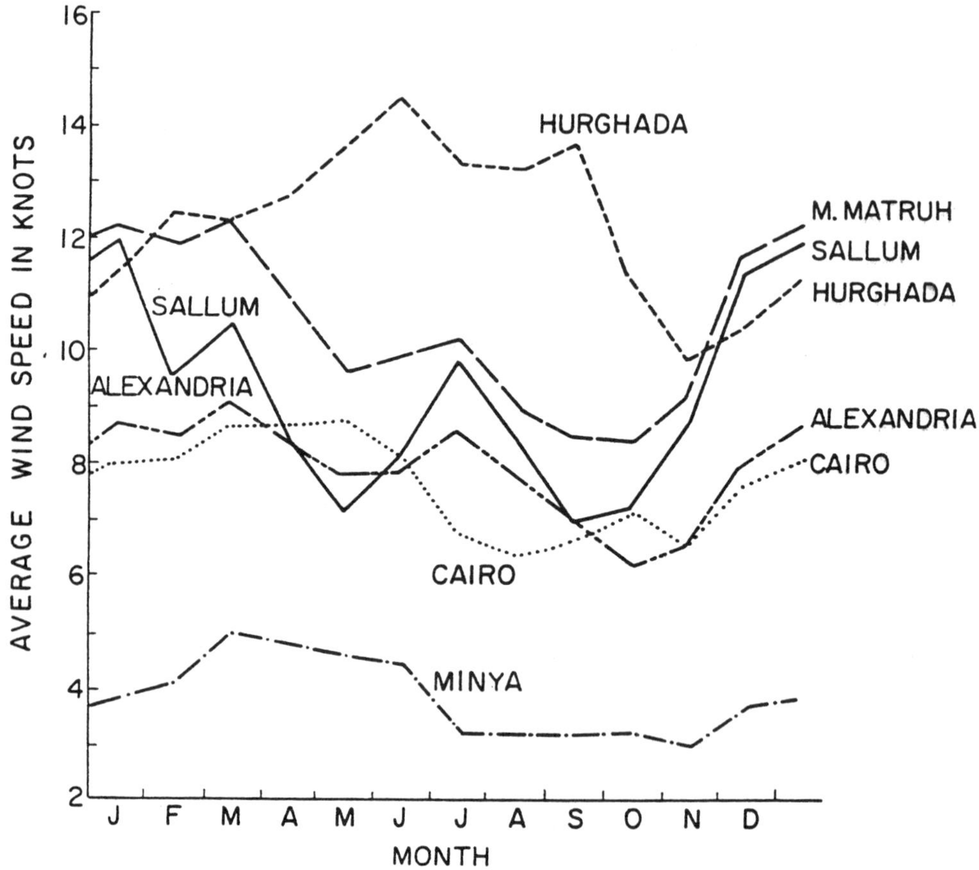


Figure 2: Average Monthly Wind Speeds in Knots

This will be disputed by many (especially those who have not made long term measurements) that they know of places where the wind blows all the time, or that the average wind is twenty-five knots per hour, or some such similar statement. The author knows of no continuous (meaning over many years) measurements indicating where such conditions exist. If they do exist, they are extremely rare. In all probability, if wind is to be a significant energy contributor, to the future, it must be on the basis of average winds of nine to thirteen knots per hour with only occasional winds as high as twenty to twenty-five knots.

In preparing this document, the author was asked to comment on operational experience of windpower units in sandstorms, flash floods, and similar occurrences in desert areas.

Some twenty-five years ago, Dr. Ibrahim Sakr, now of the Egyptian Academy of Sciences in Cario, designed a mechanical pumping windmill which was subsequently manufactured inthe military factories of Egypt and dispersed in hundreds of locations along the North Coast. The author has personally traveled along that

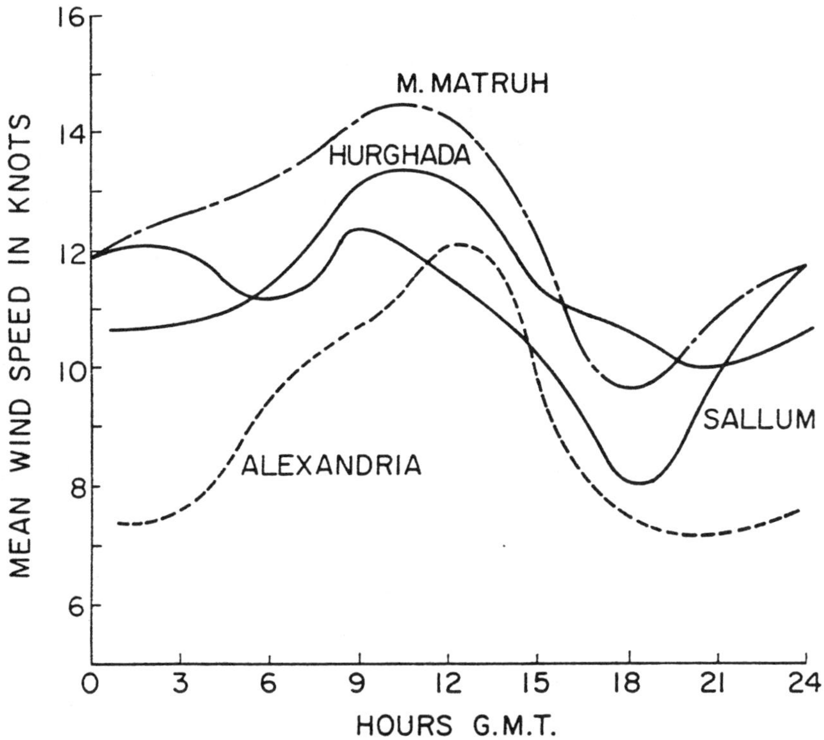


Figure 3: Diurnal Variation of Wind Speed in January

coast (from Alexandria to El Alemain) on a number of occasions and observed Dr. Sakr's windmills. Many of them are still operating and in good repair. In conversations with Dr. Sakr, he stated that the fine dust storms did indeed give severe bearing problems, and that various schemes for shielding bearings were only partially effective. A more severe problem, however, in the authors opinion at least, is the development of schemes to automatically protect the windmill (by turning it out of the wind) when winds of extremely high velocity occur. Such high velocity winds, which can easily damage or destroy a windmill, occur almost everywhere irrespective of whether the average wind velocity is high or low. The author has at this writing, designed and built a number of different wind electric generation and deep well pumping systems and has experienced some field failures. It is his belief that bearings can be made to reasonably withstand the worst sand and grit situations although in the worst cases bearings may have to be replaced every two or three years. The most severe problem, however, is to protect the windmill and associated equipment (which is designed to operate in about a twelve to fourteen knot/hour wind) from an eighty to ninety knot/hour wind. In observations the author has made on Dr. Sakr's windmills on the North African coast, on windmills in the U.S., and on systems of his own design, this is a uniform problem. The problem is not just with the wind turbine. Actually, damage to the mechanical and/or electrical systems associated with the windmill is apt to be equally serious.

The author has had no experience with problems of flash flooding as far as windmills are concerned. However, in any local situation, the areas vulnerable to flash flooding are generally known. Presumably those people erecting the windmill can be expected to exercise some common sense in such matters.

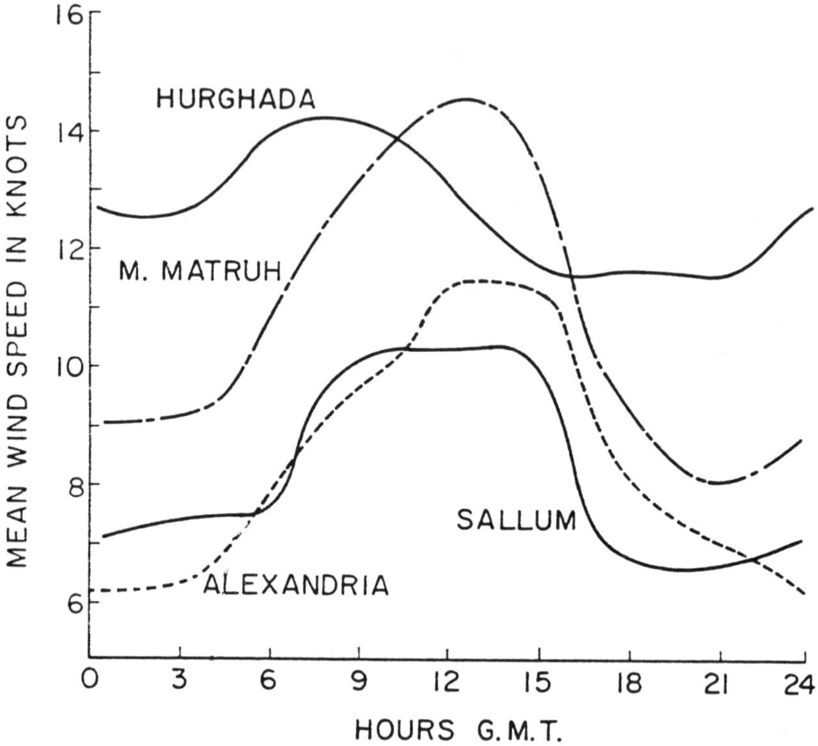


Figure 4: Diurnal Variation of Wind Speed in April

Finally, the author must acknowledge that all of the detailed information about wind in Egypt and the North African coast was prepared by Engineers of the Ministry of Electricity of the Egyptian Arab Republic.

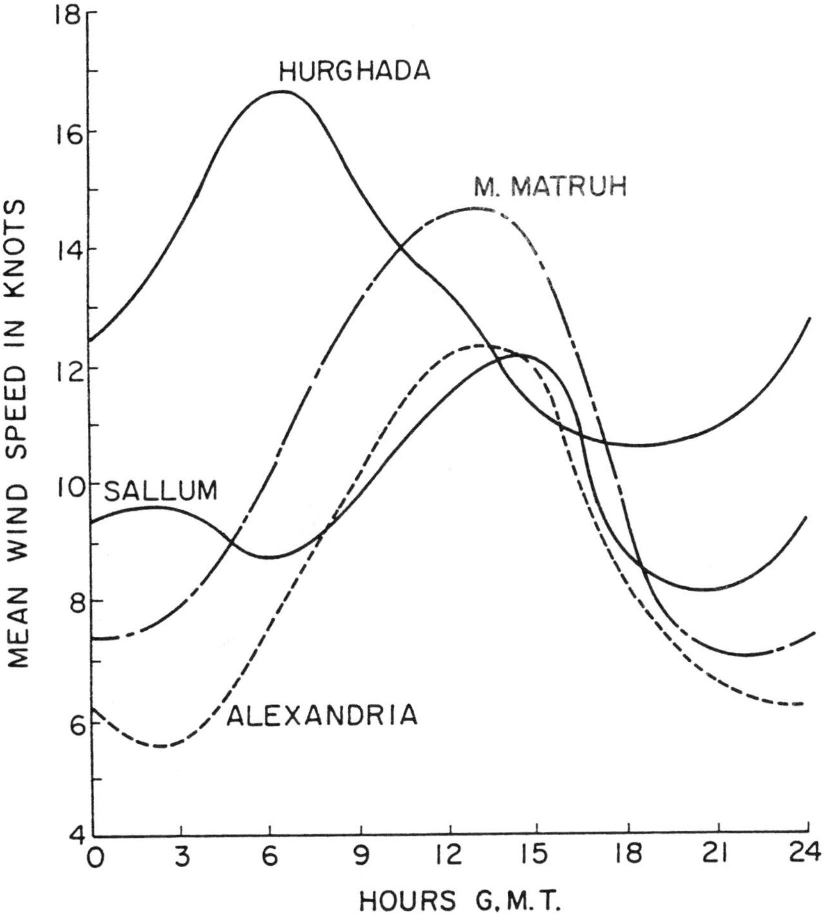


Figure 5: Diurnal Variation of Wind in July

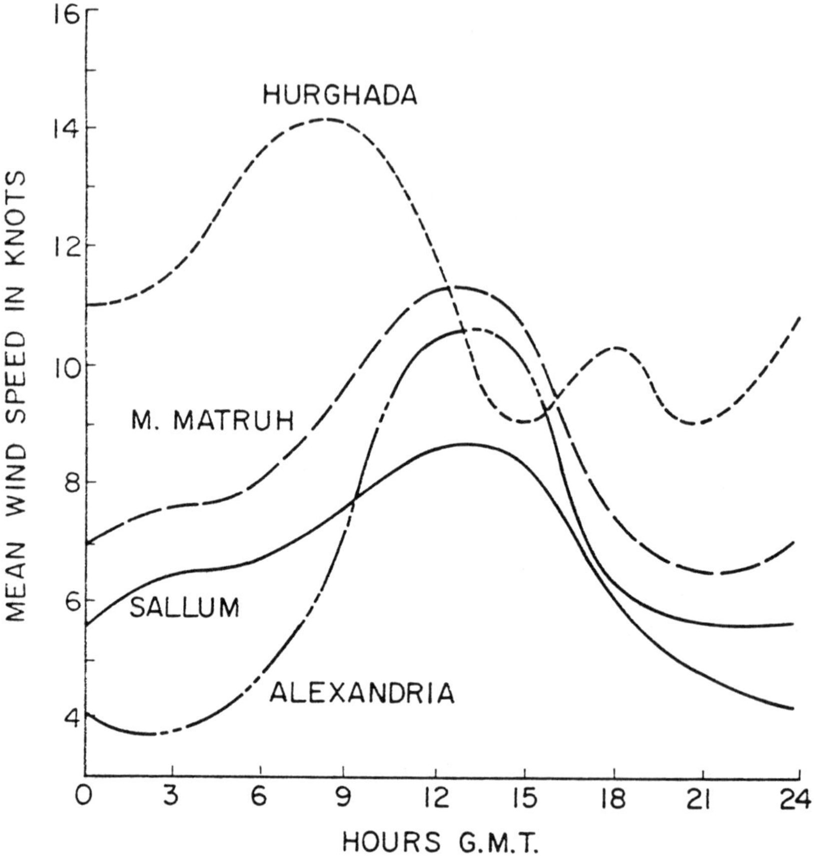


Figure 6: Diurnal Variation of Wind Speed in October.

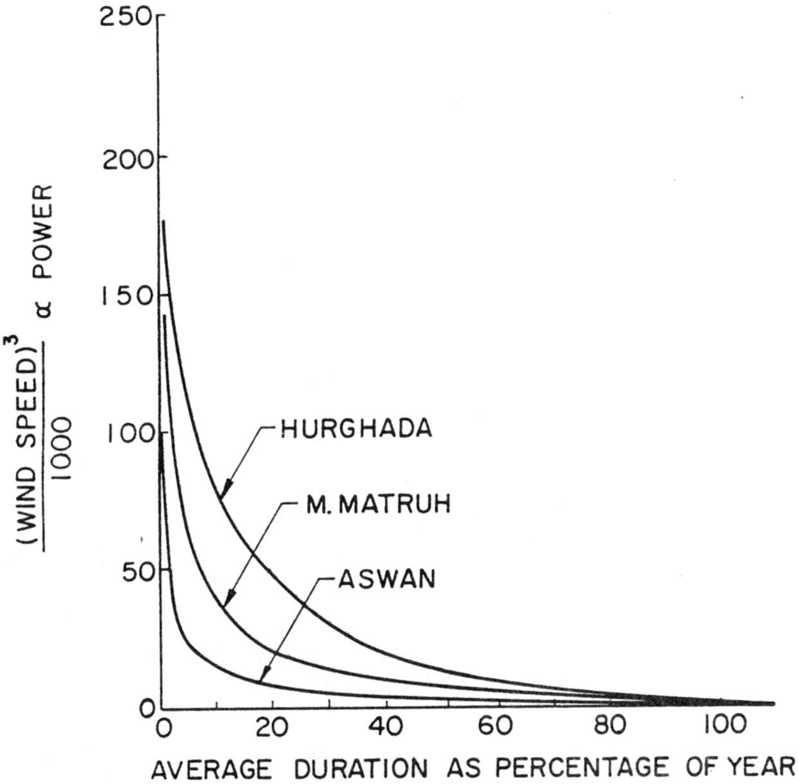


Figure 7: Power Duration Curve

# GEOTHERMAL RESOURCES IN ARID AND DESERT AREAS

By Joseph Barnea and James H. McNitt

## The Six Types of Geothermal Resources

We know today of six types of geothermal resources, namely, dry steam fields, hot water fields, low enthalpy fields, geopressure zones, dry hot rock areas, and magma energy. Dry steam fields are fields in which steam only is obtained through drill holes. The steam can be directly transported to a nearby power station and can be used to generate electricity. Such dry steam fields are found in Northern California in the Geyser field and in the Lardarello field in Italy, the first dry steam field to be developed. Dry steam fields are comparatively rare. Much more frequent are hot water fields in which water at high temperature, usually between 200 to 350° centigrade, is found at moderate depth. A drill hole into a hot water field produces a mixture of hot water and steam which usually is separated at the well head into steam and hot water. The steam can be used for power generation or other purposes and the hot water can be used for non-electric applications. Hot water fields can produce under favorable conditions 200 to 300 tons of water per hour per well and, thus, have a substantial water production capacity. Low enthalpy fields are fields which can produce water at moderate temperatures of between 40° centigrade to 150° centigrade. They are widespread and are used for their heat content but could be significant in desert areas, in addition, as a water source. Some of the low enthalpy fields require pumping. Geopressure zones are zones in sedimentary areas, usually quite deep, where water is found under very high pressure so that, in effect, geopressure zones represent three sources of energy: the mechanical energy of the water under pressure which considerably exceeds hydrostatic pressure, the heat of the water and methane in solution in the hot pressurized water. Geopressure zones in arid and desert zones may present large scale resources including very large amounts of water, some of it at high salinity. However, the heat of the water will allow self-desalination or desalination without any outside energy input. A special paper by Dr. Jones will deal with the presently known potential of geopressure zones in arid and desert areas. Dry hot rock areas and magma energy are two types of geothermal energy for which extraction technology has not yet been developed and, therefore, they are not considered further in this paper.

## The Potential Role of Geothermal Resources in Desert Areas

All the geothermal resources we are considering here produce heat and water and, thus, they are more significant in arid and desert zones than in other areas where usually use is made only of the heat content of the geothermal resource and not of the water content. Where the temperature exceeds 180° centigrade, geothermal resources can be used to produce electricity and this is now a standard technique, having been used now for about 50 years. The cost of geothermal power, expressed in dollars per kwh, is low provided the temperature required for conventional geothermal electricity generation is available. Considerable research is being undertaken to produce electricity from binary and other systems which allow the use of lower temperature geothermal water for electricity generation and recently one firm has brought on the market a small unit for the generation of electricity utilizing temperatures as low as 75° centigrade. The smaller and lower temperature systems for electricity generation are considerably higher in capital cost per kw of capacity than more conventional systems but in isolated arid or desert zones they may be useful in the future as a first stage development of reliable electricity generation based on local resources.

It should be noted that all geothermal electricity generating systems use geothermal water to provide their own cooling system; consequently, a geothermal electricity system does not require any additional water supply for cooling.

Geothermal resources lower than 180° centigrade but exceeding 120° centigrade can be used for geothermal air conditioning, commercial freezing and many industrial applications. Geothermal resources between 60° and 120° can be used for many agricultural and industrial applications, and geothermal water of 40° to 60° can be used for winter farming in desert areas. Where the geothermal water is not too saline, sequential use of the resource is possible by using geothermal water, first as a heat source and then as water for agricultural and industrial purposes, including cooling water.

In arid or desert areas geothermal water has an important function as a water source. Wherever the geothermal water is higher in temperature than 110° centigrade, the heat content of the water is high enough to allow self-desalination or desalination without any additional energy input. Two pilot units for geothermal desalination are now operative in desert areas, one in the Imperial Valley of Southern California and the other near the El Tatio field in Northern Chile. It is likely that with the development of geothermal resources in desert areas, geothermal desalination will in the future become a significant utilization of geothermal resources. It should be added that geothermal desalination, requiring no energy input, is probably the lowest-cost type of water desalination.

Geothermal resources can not only make a contribution as both an energy and a water source of human settlement in the desert, but geothermal resources may also make significant contributions to wild life preservation in the desert. Furthermore, certain types of geothermal resources may have medical signficance and may enhance health facilities in desert and arid zones.

One significant limitation of geothermal resources, which must be recognized, is that they are location-bound and, thus, as a rule, geothermal resources have to be developed where they are found.

Some hot water fields may carry significant minerals and gases and, thus, in certain cases the development of geothermal resources may also provide new sources of minerals and gases, such as magnesium, gold, potash, mercury, lead, zinc and lithium and gases, such as $CO^2$, helium and argon.

## Potential Problems of Geothermal Resources in Desert Areas

In certain cases geothermal resources may carry toxic elements, such as arsenic, and toxic gases, such as hydrogen sulphide. However, each geothermal field has its own characteristics and only detailed field measurements will provide the precise knowledge of the resource characteristics. There are established methods to deal with the toxic elements in the water and in the steam and, thus, they provide no insurmountable obstacle to geothermal development.

It should be noted at the outset that there is comparatively little practical experience in the management of geothermal resources in desert or arid areas. The first question which arises in desert and arid zones is whether the geothermal field has a source of water recharge. In the case of the Imperial Valley in California, apparently there is both heat and water recharge. There is also a source of water recharge in the El Tatio field in Northern Chile and one might assume, therefore, that wherever geothermal resources are not too far away from mountain ranges, on which there is significant rainfall or snow cover, or are not too far away from rivers or oceans, water recharge exists.

The largest present geothermal development in a desert area is the Cerro Prieto field in Mexico which has an installed electric generating capacity of 75 MW and will be expanded in the near future to 150 MW. It is a hot water field and has a water recharge.

It appears to us that in planning large-scale geothermal development, the first task is to investigate the amount of possible water recharge. Such investigations are complex and may take a number of years. If a limited development of a geothermal resource is started, the investigation of water recharge may be carried out during the first years of utilization of the geothermal resource with an expansion of the production once the amount of recharge has been established.

If in a given geothermal resource area in the desert, investigation should establish that there is only limited or no water recharge, it may become necessary to use the geothermal water in a closed system so that the hot water or steam after cooling is reinjected in the formation. There is also the possibility of using otherwise unusable water, such as sewage or saline water, for reinjection into geothermal fields. Sewage water if reinjected into a hot water field would become, after underground cooking, safe water and, thus, later through geothermal production available for reuse.

## Conclusions

From the attached appendix, which lists presently known geothermal indications in desert or arid zones, it becomes clear that indications of geothermal resources are found in 80 percent of the desert and arid zone areas. But, by-and-large, with only two or three exceptions, none of the indications have been investigated in a systematic or detailed manner. It is recommended, therefore, that in all the desert and arid zones where geothermal indications are found and where the utilization of geothermal resources may be economically attractive, such investigations should be undertaken as soon as possible.

It should be realized that geothermal exploration in desert areas is much easier than in densely populated or forested jungle or swampy areas and, as a rule, such exploration is cheaper than in most other areas. Geothermal resources in desert and arid areas are attractive and may provide both energy and water at a low cost.

## Annex
### Indications of Geothermal Resources in Desert and Arid Areas (x) (XX)

There is geologic evidence to indicate that 35 countries, located in the desert and arid areas of Africa, the Middle East, Asia and South America, North America and Australia have a geothermal resource potential. One of these countries, Mexico, is already generating 75 megawatts of geothermal power and deep exploratory drilling has confirmed the presence of high temperature geothermal reservoirs in four more: U.S.A., FTAI (French Territory of the Afars and the Issas), Kenya and Chile. Although they have not reached the drilling stage as yet, four additional arid countries have active exploration programmes: Ethiopia, Jordan, Israel and Iran. Many of the remaining 26 countries are aware of their potential, but to our knowledge, they have not yet initiated exploration programmes.

The 35 countries are listed below under their six geographic regions: In the listings for each geographic group, a distinction is made between those countries which appear to have a potential for both high and low enthalpy resources, (21 countries) and those in which it is more probable that only low enthalpy resources will be found (14 countries). The criterion for making this distinction is the presence or absence of Quarternary volcanism. With few exceptions the high enthalpy fields found thus far on a global basis are associated with Quarternary volcanism, whereas low enthalpy fields appear to be generally associated with areas of deep ground water circulation and above average regional heat flow.

I. Africa

a) Greater probability for high enthalpy fields:

| | |
|---|---|
| FTAI (French Territory of the Afars and the Issas) | Morocco |
| Ethiopia | Chad |
| Kenya | Malagasy Rep. |

(x) This list does not include geopressure zones. For these, see the paper by Dr. P. Jones.

(xx) No indications were found in the literature on geothermal resources in the desert areas of USSR and the People's Republic of China.

b) Greater probability for low enthalpy fields:

| | |
|---|---|
| Egypt | Niger |
| Sudan | Somalia |
| Libya | Angola |
| Tunisia | Namibia |
| Algeria | Botswana |

II. Middle East

Most thermal areas in this region are associated with Quarternary volcanism and therefore, all the countries are included in the probable high enthalpy category:

| | |
|---|---|
| Yemen | Jordan |
| Democratic Yemen | Israel |
| Saudi Arabia | Syria |

III. Asia

a) Greater probability for higher enthalpy fields:

Iran

Mongolia

b) Greater probability for low enthalpy fields:

Afghanistan

Pakistan

IV. South America

All countries in this group have a potential for high enthalpy fields:

| | |
|---|---|
| Ecuador | Bolivia |
| Peru | Chile |
| | Argentina |

V. North America (Greater probability for higher enthalpy fields)

Mexico

U.S.A.

VI. Australia (low enthalpy)

# Geopressured Geothermal Ground-Water Resources in Desert Regions

Paul H. Jones

## INTRODUCTION

Desert regions underlain by sedimentary basins ranging in depth from 5 to 15 kilometers or more have ready access through drilled wells to vast amounts of underground water. Unfortunately, most of the ground water is too saline for human consumption or agricultural use. In many places, however, such saline ground waters occur in extensive aquifers at very high reservoir pressures, and are hot enough to be self distilling, at least in part, when discharged at the land surface by wells. Such wells could flow at rates up to 16,000 $m^3$/day; closed-in pressures could reflect depth-pressure gradients of 200 g/$cm^2$/m or more.

Such abnormally high fluid pressures are called "geopressures" (Stuart, 1972), defined as pore fluid pressure that reflects some part, or all of the weight of superincumbent rocks. Normal fluid pressures, called "hydropressures" reflect only the weight of the superincumbent fluid column, amounting to about 107 g/$cm^2$/m (0.465 psi/ft) of depth. This depth-pressure gradient would result from a water column having a dissolved solids content of 80,000 mg/l (Dickinson, 1953).

## PRINCIPLES OF OCCURRENCE

The occurrences of geopressure may be classified as (1) depositional or (2) tectonic (Jones, 1977). The first is common in rapidly-filled young sedimentary basins in which subsidence has kept pace with progradational deltaic deposition, in a marine marginal setting. The second is common in active tectonic belts associated with crustal plate movement or mountain building, in deposits of any age, and at any depth. In the first, reservoir rocks are commonly highly porous (15 to 30 percent by volume), moderately to highly permeable (10 to 1,000 md or more), and areally extensive (10 to 100 $km^2$ or more). In the second, reservoir rocks are generally of metamorphic types, having low porosity (a few percent), low permeability (a few md, or less), and of small areal extent except where they consist of fault-shattered carbonate rocks in fold belts. From the standpoint of supply development, only the depositional geopressure zones may be considered promising.

The water in depositional geopressured zones is confined by an overlying bed of low permeability, which forms the "seal". It may be a transgressive marine shale a few tens of meters in thickness, or more; or it may be an evaporite bed (halite or anhydrite) of large areal extent. In basins of Mesozoic age, such as the South Mississippi Salt Dome basin, geopressured fluids may occupy secondary porosity, and indurated beds of any kind may form the seal (Parker, 1973). Laterally, fluid pressure may be confined by fault offset of aquifers or sand-bed pinch-out in shale; and below, massive shale deposits may grade into dense carbonates or into metamorphic rocks.

## ORIGIN OF GEOPRESSURE

Geologic conditions leading to the origin of geopressure include the following:

1. Very rapid deposition of noncarbonate clastic rock (sand and clay) in which clay predominates,
2. Faulting contemporaneous with deposition (growth faulting) the fault trends cutting off sand-bed aquifers from their landward extensions,
3. Rapidly deepening burial and subsidence, with loading at rates greater than water escape from clay beds can occur with compaction stress,
4. Progressive geothermal heating of the subsiding sedimentary mass, with thermal expansion of confined water and sediment minerals,
5. Clay mineral conversion (thermal diagenesis) releasing bound and intracrystalline water as free pore water, equal to half the volume of clay mineral so converted (Burst, 1969).

Geopressured reservoirs formed by these processes may be classed as "depositional" type; another type of geopressured reservoir is produced by mountain-building tectonics related to crustal plate movement. "Tectonic" reservoirs are characterized by pressures considerably greater than those attributable to overburden load, but their water yield is likely to be much less than from depositional type reservoirs.

## GEOPRESSURE-GEOTEMPERATURE RELATIONSHIPS

Geopressured zones formed by these processes may have very great areal extent (375,000 $km^2$ in the northern Gulf of Mexico basin). The natural "pressure vessel" so formed has marked effect upon geothermal heat flow, mainly as a consequence of the thermal properties of water. The specific heat of water is about 5 times greater than that of the sediment grains it envelops, and its thermal conductivity is only about 1/5 that of the sediment grains; consequently, water has a dominant effect on geothermal heat flow (Jones, 1975). Where water is trapped in the geopressure zone, the deposits in which it occurs form a very efficient thermal insulator, restricting heat flow and producing high-temperature reservoirs. Where water moves freely upward, as in the hydropressure zone above the geopressure zone, upward transfer of geothermal heat is so effective that geothermal gradients are very low, and ground-water temperatures are generally less than 100°C.--except where leakage from the underlying geopressure zone is in progress.

Both "depositional" and "tectonic" geopressured geothermal systems derive their heat by mass transfer; crustal rupture exposes sediments to igneous intrusion and high heat flow from the upper mantle. Both systems are therefore geographically related to continental margins, past or present.

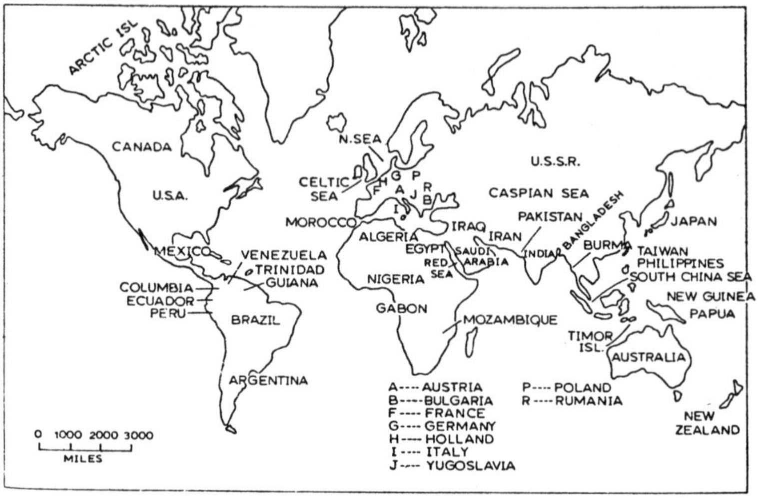


Figure 1. Worldwide occurrence of geopressured zones (from Fertl, 1976).

## WORLDWIDE OCCURRENCE OF GEOPRESSURE ZONES

The worldwide occurrence of geopressure zones is shown on figure 1; each of the countries identified on the map includes deep sedimentary basins or tectonic belts in which geopressured reservoirs are known to occur. The belt of desert countries from Morocco to India follows one of the earth's major tectonic belts (the ancient Tethys seaway), which separated the principal continental plates of the earth for more than 600 million years. Structural deformation, downwarp and uplift, and intense igneous activity have marked this belt for the past 15 million years. High rates of geothermal heat flow are evidenced widely along this belt, and at its intersections with young crustal rift zones.

## CONCLUSIONS

Development of fresh water supplies from superheated, geopressured saline ground waters, as well as by-product heat and kinetic (flow) energy from deep wells, is possible in large parts of the world's desert regions. Knowledge of this resource in desert areas is scattered and incomplete, and no realistic estimates can be made now of the size and importance of fresh water supplies that might be developed. However, the yields of wells and reservoir conditions are known for similar geopressured reservoirs in other parts of the world, and this knowledge suggests that studies of geopressured zones in desert regions is of high priority.

Access to resource information on geopressured zones in desert regions is difficult to obtain, both logistically and politically, and studies of this resource seem most appropriate for United Nations' consideration. Knowledge of this resource could be very important in considering alternate strategies for desert development.

## REFERENCES

Burst, J. F., Diagenesis of Gulf Coast clayey sediments and its possible relation to petroleum migration: Am. Assoc. Petroleum Geologists Bull. v. 53 (1969), 73-93.

Dickinson, George, 1953, Reservoir pressures in Gulf Coast Louisiana: Am. Assoc. Petroleum Geologists Bull. v 37 (1953), 410-432.

Fertl, W. H., Abnormal Formation Pressures: Elsevier Scientific Publishing Co., New York, 1976.

Jones, P. H., Geothermal and hydrocarbon regimes, northern Gulf of Mexico basin, in M. H. Dorfman and R. W. Deller, eds., Proceedings, First Geopressured Geothermal Energy Conference, the University of Texas at Austin (1975), 15-89.

__________ Gas in geopressure zones, in The Future Supply of Nature-Made Petroleum and Gas: Technical reports presented to the Second IIASA Conference on Energy Resources, sponsored by UNITAR and IIASA, July 5-16, 1976, Schloss Laxenburg, Austria; Pergamon Press, Inc., Elmsford, New York, 1977.

Parker, C. A., Geopressures in the deep Smackover of Mississippi: Jour. Petroleum Technology, v. 25 (1973), 971-979.

Stuart, C. A., Geopressures, in Proceedings of the Second Symposium on Abnormal Subsurface Pressure, Louisiana State University, Baton Rouge, La., 1972.

UTILIZATION OF GEOTHERMAL BRACKISH WATER FOR THE PRODUCTION OF OUT-OF-SEASON CROPS IN THE NEGEV DESERT[1]

D. Pasternak, E. Rappaport, M. Twersky and I. Borovic
Research & Development Authority, Ben-Gurion University of the Negev
Y. De Malach
Ramat Negev Regional Council

## ABSTRACT

A technology for the utilization of low temperature (25-50°C) geothermal water for soil and space heating is described. Soil heating is carried out by passing warm water through polyethylene pipes buried in the soil. Greenhouse space heating is achieved by running a film of warm water between the two clear layers which comprise the greenhouse roof. In both cases temperature is regulated by the flow of water. A system for growing out-of-season crops in the field or in greenhouses with the aid of soil and air heating is described. Raising soil temperatures from 14°C to 30°C increased cucumber yields from 6.4 to 15.3 kg $m^{-2}$. Utilization of geothermally heated greenhouses made possible the production of winter cucumbers and melons in the Central Negev.

## INTRODUCTION

Two brackish water aquifers with a temperature of 35-40°C underlie most of the Negev desert. Field experiments which have been conducted since 1970 have demonstrated that this brackish water, which contains approximately 2500 ppm total dissolved solids (TDS), can be successfully used for irrigation of agricultural crops such as cotton, wheat, vegetables etc. (Twersky, et al., 1974; Pasternak, et al., 1973; Pasternak et al., 1975).

The winter season in the Negev is characterized by cloudless days and mild night temperatures in the coastal area and in the inland valleys. As a result of this climate, the Negev is now Israel's major producing area for out-of-season crops a large proportion of which is exported. Utilization of geothermal energy for soil warming of fields or for heating of greenhouse space could further increase this relative advantage of the desert.

---

[1]Presented by J. Schechter

The effect of soil temperatures on plant growth and yield has been reviewed by Richards, et al., (1952); Nielsen and Humphries, (1966) and more recently by Cooper, (1975). Yields can be increased by raising root temperatures up to the optimal temperature for each plant species. For growth of summer crops like beans, tomatoes, maize, etc. optimal root temperatures range from 25-35°C. Rykbost, et al., (1975a, 1975b) showed that soil warming resulted in higher yields under field conditions.

Geothermal water of low temperatures can also be used to regulate the temperature of greenhouse space, on the condition that economically feasible technology for the distribution of heat in the greenhouse can be developed.

## EXPERIMENTAL

### Soil Heating in Fields

A system for soil heating and for crop irrigation with brackish water has been developed over three seasons of experiments. In the first season, the trials were carried out in the field, and brackish (2300 ppm TDS) warm water was used. In the second and third seasons, the experiments were carried out in the grounds of the Research and Development Authority with better quality water (600 ppm TDS) artifically heated to 40°C.

The integrated system comprised four components: a well, soil-heated field space, a storage pond and irrigated fields (Fig. 1). A network of loop-shaped 12 mm plastic tubes was buried in the ground at a depth of 30 cm. The distance between the two arms of each loop was 40 cm and between the centers of two adjacent loops - 180 cm. Warm water from the well was pumped through each tube, exiting into a central collecting pipe, through which it flowed, "cooled", to a storage pond. It could then be utilized again, for conventional irrigation.

Cucumbers (Cucumis sativa) were planted in all three seasons and served as a test crop. The growing system for the out-of-season cucumbers consisted of: a) buried plastic lines for soil warming; b) trickle irrigation lines; c) plastic mulch, and d) polyethylene tunnels. Table 1 demonstrates the importance of mulching in the system.

Table 1. Effect of Polyethylene Mulch on Cucumber Yields in Heated and Non-Heated Plots - Ninety Days After Planting (Average rate of warm water supply - 2.5 liters $m^{-2}$ $hr^{-1}$)

| Treatment | Cumulative yield (kg $m^{-2}$) | |
|---|---|---|
| | Non-Heated | Heated |
| Clear mulch | 1.5 | 3.0 |
| No mulch | 0.5 | 1.8 |

Plastic tunnels were necessary. Cucumbers planted on soil-heated beds with plastic mulch but without tunnels did not reach maturity.

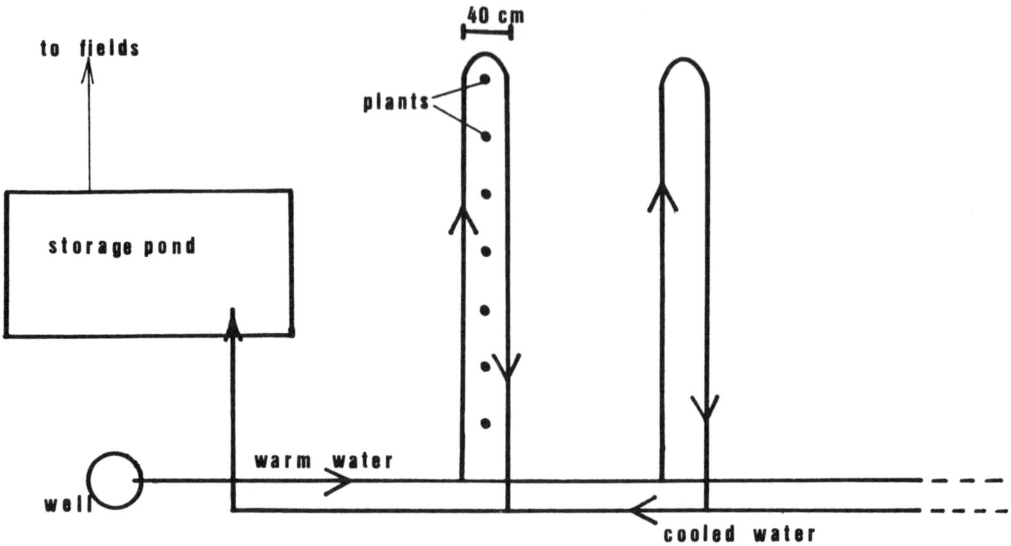


Figure 1: A schematic presentation of the system for the utilization of the geothermal energy of brackish water for soil heating combined with the use of this water for irrigation of conventional crops.

## Effects on Soil Temperature

Soil temperatures, measured at a depth of 30 cm, were stable during the day with a deviation of ± 0.5°C. Table 2 gives the relationship between the rate of water supply and the 30 cm temperatures achieved during January-February 1975.

Table 2. Effect of the Rate of Water Supply on Soil Temperatures at a Depth of 30 cm in Mulched Plastic Tunnels
(Inlet Temperature - 40°C)
(Soil temperature: bare soil 10°C. Non-heated tunnel - 14°C)

| Water discharge (D) (liter $m^{-2}$ $hr^{-1}$) | Soil temperature (daily average in °C) | Inlet-outlet temperature difference ( t in °C) | Energy requirement ( t x D Kcal $m^2$ $hr^{-1}$) |
|---|---|---|---|
| 1.8 | 25.0 | 9 | 16.2 |
| 2.7 | 26.0 | 7.2 | 19.4 |
| 5.4 | 27.0 | 5.2 | 28.1 |

Soil temperatures can be effectively regulated by adjusting flow rates. The system is more efficient at low soil temperatures. A water supply rate of approximately 2.0 $l \cdot m^{-2} \cdot hr^{-1}$ was enough to raise soil temperature to 25°C, the optimal temperature for many crop species (Cooper, 1975). The brackish water wells of the Central Negev have a capacity for discharging 300-400 $m^3\ hr^{-1}$. Such a well can therefore heat an area of 15 to 20 hectares.

## Effects on Yield

Soil warming resulted in a marked increase in cucumber yields (Fig. 2). Yields were 6.4 kg $m^{-2}$ in the control plots and 15.3 kg $m^{-2}$ in the 30°C plots. Even more important was the fact that heated plots yielded earlier. On the 130th day after planting cumulative yield of the 30°C plots was 6.2 kg.$m^{-2}$ whereas that of the non-heated plots was only 1.5 kg.$m^{-2}$.

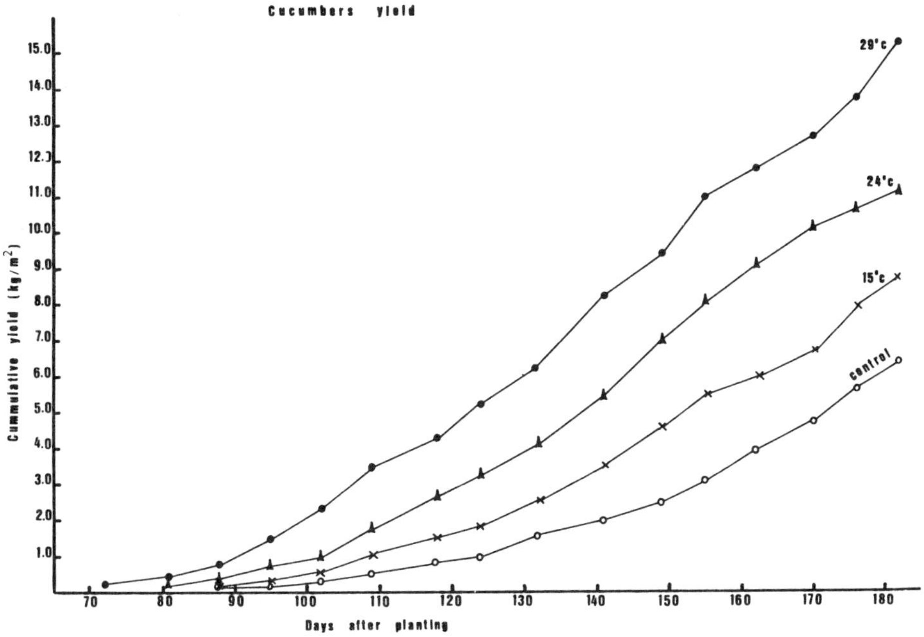


Figure 2: Cumulative yield of cucumbers grown during the winter months at three soil temperatures and in a non-heated plot. Soil temperatures at a depth of 30 cm were 29°C (●), 24°C (▲), 15°C (x) and the temperatures in a non-heated plot (o) varied between 12°-15°C according to season.

## Greenhouse Heating

A system for heating and cooling greenhouses through the utilization of warm well water and cold pond water has been developed. The system comprises a double-layer greenhouse roof, a network of buried polyethylene pipes, circulating pumps and thermostats (Fig. 3). When the greenhouse nighttime temperature drops below a predetermined level, warm water is pumped between the two clear layers of the roof. The heat of the water is radiated into the greenhouse and the relatively warm roof serves as a barrier against further heat loss to the atmosphere. When daytime temperatures in the greenhouse rise above a desired level (approximately 25°C), the relatively cool water from the storage pond is pumped between the two roof layers. The cold water absorbs some of the sun's long wave radiation and

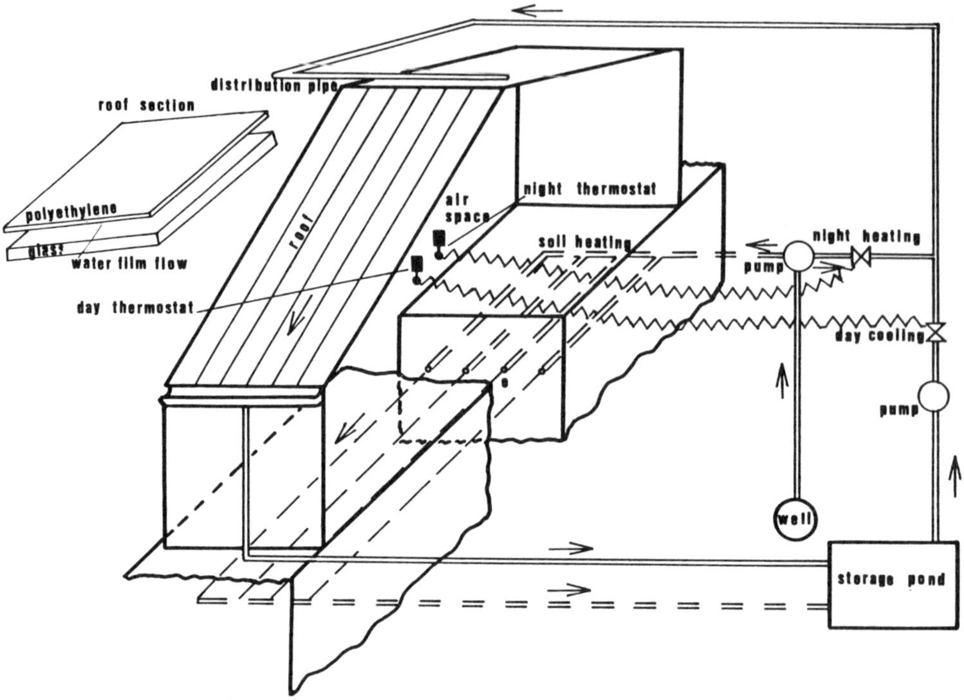


Figure 3: A system for heating and cooling greenhouse space and for heating greenhouse soil using warm well water and cool storage-pond water.

some of the greenhouse radiated and convected heat. As a result, air temperature in the greenhouse drops, and it is possible to keep the greenhouse closed during the winter months without the need of ventilation. (In the Negev, it is usually necessary to ventilate greenhouses on sunny winter days in order to dissipate excess heat). Consequently water consumption is markedly reduced. A closed greenhouse permits the artificial application of $CO_2$ leading to higher and better quality yields.

Soil heating is carried out in a similar manner as in the field. Warm geothermal water runs continuously through the plastic pipes buried in the soil. Both soil and air temperatures in the system can easily be regulated by changing water flow rates.

Energy costs in this system, as in the case of field soil heating, are very low because the water after dissipation of the heat is used for irrigation of field crops (Fig. 4). Figures 5 and 6 show twenty-four hour variations in air and soil temperature in a geothermally heated (and cooled) plastic greenhouse as compared with the variation inside an identical but unheated greenhouse. It can be seen that the heating-cooling system effectively controls the temperature by day and night during a typical sunny winter day. Thus 15-20 $kgs \cdot m^{-2}$ fuel are saved per annum. Plate 1 shows the development of tomato plants in the "geothermal" and the "control" greenhouses and Table 3 gives the tomatoes, melon and cucumber yields in the two greenhouses.

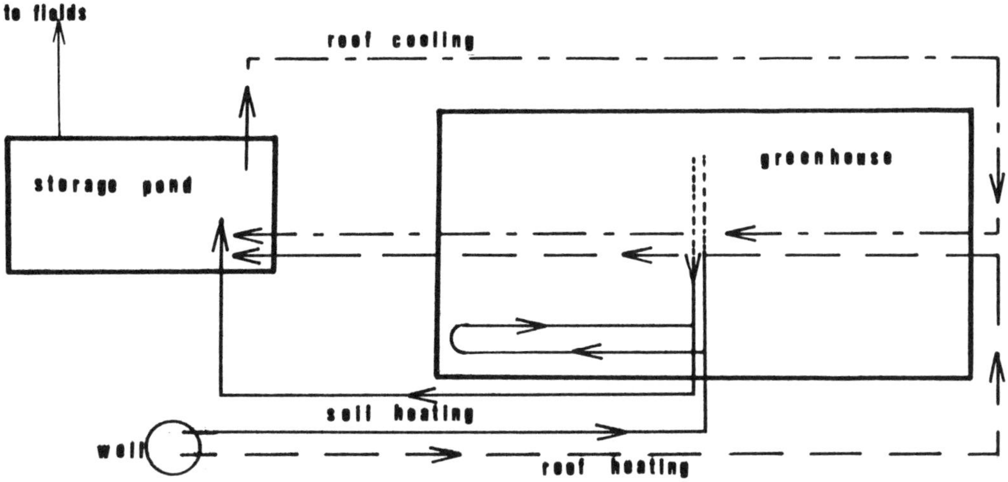

Figure 4: A schematic presentation of the system for the utilization of the geothermal energy of brackish water for greenhouse heating, of storage pond water for greenhouse cooling and of the water for irrigation of conventional crops.

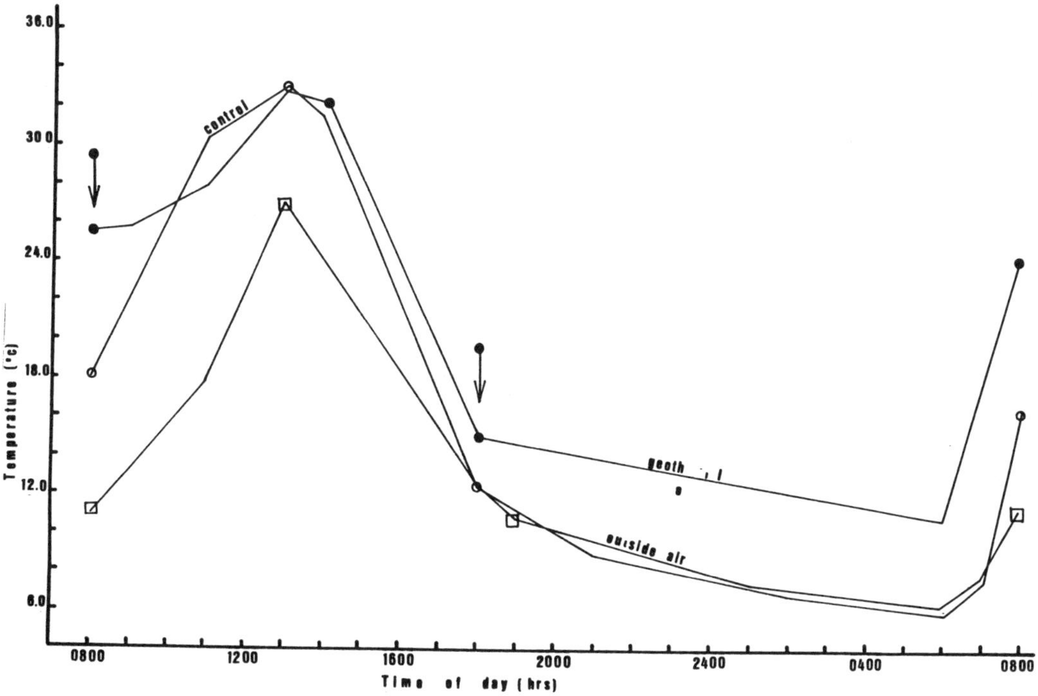


Figure 5: Twenty-four hour variation in air temperature inside a geothermally-heated greenhouse (●), in a control non-heated greenhouse (o), and in the open air (□) during a sunny winter day. Arrows indicate beginning of pump operation.

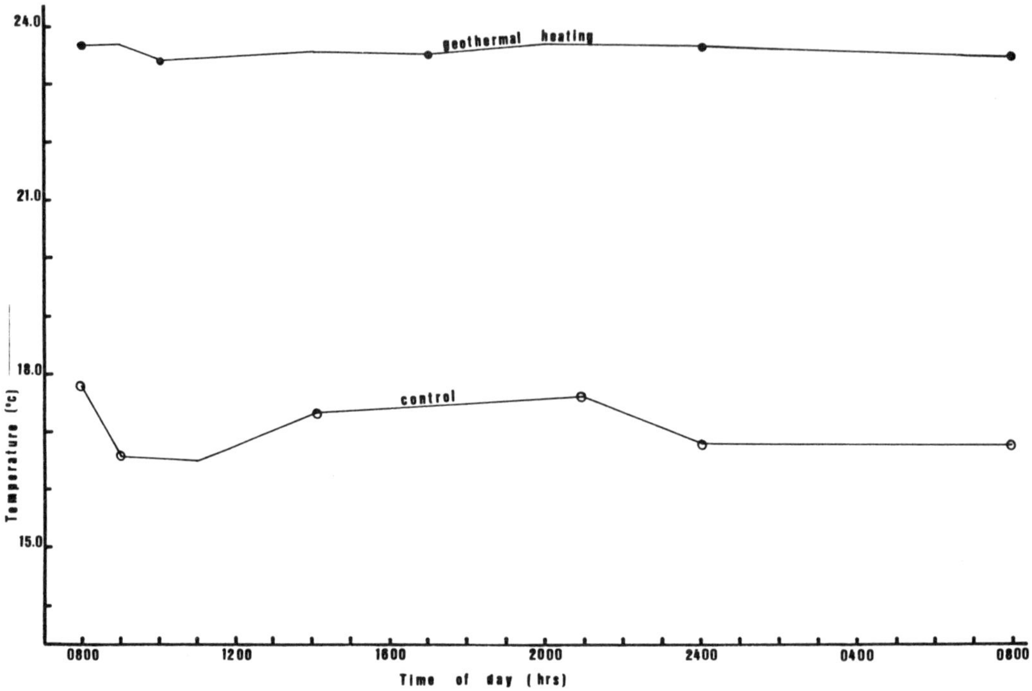


Figure 6: Twenty-four hour variation in soil temperatures at a depth of 40 cm inside a geothermally heated greenhouse (●) and a control, non-heated greenhouse (o).

## CONCLUSIONS

Cheap and reliable technologies have been developed for greenhouse space heating and for open field soil heating, using a warm (25-50°C) water source.

Geothermal brackish water can be used in an integrated system in which the water is utilized first for soil and space heating and then for conventional irrigation of fields. Energy costs in such a system are very low.

The raising of soil and air temperatures safeguards, increases and results in earlier yields.

### Acknowledgements

This work has been supported by the Israeli Ministry of Commerce and Industry.

Plate 1: Tomato plants grown in a geothermally-heated greenhouse and in a non-heated but otherwise identical greenhouse (control). Both plants are 112 days old. Note the differences in size and in fruit load between the two plants.

Table 3. Cumulative Yields (kg.m$^{-2}$) of Tomatoes, Melons and Cucumbers in a Geothermally Heated, and in a Control, Non Heated Greenhouse

| Crop | Greenhouse Type | Jan. '77 | Feb. | Mar. | Apr. | May 1st-19th | June '77 |
|---|---|---|---|---|---|---|---|
| Tomatoes | Heated | - | - | 1.28 | 9.27 | 13.75 | - - - - - |
| (<u>Lycopersicon esculentum</u>) | Control | - | - | 0.10 | 3.52 | 7.24 | - - - - - |
| Melons | Heated | - | 0.95 | 3.28 | 4.07 | 5.20 | - - - - - |
| (<u>Cucumis melo</u>) | Control | - | - | - | 0.07 | 0.36 | - - - - - |
| Cucumbers | Heated | 2.45 | 7.40 | 12.18 | 16.43 | 19.03 | - - - - |
| (<u>Cucumis sativa</u>) | Control | 0.38 | 0.73 | 1.24 | 2.10 | 3.21 | - - - - - |

*Preliminary data

Date of sowing: Tomatoes - 2 Nov. 1976; melons and cucumbers 11 Oct. 1976

REFERENCES

(1) Cooper, A.J. 1975. "Root temperatures and plant growth" Commonwealth Bureau of Horticulture and Plantations Crops, East Molling, Maidstone, Kent. Research Review No. 4.

(2) Nielsen, K.R. and E.C. Humphries 1966. Effects of root temperature on plant growth. Soil and Fertilizers, 29:1-7.

(3) Pasternak, D., M. Twersky and I. Borovic 1973. Irrigation of wheat with brackish water in Ramat Hanegev. Hassadeh, 53:381-384. (In Hebrew).

(4) Pasternak, D., M. Twersky, I. Borovic, Y. De Malach and E. Rappaport 1975. Use of geothermal brackish water for soil warming and vegetable irrigation. In: "Brackish Water as a Factor in Development", A. Issar ed. Proceedings of an International Symposium, Beer-Sheva, Israel.

(5) Richards, S.J., R.M. Hagan and T.M. McCalla 1952. Soil temperature and plant growth, p. 303-480. In: "Soil Physical Conditions and Plant Growth." B.T. Shaw ed. Academic Press, N.Y.

(6) Rykbost, K.A., L. Boersma, H.J. Mack and W.E. Schmisseur 1975a. Yield response to soil warming: agronomic crops. Agron. J. 67:733-738.

(7) Rykbost, K.A., L. Boersma, H.J. Mack and W.E. Schmisseur 1975b. Yield response to soil warming: vegetable crops. Agron. J. 67:738-743.

(8) Twersky, M., D. Pasternak and I. Borovic 1975. Effect of brackish water on yield and development of cotton. In: "Brackish Water as a Factor in Development". A. Issar ed. Proceedings of an International Symposium, Beer-Sheva, Israel.

# GEOTHERMAL ENERGY DEVELOPMENT IN THE DESERT: THE CASE OF IMPERIAL VALLEY*

Stahrl W. Edmunds
Director, Dry Lands Research Institute,
University of California, Riverside

## ABSTRACT

Imperial Valley in California has the potential to develop about 4,500 megawatts of power over the next 40 years, which will result in substantial fiscal and industrial development benefits while still not affecting the present rich agricultural base. The development planning process being used in Imperial Valley is described in this paper and may be utilized by other desert regions to develop their own geothermal resources. The total potential geothermal resources in the world are estimated at 1 billion megawatts, about equal to all the electrical energy installations now in place in the world. Much of the geothermal resources are located in the deserts of the world, making it possible for them to become major sources of energy in the future. In the process, the desert countries have the opportunity for industrial development and economic growth, similar to that described for Imperial Valley. The purpose of this paper, then, is to aid the desert regions of the world to plan the utilization of their geothermal resources and to describe the economic benefits which such development can bring.

## INTRODUCTION

Geothermal energy is likely to be important to the future of desert areas throughout the world, because the deserts of the world are well endowed with potential geothermal resources. The utilization of these geothermal resources would be a stimulus to general economic development, and thus bring new sources of livelihood to regions which have been traditionally poor.

Imperial Valley in the United States is one of the first deserts to become the site of geothermal energy development; and fortunately the development process is taking place through a highly visible form of planning which can be utilized by

---

*The data presented in this report were made possible by a grant from Imperial County under contract CI 75/76-1 and from NSF/ERDA Grant AER75-0893.

other desert countries to plan their own development. This paper is concerned, then, with describing the planning process and its research findings, so that other nations may adapt the approach to their own situations.

## WORLD AND DESERT GEOTHERMAL RESOURCES

First, let us take a look at the world energy situation which is heavily dependent upon oil that may become exhausted in 30 years at present consumption and growth rates. Petroleum is not only a principal source of automotive fuel but also of electrical energy generation.

The world has about 1¼ billion megawatts of installed electrical generating capacity, much of which will have to depend upon new fuel sources other than petroleum after the turn of the century (1). What are the options in new fuel sources? The United States is planning heavy development of its coal reserves for electrical energy conversion. Nuclear power is another possible source, but environmental contamination remains a serious barrier to its full development. Geothermal energy is a third option, which is emerging as a major potential source.

Worldwide geothermal energy resources have been estimated by Rex at a potential of 1 billion megawatts of power, about equal to all the installed electrical generating capacity in the world today (2). Banwell and Meidav characterize geothermal energy as "orders of magnitude greater than other sources of fossil energy." (3).

Equally important is the fact that much of the geothermal energy potential is located in desert regions of the world, which have so long been underutilized and underdeveloped. Table 1 gives a list of selected desert countries with potential geothermal resources. The table indicates that all continents are represented throughout the arid belts running worldwide between 10 and 30 degrees North latitude.

The fact that geothermal resources are so widely distributed throughout the world, and so generously in desert areas needing development, presents the possibility that economic development, growth, wealth, and income distribution could be more widely and equally shared in the future than in the past. However, this promise will only be realized if the desert nations with geothermal potential undertake developmental planning for their own futures. This is where the Imperial Valley case becomes of some interest, because it may be viewed as an experiment in developmental planning of geothermal resources which may be applicable to desert nations everywhere.

## THE IMPERIAL VALLEY CASE

The geothermal planning process in Imperial County has multiple origins and stimuli. These are:

1. The federal government, especially through the Energy Research and Development Agency and the National Science Foundation, has funded research to establish baseline data on the nature and use of the geothermal resource.

2. The geothermal industry, especially oil companies, has financed exploratory wells.

Table 1. Selected Arid Areas and Geothermal Resources
(X indicates probable presence of resource)

| Area | Tect | Volc | HS | GP |
|---|---|---|---|---|
| Mali | | | | |
| Mauritania | | | | |
| Niger | | | | |
| Chad | X | | | |
| Sudan | | X | | |
| Namibia | | | | |
| Somalia | X | | | |
| Mongolia | | | | |
| North Ethiopia | X | | | |
| Botswana | ? | | X | |
| Algeria | X | X | | X |
| Libya | | X | | |
| Egypt | | | | |
| Morocco | X | | | X |
| Tunisia | X | | | |
| Yemen | X | X | | |
| India | X | X | X | X |
| Iran | X | X | X | X |
| Saudi Arabia | X | X | X | X |
| Iraq | | | | X |
| Turkey | X | X | X | X |
| Chile | X | X | X | |
| Peru | X | X | X | |
| Bolivia Altiplano | X | X | X | |
| Pakistan | | | | X |
| Nigeria | | | | X |

Classification of Sources of Geothermal Energy:

Tect--Tectonic belts rift zones
Volc--Quaternary and rent volcanism
HS--Hypernormal gradient
GP--Geopressure zone

Source: Banwell

3. The electric power companies have assessed engineering alternatives and developed possible electric generating plans.

4. The Electric Power Research Institute, which is privately financed by power companies, has supported designs of new conversion technologies.

5. The California State government has cooperated by setting flexible initial standards for well drilling and testing in its role of monitoring exploration and production.

6. The County of Imperial has issued well drilling permits and initiated a geothermal element to its general plan which will foster energy development while protecting existing agriculture and the environment.

7. Research institutions have been utilized to develop data to assess the resource, its use and its impacts. Among these are the Livermore National Laboratory which is establishing environmental baseline data. Also, the Dry Lands Research Institute (University of California, Riverside) and the Environmental Quality Laboratory (Cal Tech) have jointly supplied assessments to Imperial County for its geothermal element to the general plan, and the findings which follow are summarized from this research.

Imperial County lies east of the coastal range between San Diego, California, and Yuma, Arizona. The area is a desert converted to agricultural production by the allotment of 3 million acre-feet of water annually from the Colorado River. Agricultural output is valued at about $500 million per year; and the valley supplies much of the lettuce, asparagus and winter vegetables for the whole nation, as well as substantial amounts of livestock, cotton, and field crops for regional consumption. Naturally, Imperial County wishes to maintain its rich agricultural base. Therefore, geothermal energy must be developed compatibly with agriculture.

## RESOURCE ASSESSMENT

Imperial Valley is located in the Salton rift where tectonic plate movement is causing the valley to spread at the rate of 2 centimeters per year, thus opening the basement rock, at depths of 5 to 6 kilometers, to the intrusion of hot magma as a heat source. The sedimentary overburden consists of aggregate, clay, and silicic material formed into sandy aquifers separated horizontally by impermeable strata and fractured irregularly along fault lines which radiate from the tectonic movement below. The fracturing and impermeable layers cause the geothermal reservoirs to be ambiguous and difficult to define, except by exploratory drilling.

The most active area of tectonic movement is in the juncture of the Imperial and Brawley faults. This strike-slip fault is rotating to the southwest, forming a trapezoidal block between the ends of the two faults, roughly defined by the Salton Sea, Brawley, Heber, and Westmoreland.

The geology assessment correlated data from well logs, thermal gradient and magnetic anomaly measurements, as well as studies of geophysical structure. The new contribution of the geologic assessment was to introduce and correlate seismic and gravity anomaly measurements with previous data.

The estimate of potential energy in the valley, from these measures, is about 10,000 MW (megawatts) if heat is extracted only from fluids, and about 20,000 MW-plus if heat is extracted from both rock and fluid. About half the available energy is in the Salton Sea KGRA, where total dissolved minerals (at 350,000 ppm) make fluid handling difficult with present technology, although experiments are under way to use binary cycle conversion or a new total flow process.

The Brawley, Heber, and East Mesa KGRA's have fluids with an order of magnitude less salinity, and are thus more likely for early development. The fluids, at around 200°C, are somewhat cooler than in the Salton Sea area, but they appear capable of economic development by either flash-steam or binary processes. Brawley has a flowing well under test by Union Oil. Chevron and San Diego Gas & Electric are in the development and decision stage at Heber. Magma, Republic, and the U.S. Bureau of Reclamation have wells in the East Mesa KGRA.

Perhaps the most interesting new hypothesis is that substantial new geothermal reservoirs may lie to the west of the present KGRA's, roughly along a line west of Westmoreland and Heber. The basis for this hypothesis is from data on an exploratory well brought in by Republic near Westmoreland with high heat and low salinity, plus measurements from seismic arrays in the west valley. This seismic work, by Dr. Shawn Biehler, suggests that current tectonic activity is opening new heat sources to the west of present KGRA's, and recent infiltration of water accounts for the low salinity.

The newly recognized reservoirs, however, are too young to have set up convection-cell deposition of minerals in porous rock; thus they have no gravity anomaly signature. The new reservoirs are detectable mainly by seismic arrays to locate the general heat source, helium detection to locate the fractures along which the gas escapes, and drilling to locate the fluid in fracture zones.

Thus, these findings suggest that the geothermal energy resource in Imperial Valley is not only substantial but possibly expanding.

## ENGINEERING ASSESSMENT

The surface area requirement for a geothermal well is one acre, of which half can be returned to agriculture or other uses after well and transmission lines are installed. By use of slant drilling, a single drilling location can cover about 300 subsurface acres. Thus, a single drilling platform could supply a 50 MW generating plant.

Flash steam plants appear to be the most feasible at present technology; but experimental binary cycle systems may be tested to provide closed loop handling of geothermal fluids. Flash steam plants in Imperial Valley will operate at relatively low temperatures and steam pressures, which means larger steam exit sections and turbines, as well as higher plant costs.

A 20-25 percent deficit in the volume of water available for reinjection may occur through evaporation in a steam flash plant. Surface water, irrigation drainage, and Salton Sea fluids may be sources of water supplements, provided they are compatible with the reservoir water upon reinjection.

Water will also be required for the use of evaporative cooling towers at about 6,000 acre-feet per year for each 100 MW of power for a flash steam plant at 200°C. Imperial County has sufficient agricultural drainage waters for about 10,000 MW of plant capacity. However, drainage water is likely to need treatment for organic substance and nutrients, possibly at a cost up to .4 mil per kilowatt hour.

## GEOGRAPHY AND ENVIRONMENTAL IMPLICATIONS

The impact of geothermal energy development upon agricultural resources in Imperial Valley appears to be minor. The land area requirements would displace perhaps 1 percent of the present agricultural land under production, and most of this production could be restored by applying water to other acreage within the boundaries of the irrigation districts. Within KGRA's about 44 percent of the acreages are solely field crops; about 30 percent are rotated predominantly with field crops; about 15 percent is in mixed vegetable crops; 2 percent is in asparagus; and 1 percent each is in tree crops and feed lots. The vegetable and fruit crops have high value, but not nearly as high as geothermal energy usage. The possibility of drift of noncondensible gases was considered, but the concentrations of emissions appeared insignificant compared with the salts already being deposited in the soils from irrigation.

Geothermal experience in other countries of the world indicates that the main areas of environmental concern are subsidence, seismicity, blowouts, and water availability or quality. Subsidence can be controlled by reinjection to maintain reservoir volume. Environmental impacts may be abated by regulation or control equipment for air pollution, noise, waste water, disposal of drilling muds with detrimental trace elements, disposal of solid wastes from blow-down, and prevention of hazards from transmission lines or transmission pipes. An ambient standard of 0.03 parts per million of hydrogen sulphide within one mile of the geothermal plant would prevent adverse impacts on crops or plants.

The County should encourage initial geothermal operations so tests may be made for such things as subsidence, the workability of reinjection, and measurements of environmental effects. Since electrical generation will require mainly high temperature fluids, serious consideration should be given to using lower temperature fluids, which exist in the reservoir in large volumes, for nonelectrical uses.

## SOCIOLOGY

Imperial County's population in 1975 is estimated at 83,000, which represents an increase from the 1970 census figure of 75,000. Between 1940 and 1970, the population of Imperial County grew slowly, due to the high outmigration of younger age groups. About 40 percent of the population is of Spanish origin or descent, and this portion of the population has high fertility rates and relatively young age distribution. The increase in population in 1970 to 1975 suggests that outmigration may be diminishing.

Projections of future population depend heavily on migration assumptions. Assuming no migration in or out and constant mortality rates, the present fertility and age distribution of the population would indicate a population growth by the year 2000 to 113,000 persons. If one assumes that net migration rates are directly related to the installation of new geothermal energy installations, the population might be expected to grow to 120,000 by the year 2000. The important point is that Imperial County has the basis for an expanding labor force by training the existing young population for higher skilled employments, thus checking the outmigration, plus attracting new population through the expansion of geothermal energy development.

A public opinion survey was undertaken to ascertain attitudes toward geothermal development. About 90 percent of the population reported themselves in favor of geothermal development. About three out of four felt that the geothermal development should be regulated to make it compatible with agriculture and the environment. There was some expectation that geothermal development would result in cheaper source of energy, more power available to Imperial County, or benefits in the way of employment or tax revenues. A portion of the population felt that they were not well informed on geothermal energy, and most of the information came from news reports rather than through meetings or other media.

## ECONOMICS

The cost of field development and plant investment is expected to be in the range from \$500 to \$750 per kilowatt of installed capacity, although some industry sources feel the cost may run as high as \$1,000 per kilowatt. The higher estimates contemplate extensive gathering lines for transmission of geothermal fluid, environmental abatement costs, and new engineering design problems to handle geothermal fluids, plus the general inflation in construction costs. The cost of electric power under the midrange assumptions would be about 30 to 35 mills per kilowatt hour, which would be competitive with fossils and coal, perhaps slightly less than nuclear power.

The principal benefits to Imperial County from geothermal development would be a moderate increase in the employment base, and an increase in property tax revenues. At 4,500 MW of installed capacity in the years 2000-2020, the direct employment for operation of geothermal plants would add about 500 new jobs, and the direct and secondary effects would add about 5,000 jobs. The secondary effects would be largely services to the geothermal industry, such as repairs of materials, development of related industries, plus the construction and trade to support a larger working population.

The fiscal benefit to the County would be of the order of \$5,000 per megawatt of installed capacity. Per capita income is projected to increase from \$5,000 per year in 1975 to about \$7,500 per year in the year 2000. The main areas of economic expansion would be in agriculture, food processing, agricultural chemicals, gypsum and mineral products, geothermal energy, transportation, wholesale and retail trade, personal services, and local government. An existing water supply can support the projected economic and population expansion, provided there is a 30 percent conservation in the usage of water through such practices as lining canals and more efficient irrigation methods.

## JURISDICTION AND LEGAL IMPLICATIONS

While the state government has jurisdiction to regulate well drilling, reservoir management, and power plant siting, Imperial County has sufficient authority under its police powers and land use control to regulate the environmental effects of geothermal development. Development may be encouraged, while still protecting environmental amenities, by initiating a well-articulated land use plan, providing for the special needs of geothermal development by floating zoning and performance standards for operations. Imperial County currently has a planning structure within which to integrate a geothermal element. The County, moreover, has management competence in local government to provide encouragement and guidance to geothermal development. The attitude of voters is generally similar on political issues to other portions of the state, with some tendency to be more inclined towards local control and towards environmental protection. To facilitate geothermal development, the county could develop a generic environmental report, and subsequently require developers to show specific impacts and performance criteria for development. To reduce the initial burden on county governments and taxpayers, the developers could be assessed for direct development costs. A citizen role in advising on geothermal development policy could help resolve environmental and economic trade-offs.

## RECAPITULATION OF THE PLANNING FUNCTIONS FOR GEOTHERMAL DEVELOPMENT

The major findings of the research and planning process for geothermal development in Imperial County have been outlined briefly above. (The total report contains eight documents of 900 pages.) More important than the findings, to other desert countries, which may wish to pursue their own geothermal developmental planning, is the form and content of the planning process itself. The following section summarizes the structure and methodology of the planning program.

Imperial County charged its Director of Public Works and its Planning Director with the responsibility to develop a geothermal element to its general plan. The Planning Director reports to the Planning Commission, which must approve a new geothermal planning element, before it goes to the County Board of Supervisors which has final approval. The County Board of Supervisors is a legislative body which implements the geothermal plan by ordinances governing land use, geothermal permits, conditions of operation, and taxation.

The Director of Public Works, with his Planning Director, established a research group by a grant to the Dry Lands Research Institute at the University of California, Riverside, and the Environmental Quality Laboratory at the California Institute of Technology. The progress of the research group and its findings were reviewed by three advisory committees.

1. The Industry Advisory Committee was made up of geothermal developers, mainly oil companies and electric utilities.

2. The External Management Review Committee was made up of city, country, state and federal officials who might utilize or influence the geothermal development plan.

3. The Internal Management Review Committee was made up of County department officials who would have to utilize the planning data to implement a geothermal element to the general plan.

Imperial County also set up a Project Director within the Public Works Department to work closely with the universities in the execution of the research program. The Director of the Dry Lands Research Institute was the research director coordinating the tasks for the whole program. The research program was defined into several hundred tasks and activities which were assigned to principal research investigators among the faculty in the two universities. These many tasks, and the methodologies involved, are summarized in the following outline:

I. Management of the research program

   A. Establish research goals consistent with needs of the geothermal element of the general plan
   B. Develop a system analysis of the flow of tasks to be performed, and convert this network into a system management program plan
   C. Assign the research tasks and budgets to interdisciplinary faculty groups for implementation
   D. Hold regular research reviews of progress and findings
   E. Monitor schedules and performance
   F. Integrate results

II. Resource assessment

A. Analyze the geophysical structure of the Imperial Valley
B. Estimate the total heat in storage from well log data supplied by government and oil company drillings, plus data from bore-holes, thermal gradients, measures, gravity anomalies, and seismic arrays
C. Analyze the geothermal fluids, compositions, temperatures, and reservoir characteristics
D. Estimate reservoir hydraulics and dynamics, convection, porosity, and sealing characteristics
E. Examine possibilities of subsidence, reinjection, and seismic activity
F. Estimate the recoverable heat

III. Engineering assessment

A. Make a technology assessment of the state of the art engineering designs applicable to geothermal energy recovery, particularly flash steam and binary cycle processes
B. Analyze schematic designs of alternate technologies for energy conversions with trade-offs and economic costs
C. Assess drilling alternatives and land use requirements
D. Examine fluid transmission designs
E. Analyze cooling tower requirements and alternative sources of cooling water
F. Examine nonelectrical or process heat possibilities
G. Analyze the possibility of potable water production from geothermal fluids

IV. Geographical and environmental assessment

A. Provide an agriculture data base with aerial survey maps of Imperial Valley showing present land use by parcel, with type of crops, soils, and irrigation lines
B. Estimate any crop shifting or displacement, in land area and economic value, which may occur from geothermal installations
C. Assess the environmental impacts of geothermal development in terms of air and water quality, subsidence, crop effects, or changes in natural habitats
D. Define electrical transmission corridors together with land requirements and use conditions which should apply
E. Suggest alternative land use patterns

V. Social impacts

A. Study the social structure, demography, employment and unemployment patterns in Imperial County
B. Estimate future population growth, labor force, and employment effects of geothermal development
C. Survey public attitudes towards geothermal development
D. Conduct a leadership survey of decision factors affecting geothermal development
E. Evaluate the net social effects upon the area

VI. Economic assessment

A. Analyze the capital equipment and operating costs of geothermal development, together with the feasibility of investment, risks, rate of return, tax yield, and selling price of electricity
B. Develop an economic sector analysis of the Imperial County economy based upon an 80 sector input-output matrix to show the interaction of industries (including geothermal) in terms of output, employment, value added by manufacture, and income
C. Project the economic benefits and impacts of geothermal development, through the input-output model, by introducing the capital and operating inputs in the form of employment and expenditure matrixes
D. Estimate the net benefits to Imperial County from primary geothermal development, secondary industrial development, employment effects, fiscal effects, and income distribution effects
E. Examine possible external environmental costs

VII. Jurisdictional, political and legal assessment

A. Analyze the political decision structure affecting geothermal development decisions, together with electoral voting patterns which may affect development
B. Characterize the powers and jurisdictions of the local, state, and federal agencies which have regulatory powers over geothermal development
C. Describe the regulatory options and land use planning alternatives which Imperial County may use to encourage geothermal development consistent with maintaining agriculture and environmental quality
D. Outline the planning characteristics which would contribute to a feasible and effective geothermal element to the general plan

## CONCLUSION

Imperial County is likely to develop about 4,500 megawatts of electrical energy from geothermal resources during the next 40 years with substantial fiscal benefits, moderate employment and industrial development effects, and minimal environmental impacts.

The developmental planning process used by Imperial County, which has been outlined in the previous section, is broadly applicable to geothermal development throughout the world.

The world geothermal energy resources are very large, about equal to all the electrical generating capacity now in existence; and the deserts of the world are well endowed with potential geothermal reserves.

Desert nations may find a substantial basis for their own economic development in the future by utilizing appropriate portions of the geothermal developmental planning process outlined above in their own countries.

## FOOTNOTES AND REFERENCES

(1) Robert Rex, "Geothermal Energy in the United States," The Institute of Geophysics and Planetary Physics, University of California, Riverside, 1972.

(2) J. Banwell and Tsvi Meidav, "Geothermal Energy in the Future," Geothermal Energy Magazine, December, 1974.

(3) United Nations, "World Energy Supplies," Statistical Papers, Series J, No. 17. Department of Economic and Social Affairs, N.Y., 1974.

# BENEFITS AND CONSTRAINTS OF OIL-SHALE AND TAR-SAND DEVELOPMENT IN ARID AREAS

John R. Donnell

## ABSTRACT

Desert areas of the world in the past have not fully benefited from the advances of modern science and technology. Discovery and development of the immense deposits of oil in the desert areas of the Middle East have led to the investigation of the possible existence of other commodities that might prove equally beneficial. Two of these commodities are tar sands and oil shale. The known world-wide resource potential of oil shale and tar sands is more than double that of petroleum. Although at present, oil is being produced commercially from tar sands only in Canada, and from oil shales only in the U.S.S.R. and the Peoples Republic of China, the increasing demand for petroleum products and limited supply will probably necessitate the development of a synthetic fuels industry supplying oil from oil shale and tar sands.

The large capital investment for a synthetic fuel industry requires a long period of time (20 or more years) to amortize the investment. This in turn requires a large in place resource to sustain an industry over the long term. In arid areas, only the United States contains a large enough tar-sands resource, and the United States and Morocco contain a large enough oil-shale resource to sustain the minimum size 50,000 barrel (7950 $m^3$) per day commercial development projected by U.S. and Canadian industry. However, if a smaller scale (5,000 to 10,000 barrels (795 to 1590 $m^3$) per day) industry is deemed to be economically feasible then arid areas of Israel, Jordan, Chile, and the U.S.S.R. contain sufficient in place resources.

Many of the constraints on an oil-shale or tar-sand industry are the tremendous initial capital investment, solid, liquid, and gaseous waste disposal problems, availability of water, availability of equipment, and availability of adequately trained personnel. The benefits that may accrue include a secure source of petroleum and other mineral by products, improving the country's balance of trade position, increasing employment, and, by revegetation, possibly increasing the productivity of the land disturbed by industry.

## INTRODUCTION

Immense quantities of oil are now available from existing fields utilizing primary production methods, and an equal amount may be available through enhanced recovery methods. Additional significant discoveries of crude oil will be made in the future. However, world energy demands during the next 25 years will probably necessitate supplemental production of considerable quantities of oil from tar sands and synthetic oil from oil shale.

Two trillion barrels (318 billion $m^3$) is the consensus figure in recent estimates of the world resources of crude oil from conventional sources (Warman, 1973). King Hubbert (1972) estimates that shortly before the year 2000, we will have consumed one-half of the recoverable petroleum resources of the world. Most of the easily discovered and easily producible petroleum has already been found; therefore, future production from as yet undiscovered fields and additional recovery from existing fields using enhanced recovery methods will be increasingly expensive. As the cost of finding and producing crude oil becomes increasingly expensive, the production of oil from tar sands and synthetic oil from oil shale becomes more attractive. A total known resource of about 3 trillion barrels (477 billion $m^3$) of oil is in place in the oil shales of the world that will yield an average of 10 or more gallons (38 liters) per ton, and a slightly smaller amount of oil, 2.3 trillion barrels (366 billion $m^3$), is contained in the extensive tar sands of the world.

## DISTRIBUTION OF KNOWN OIL SHALE AND TAR SAND DEPOSITS

Oil shale and tar sand deposits are widely distributed in the world (figures 1 and 2). It is apparent from the distribution shown on these two figures that the greatest number of deposits and the greatest known resource is in the Northern Hemisphere. This possibly is related to a much greater exploratory effort by the countries in the Northern Hemisphere. The greatest resource potential of both oil shale and tar sands is on the North American Continent, in contrast to the overwhelming dominance of Asia in the present production of and future resource potential for crude oil.

## OIL SHALE AND TAR SAND RESOURCES

About one trillion barrels (1959 billion $m^3$) of oil are contained in the Canadian tar sands, and of this about 860 billion barrels (137 billion $m^3$) are in the province of Alberta. Two-thirds of the known oil-shale resources of the world are in the Eocene Green River Formation in the States of Colorado, Utah, and Wyoming.

Bowman and Carrigy (Alberta Oil Sands Technology and Research Authority) (1976) state that based on the experience gained from present oil from tar sand extraction operations in Canada, the minimum economic size for an oilsands plant is one that produces 50,000 barrels (7950 $m^3$) of oil per day over a projected period of 30 years. They postulated that this would require a proven in place reserve of about one billion barrels (1959 million $m^3$) of oil. At present there is no commercial extraction of shale oil from the Green River oil shales. However, it is estimated that the optimum size operation in the United States would entail a 50,000 barrel (7950 $m^3$) a day operation over a period of 20 years. This would require an in place resource of about three quarters of a billion barrels (119 million $m^3$) of oil assuming a recovery of 50 percent. Economic parameters vary

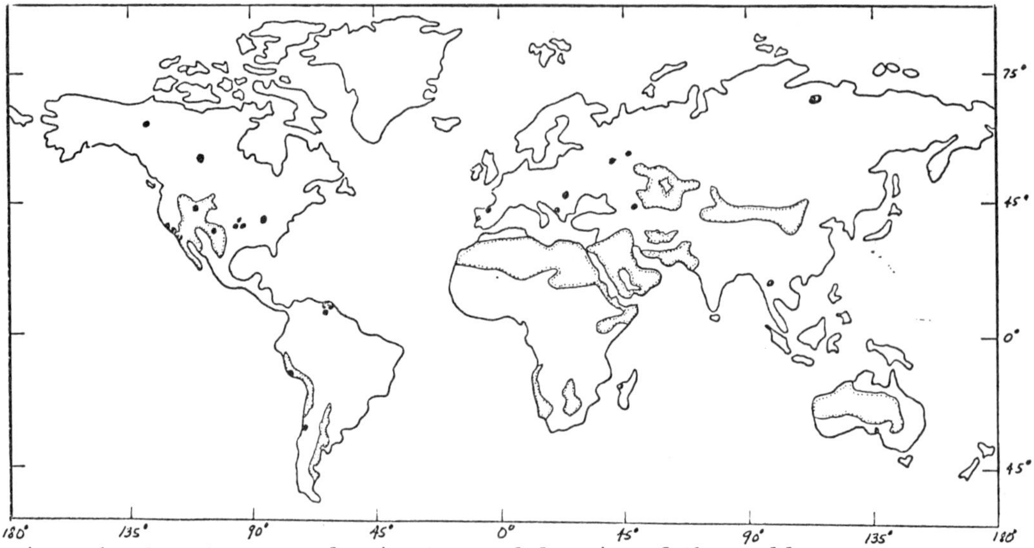


Figure 1. Desert areas and major tar sand deposits of the world.

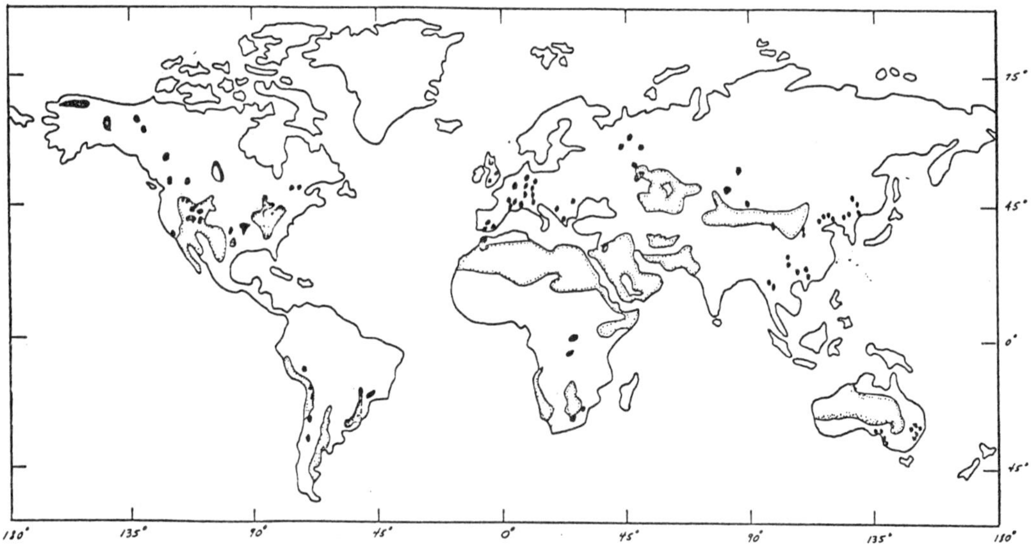


Figure 2. Oil-shale deposits and desert areas of the world.

from country to country but assuming that an initial oil-shale or tar-sand development would require the same resource base as that postulated for the Canadian and United States deposits then the following lists indicate the areas which are believed to meet the minimum resource requirements.

TAR SANDS, OIL SHALES, AND DESERT AREAS

Deserts are defined, for the purposes of this paper, as areas that receive less than 10 inches (25.4 cm) of precipitation a year, and associated semi-arid areas receive less than 15 inches (38.1 cm ) of precipitation a year. The distribution of tar-sand and oil-shale deposits, and their location with respect to modern major desert areas of the world, as depicted on figures 1 and 2, indicate that in general there is an inverse relationship. This inverse relationship essentially holds true for the geologic past as well as for the present. Tar-sand deposits contain the residual hydrocarbons remaining in oil fields whose lighter fractions have escaped into the atmosphere. Most of the oil in these parent fields was formed from organic materials deposited in a stable marine environment whose only relationship with deserts was incidental geographic proximity. Some small oil-shale deposits were formed in playa lakes that existed in a desert environment; however, the major oil-shale deposits of the world were deposited in shallow marine platform areas marginal to intercratonal geosynclines or in lakes of large areal extent and long duration. The oil shales of the Green River Formation were formed in several lakes that, in general, existed as permanent bodies of water for several million years. Well preserved leaves and pollen in the lake sediments indicate that the climate in the area throughout most of the life of the lake was humid subtropical. The temperature in the area of the lakes fluctuated widely from a mean annual temperature of 65°F (15.6°C) and a mean annual precipitation of 30 to 40 inches (76.2 to 101.6 cm) (Bradley, 1929). During some phases of the existence of the lakes, commercial deposits of sodium minerals were precipitated from the highly saline lake waters. This was a result of supersaturation of the salts in the mix as a result of temporary excess of evaporation over precipitation. This also resulted in minor constrictions of the lake but not evaporation to total dryness.

A cursory examination of figure 1 and table 1 shows that the only tar-sands deposits in arid and semi-arid areas with the requisite one billion barrels (119 million $m^3$) of in place reserves are those in Utah in the northern part of the Great American Desert. Table 2 and figure 2 indicate that in arid and semi-arid regions, only the Green River Formation in the northern part of the Great American Desert, and possibly the oil shales of the Middle Atlas region of central Morocco, contain a sufficient resource to be of interest. The precipitation in these prospectively economic tar-sand and oil-shale areas range from less than 10 inches (25.4 cm) to slightly more than 15 inches (38.1 cm) a year.

Some American oil companies are investigating a modular approach to oil-shale development that involves as a first step the mining, crushing, and processing of sufficient oil shale for a single commercial size retort producing 5 to 10 thousand barrels (795 to 1590 $m^3$) of oil per day. The one module approach would require a much smaller resource to sustain commercial development and enable countries such as Israel, Jordan, Chile, and Russia to develop smaller oil-shale deposits in arid areas.

The early history of oil-shale development is replete with examples of many years of continuous production of small quantities of oil. Two of these examples are the Scottish and Swedish industries. Production started in Scotland in 1851 and continued until 1962. Shale oil was produced in large banks of 50 or 60 retorts with an individual throughput of 5 to 15 tons of shale per retort (Stewart, 1968). Sweden, during World War II, was cut off from essentially all sources of imported oil and forced to rely on synthetic oil from shale. Production during the war totaled 2.2 million barrels (350 thousand $m^3$) from large banks of retorts containing a total of approximately 4,000 small capacity retorts. Production continued until 1966 when rapidly rising labor costs and the availability of low

Table 1. Tar Sands

| Deposit | Country | In place resource | |
|---|---|---|---|
| | | Billions of Bbls | (Billions of $m^3$) |
| Orinoco | Venezuela | 700.00 | (111.3) |
| Athabasca | Canada | 600.00 | ( 95.4) |
| Olenek | U.S.S.R. | 600.00 | ( 95.4) |
| Cold Lake | Canada | 160.00 | ( 25.4) |
| Wabasca | Canada | 85.00 | ( 13.5) |
| Peace River | Canada | 75.00 | ( 11.9) |
| Tar Sand Triangle | U.S.A | 18.00 | ( 2.9) |
| P. R. Springs | U.S.A | 4.00 | ( 0.6) |
| Sunnyside | U.S.A. | 3.00 | ( 0.5) |
| Bemolanga | Malagassy | 1.75 | ( 0.3) |
| Circle Cliffs | U.S.A | 1.30 | ( 0.2) |
| Asphalt Ridge | U.S.A. | 1.20 | ( 0.2) |
| North Alabama | U.S.A | 1.18 | ( 0.2) |

Table 2. Oil Shale

| Country | In place resource | |
|---|---|---|
| | Billions of Bbls | (Billions of $m^3$) |
| U.S.A. | 2,000.22 | (318.0) |
| Brazil | 800.84 | (127.3) |
| U.S.S.R. | 112.59 | ( 17.9) |
| Zaire | 100.64 | ( 16.0) |
| Canada | 44.03 | ( 7.0) |
| Italy (Sicily) | 35.22 | ( 5.6) |
| Peoples Rep. of China | 27.86 | ( 4.4) |
| Morocco | 4.00 | ( 0.6) |
| Sweden | 2.52 | ( 0.4) |
| Burma | 2.01 | ( 0.3) |
| West Germany | 2.01 | ( 0.3) |
| Great Britain | 1.01 | ( 0.2) |
| Thailand | .82 | ( 0.1) |

cost oil forced the discontinuance of the oil-shale industry (Gejort, 1968). It is quite possible that today similar production methods may be utilized to produce small quantities of oil economically from small oil-shale deposits in under-developed countries where there is a growing demand for petroleum and an available supply of low cost labor.

## CONSTRAINTS TO DEVELOPMENT

### Financial

Thus far, 8 years of experience gained by Great Canadian Oil Sands Limited from a tar sands operation with a rated capacity of approximately 50,000 barrels (7950 $m^3$) of oil per day has indicated that a large initial capital expenditure is required for tar sands production. R. C. Osborne, et al. (1976) estimated that the initial investment required for a 100 thousand barrel (15900 $m^3$) a day tar sands project is $19,000 per daily barrel of rated capacity in 1976 dollars. A project this size could cost $2.4 billion if construction started in mid 1976. Additional annual capital costs are estimated to be $10 million to $40 million for either mining and surface processing or in-situ recovery methods.

Figures are not readily available on either capital costs or operational costs for shale-oil extraction in the U.S.S.R. or the Peoples Republic of China where oil-shale deposits are currently being exploited. Paper studies have been made on the capital costs required to produce oil from the Green River oil shales. These are based on pilot and semi-works studies of various extraction methods and range from a high of approximately one billion dollars for conventional underground mining and surface retorting to a low of about $443 million (Ashland Oil Inc. and Occidental Oil Shale Inc., 1977) for an operation producing approximately 50,000 barrels (7950 $m^3$) of oil per day utilizing a modified in-situ recovery method.

Capital expenses for a single module of the Paraho retort scaled up for production of 11,500 barrels (1829 $m^3$) of oil per day are estimated to be between $75 million and $100 million.

A plant consisting of 104 retorts handling a total of 1,525 tonnes of shale per day was constructed in Scotland in 1940 for a capital cost of approximately one million pounds. A Swedish plant, constructed in 1960, with a yearly retorting capacity of 2,047,400 tonnes (approximately 5,500 tonnes per day) cost 200 million crowns.

### Solid Waste Disposal

Mined rock occupies 1.3 to 1.5 times the space of the unmined rock; therefore even after extraction of the organic fraction (27 percent by volume for 30 gallon (114 liter) shale), more space is required for solid waste disposal than is created by mining. A 50,000 barrel (7950 $m^3$) a day oil-shale industry may require disposal of more than 36,000 $m^3$ of retorted shale per day, and of this total volume possibly 27,000 $m^3$ may be used as backfill in the mine and the remaining 9,000 $m^3$ disposed of on the surface. This large quantity of solid waste generated over the expected life of the industry will cover an extensive surface area creating different land forms and changing former land use patterns. It will be necessary to revegetate the disturbed surface to prevent extensive wind and water erosion and, if possible, to put the waste disposal area into productive use.

### Technical Constraints

Technology has been developed on a commercial scale for the extraction of oil from tar sands in Canada and for the extraction of oil from oil shale in the Peoples Republic of China and the U.S.S.R. Numerous retorts have been tested successfully for the processing of up to 1,000 tons of Green River oil shale per day. Processing characteristics may vary with individual deposits of oil shale and tar sand; therefore difficulties may be encountered in adapting existing technology to other deposits.

Equipment, other than that needed for large scale mining operations, cannot be obtained from manufacturer's existing stock but in many instances must be specially designed and ordered. A very great demand exists world-wide for heavy equipment, whether it be a stock item or a special order. This results in possible equipment delivery delays of as much as several years.

Several thousand people, including engineers, technicians, and skilled labor, will be employed in the construction phase of a synthetic fuels industry. A permanent force of 2,000 skilled operating and maintenance workers may be required for the commercial operation. There is now a shortage of mining and processing engineers, and very few personnel are trained in the technical skills required for a synthetic fuels industry. Recruitment of personnel for a synthetic fuels industry will be in direct competition with recruiting needs of an expanding oil, coal and uranium industry and with the constant demands for technical personnel by the metal mining industry.

## Air Quality

Initiation of a synthetic fuels industry will probably result in a significant deterioration of existing air quality. Dust will be generated by loading of the ore, transport, crushing, screening and storage of the material, ejection of solid waste from the processing facility, and finally solid waste disposal. Most of the dust can be controlled by wetting the solids throughout all stages of the operation.

Gas will be expelled into the atmosphere from equipment use, retort stacks, and finally evaporation from product storage facilities. The major concern with noxious gasses is with the sulfur and nitrogen oxides and volatile trace elements that may be released into the atmosphere as a result of processing. Sulfur and nitrogen are in most organic-rich rocks in amounts ranging from one to several percent. Removal of these two elements in processing produces considerable amounts of elemental sulfur and ammonia as valuable byproducts. Most of the volatile trace metals are in the oil shales and tar sands in amounts that are too small to be of any environmental concern.

## Water

Some of the greatest concerns associated with a large scale synthetic fuels industry are water requirements, water availability, and the probable effects of an industry on the quality of water.

### Needs

Vasts amounts of water are required for a conventional tar-sands mining and hot water extraction recovery method, and equal amounts are required when water is utilized for in-situ extraction of oil from tar sands. Other extraction methods such as fire flooding and the use of hot gas will require considerably less water. Equally large amounts of water are needed for oil-shale development. Estimates of water requirements for a conventional mining and surface retorting oil-shale operation range from a minimum of 1.0 $m^3$ to a maximum of 5.4 $m^3$ of water for each 1.0 $m^3$ of oil processed. More than 80 percent of the estimated water usage is related to processing. This includes mining and crushing, retorting, upgrading, moisturizing processed shale, power generation, and revegetation of processed shale disposal piles and other disturbed areas. The remainder of the water is designated for community use. In-situ recovery of oil from oil shale is estimated to require only 1.0 $m^3$ to 1.5 $m^3$ of water for every 1.0 $m^3$ of water produced. The in-situ process has little water requirement for mining and no requirements for crushing, retorting (unless steam is used), and processed shale moisturizing and may not require water for an upgrading process.

### Availability

The main sources of water for a synthetic fuels industry are surface water and ground water. Water utilized but not consumed in the processing is a secondary source as is a small quantity of water derived from oil shale in the retorting process. Arid and semi-arid areas, because of the paucity of precipitation, spawn few, if any, permanently flowing streams but commonly are traversed by high volume streams with headwaters at some distance in an area of high precipitation. Aquifers whose intake areas are far removed from the arid region may completely underlie deserts and contain abundant quantities of water. Even though an arid region at present has sufficient water, an oil-shale or tar-sand industry may consume more than the amount that is readily available to industry, requiring a diversion of water supplies from areas of current usage.

### Quality

Additions to the existing supply of water in permanently flowing streams in arid areas are by runoff from infrequent rains or returns from irrigated lands. These sources of water are high in salts and tend to degrade the downstream quality of water. The diversion and consumption of large quantities of water by a synthetic fuels industry may compound the water quality problem because it removes higher quality water that formerly acted as a dilutant.

Leachate from processed solid waste disposal piles, unless strictly contained, may enter the existing drainage and degrade the downstream water quality. This may be especially true in an oil-shale industry inasmuch as the solids in processing may be subjected to temperatures approaching 1000°C, thus altering the original mineral composition of the rock and rendering it more susceptible to chemical decomposition.

Poor quality water derived in processing oil shale and tar sands that otherwise might degrade the existing water supply may be utilized in dust control and in wetting the processed solid waste for compaction prior to revegetation.

## BENEFITS OF DEVELOPMENT

### Economic

Economic benefits are many and varied and range from direct financial benefits that may be realized by the sale of the product or products resulting from an industry, to an increase in the overall work force.

All of the free world, except the OPEC countries, has an shortage of petroleum supplies and a constantly increasing demand for petroleum products. A synthetic fuel industry may enable an underdeveloped country to supply its entire needs for petroleum products and become a net exporter of petroleum. The economy of developed countries is heavily dependent on energy in large part now furnished by petroleum. A significant supply of synthetic oil in these countries will help maintain a viable economy. Although synthetic oil production in developed countries may not, along with domestic crude oil supplies, fulfill all of the countries' petroleum needs, it will provide a secure supply and reduce imports. Many oil-shale and tar-sand operations could recover considerable quantities of sulfur and nitrogen as byproducts. In addition many of the oil shales contain abnormally high amounts of trace metals such as uranium and vanadium or associated saline minerals that may be recovered economically as part of the overall operation. Some of the oil from tar sands also contains valuable trace metals. All of these recoverable products, whether consumed domestically or exported, will serve to improve the country's balance of trade position.

Increased employment opportunities will be provided not only by the synthetic fuel industry but by companies providing ancillary services and increased domestic services required for industry employees and their families. Although much of the labor force will consist of technical personnel not available locally, there will also be a large demand on the local labor market for unskilled personnel.

Much of the arid to semi-arid areas of the world that formerly sustained a fair cover of vegetation have been subjected to excessive agricultural and grazing use, and as a consequence have been heavily eroded and no longer have any economic surface use. Areas of high precipitation that have been covered with mine and mill tailings have been rehabilitated and revegetated and now are more productive than they were prior to their initial disturbance. Revegetation experiments on processed oil shale have been conducted in western Colorado since 1965, initially in greenhouse experiments and from 1967 until the present (1977) in the field. Thus far, results have shown that with proper conditioning of the processed shale and initial heavy watering, a growth of native vegetation may be established and after a short time be self sustaining. Other imported plant species with a greater economic potential have established thick growth patterns on the processed shale plot. One of the problems with reestablishing native vegetative cover in arid and semi-arid areas is the availability of seed. The U.S. Soil Conservation Service has established a plant materials center in the vicinity of the Piceance Creek basin, Colorado, to provide the seed necessary for planting native species in the oil-shale area and the surrounding area now being heavily mined for coal. The results so far indicate that with proper management, an area disturbed during a synthetic fuels operation may be revegetated and prove to be more productive that it was prior to the advent of an industry.

## SUMMARY

Oil-shale and tar-sand deposits that are known throughout the world vary widely in resource potential. Few of these known deposits are associated with arid or semi-arid areas. Of these, only the tar sands of Utah and the Green River oil shales in the Rocky Mountain area of the United States, and the Moroccan oil shales contain a sufficient resource to sustain a 50,000 to 100,000 barrel (7950 $m^3$ to 15900 $m^3$) a day synthetic oil industry. If commerical development of a smaller module of 5,000 barrels to 10,000 barrels per day (795 $m^3$ to 1590 $m^3$ per day) is feasible, then oil shales of the arid areas of Israel, Jordan, Chile, and Russia may be amenable to development. Some of the many constraints on oil-shale and tar-sand development are the tremendous initial capital investment; potential solid, liquid, and gaseous waste disposal problems; and availability of water, equipment, and adequately trained personnel. The benefits that may accrue in a synthetic fuels industry are mainly economic. These benefits include a secure source of oil and any other mineral byproducts, improving the country's balance of trade position, increasing employment, and by revegetation possibly increasing the productivity of the land disturbed by industry.

## REFERENCES

Ashland Oil Inc., and Occidental Oil Shale Inc., 1977, Modification to Detailed Development Plan Oil Shale Tract C-b, Feb. 1977.

Bowman, C.W., and Carrigy, M.A., 1976, World-Wide Oil Sand Reserves Part A of Development Status of Alberta Oil Sands, 1976, Paper presented at the International Conference on the Future Supply of Nature-Made Petroleum and Gas, at Schloss Laxenburg, Austria, July 5th to 16th, 1976.

Bradley, Wilmot H., 1929, The Varves and Climate of the Green River Epoch: U.S. Geol. Survey Prof. Paper 158, p. 87-109.

Chaffin, H. S., Jr., Beavers, W. M., and Boone, P. A., 1975, Petroliferous rocks (Mississippian age) of North America: Alabama Geol. Survey Open-file rept., 33 p.

Gejort, Claes, 1968, History, Technology and Economics of Oil-Shale Utilization in Sweden, _in_ United Nations Symposium on Development and Utilization of Oil Shale Resources, sec. 2, Tallin, Estonia, U.S.S.R.: 20 p.

Hubbert, M.K., 1972, Man's Conquest of Energy, in the Environmental and Ecological forum 1970-1971, U.S. Atomic Energy Commission.

Osborne, R.C., et al., 1976, Oil Sands Development Risks, Economics and Outlook Part D of Development Status of Alberta Oil Sands - 1976, paper presented at the International Conference on the Future Supply of Nature Made Petroleum and Gas, at Schloss Laxenburg, Austria, July 5th to July 16th, 1976.

Stewart, Daniel, 1968, The History, Technology and Economics of Oil Shale Utilization in the United Kingdom, _in_ United Nations Symposium on Development and Utilization of Oil Shale Resources, sec. 3, Tallin, Estonia, U.S.S.R.: 25 p.

Warman, H., 1973, Future Availability of Oil, Paper presented at the Conference on World Energy Supplies organized by the Financial Times in Association with BOAC, London, Sept. 18-20, 1973.

# THE OCCURRENCE OF CONVENTIONAL OIL AND GAS IN DESERT AREAS

Robert E. King

## ABSTRACT

This is no direct relationship between climates of the present day and relative oil and gas resources of different regions. However, in some major oil producing regions, notably the Perisan Gulf, aridity prevailed through millions of years of geologic time, and a causal relationship probably existed. Many sedimentary basins in the interior of large continents are characterized by fluviatile and lacustrine deposits, with a deficiency in source rocks from which petroleum is generated. In some of the intracontinental sedimentary basins, there are considerable thicknesses of marine shale, sandstones, and limestones with high organic content. These are intercalated with or overlain by sedimentary rocks of predominantly terrestrial origin, and some of these have an excellent petroleum potential. In interior portions of some ancient gulfs, the sea water was brackish and the oil derived from sediments deposited in this environment is commonly very waxy, with a high pour point. This type of crude oil requires heating for pipeline and tanker transportation, and is more costly to exploit than normal crude oil. Those desert regions that formed part of the ancient Gondwanaland continent, mostly in the Southern Hemisphere, have poor petroleum prospects except along the margins of the present continents.

The petroleum prospects of some of the world's large deserts are reviewed, with emphasis on regions not presently containing large oil and gas fields. Oil has recently been found in rift basins in Chad, and similar basins in Niger and Mali, as yet little explored, may have interesting possibilities. The Hadramaut in South Yemen contains an inadequately tested deep basin. A large basin in southwestern Afghanistan is undergoing preliminary evaluation. Interior basins beneath the Iranian Plateau appear to have untested prospects. Large basins in northern and northwestern China contain some giant oil fields and they probably have a high potential for further discoveries. The coastal desert of Peru and northern Chile, though productive in the Talara district of northwestern Peru, has limited onshore prospects, but the offshore potential may be large. Many deserts, such as the Kalahari, much of interior Australia, and extensive parts of North Africa, have poor prospects, and it is unlikely that fluid oil and gas reserves of significant size will ever be found in many of these places.

## THE OCCURRENCE OF CONVENTIONAL OIL AND GAS IN DESERT AREAS

In the course of the last 450 million years of geologic time, the earth's continents have moved as immense blocks of granite, slowly creeping over a core of rock of greater density. Sometimes the continents drifted apart, and at other times they clashed together, forming mountain ranges and deep fracture zones. During this evolutionary process, the surface of the granite has moved vertically, forming plateaus where there was uplift, and basins where there was downward movement. The sea invaded the basins depressed below sea level. Clay, silt, and sand were transported by rivers and by wind into the seas from the surrounding lands, and they were deposited in the marine basins. Calcium carbonate, calcium sulfate, and sodium chloride were precipitated from sea water to form limestone, anhydrite, and rock salt in thick layers. After existing for some millions of years, most of the ancient seas vanished from the continents, but their former existence is recognized from the rocks formed by compaction and lithification of the sediments that filled the ancient basins. These sedimentary rocks are mapped in surface outcrops, and studied in wells, some of which have reached depths of 6,000 to 9,000 meters, before reaching the underlying granitic core of the continent. The crystalline rocks are referred to as "basement", being at the base of the stratified section of sedimentary rocks.

Besides the basins which were filled with sea water, many basins which resulted from depressions in the continents were never connected with the sea, and they were filled by sediments deposited in lakes or in the floodplains of rivers. These non-marine basins are particularly common in the desert interior of large continental areas. The vestiges of ancient seas, lakes, and river basins, with their shales, sandstone, limestones, and so on, are termed "sedimentary basins."

Organic matter is included in the sediments of the basins, some of it is derived from animal and plant life in the sea, and some of it is washed into the seas or lakes by rivers. If the organic matter was protected from oxidation, it was gradually altered chemically as each layer of sediment was covered by younger deposits. With increasing depth of burial, the sediments became more compact, and the temperature increased, causing alteration of some of the organic matter to oil and gas. The hydrocarbon molecules, having a low specific gravity, migrated laterally and upward through minute pores in the rock and were concentrated in underground reservoirs sealed by impermeable covers, which are referred to as "traps".

If the organic matter in the sediments was chiefly of marine origin it is proteinaceous, and the hydrocarbons generated from them are predominantly heavy molecules of liquid oil. The organic matter of terrestrial origin is chiefly humic and lignitic, being derived from land plants, and this provides a predominance of gaseous hydrocarbons, chiefly methane. The nature of the original organic matter is not the sole reason why some fields produce gas with little or no liquid petroleum. One additional reason is that a great depth of burial - 5,000-6,000 meters - and associated high temperatures, the more complex petroleum molecules crack or break up into simpler ones, eventually reducing the hydrocarbons to the simplest molecule - methane.

Among sedimentary basins there is wide variation in the amount and character of the organic matter that was buried in the rocks and later transformed to petroleum. There is generally wide variation in the porosity of the reservoir rocks and the dimensions of the traps into which the hydrocarbons migrated. The sedimentary rocks of some basins have poor porosity, low organic content, and traps of small size. Under these conditions only small oil and gas fields are present, and their economic importance is small. At the other extreme is the rich

Persian Gulf region, where the significant factors were all developed to an extraordinary degree, not only at one epoch of geologic time but throughout a time span of 250 million years, from the Permian to the Miocene, though the optimum conditions for oil generation and accumulation reached their peak in different segments of the region at different times, as the seas shifted back and forth across the basin and sediments of different kinds and varying thicknesses were deposited.

As the continents have shifted about and mountain ranges and plateaus have been formed, climatic zones also have shifted. A polar, glaciated region during one span of geologic time may later have been a nearly rainless desert or had a humid, tropical climate. The mere fact that a country or a region is a desert does not make it potentially rich in petroleum resources. It is true that much of the world's petroleum is found in deserts, such as the Persian Gulf region, North Africa, and parts of the western United States, but it is equally true that a large percentage of the earth's surface is desert. There are also rich oil and gas resources in tropical rain forests, as in Indonesia, Nigeria, northern South America, and southern Mexico. They also occur in temperate climates like the United States Gulf coast, the North Sea, and parts of the Soviet Union, and in boreal zones like Alaska and western Siberia.

Despite the diversity of climatic conditions characterizing the oil producing regions of the world, and the climatic changes that have occurred throughout geologic eras, there are interesting links between some present and ancient zones of aridity which appear to be related to the oil and gas potential of these places.

It is striking that aridity has prevailed in the Persian Gulf region since the beginning of the Cambrian, about 480 million years ago, when the first shelly invertebrates appeared abundantly in the earth. The Hormuz salt was laid down in a sea that extended from what is now central Iran to central Oman; the salt is associated with reddish colored sandstones and purple shales, all indicative of an arid environment. The Permian rocks of the Persian Gulf, which contain immense gas reserves, were deposited in an arid climate; they are characterized by dolomite and anhydrite. The Triassic is red shale and sandstone, and the Upper Jurassic, deposited about 140 million years ago, contains extensive salt and anhydrite layers, associated with the porous limestones so highly productive in Saudi Arabia. In Iran, salt was deposited in the Eocene and in the Miocene, 20-50 million years ago, while most of the rest of the Persian Gulf region was covered by a shallow sea.

Evidence of recurring aridity in this region from the Cambrian up to the present time cannot be merely coincidental to the occurrence of enormous oil and gas resources. The seas that covered the region for so many millions of years were warm and clear, and there was evidently abundant organic life, as in the modern Red Sea. Limestone banks and reefs, rich in algae and shelly invertebrates, attained great thickness in the broad, quiet seas, and lime muds rich in planktonic organic matter were simultaneously deposited in deeper water. There was little influx of sand and mud from the bordering lands, which were of low relief and very arid. Periodically, circulation of water between the Gulf basin and the deep Indian Ocean was impeded by shoals, and within the Gulf there was evaporation of trapped sea water which caused precipitation of sodium chloride and calcium sulfate, sealing the underlying porous limestone and sandstone and preventing escape of hydrocarbons.

To some degree these conditions in the Persian Gulf were duplicated in many other arid regions of the world. Large parts of the interior of continents have been rain shadow zones through tens of millions of years. Examples are the Gobi and other vast deserts of central Asia, the Iranian Plateau, the Arabian Peninsula, the Sahara and Nubian deserts, the Kalahari, Australia, and the western United States. In addition to aridity through a great span of geologic time, many of the interior regions were beyond the reach of seas which overflowed the continental margins periodically, and the basins of the interior zones were filled with sediments deposited in inland lakes, on the flood plains of rivers, or as blankets of wind-blown sands.

As a consequence of lack of marine sediments and prolonged aridity, many of the inner continental basins have a deficiency of organic matter in their sedimentary strata. In the fluviatile sands and shales oxidation destroyed much of organic molecules before hydrocarbon generation could commence. The only potential source of petroleum in these basins was in algae and other primitive organisms preserved in clays and marls deposited in ephemeral lakes. Some of these lacustrine deposits are extraordinarily rich, as in the Green River oil shales of Colorado, Utah, and Wyoming, and oil shales in various basins in China. Nevertheless, recoverable reserves of liquid and gaseous hydrocarbons in these regions is much less than on the exterior parts of the continents where sedimentary rocks of marine origin predominate.

In other interior arid regions, marine invasions took place alternately between cycles of extreme aridity. In those places, optimum conditions were present for generation and preservation of hydrocargons from the marine strata and their accumulation in porous non-marine formations associated with them. Along the margins of many ancient basins, the normally saline sea water was diluted by fresh water, and the organic matter in the sediments deposited in these brackish seas are prone to be highly waxy, with a high pour point. A notable example of this is the Serir field, deep in the interior of Libya. Oil of this type requires special facilities for pipeline and tanker transportation, and it is more costly to exploit than more normal crude oil.

In the Southern Hemisphere and Peninsular India, a special circumstance has had a profound affect on the oil and gas potential of vast continental areas, many of them deserts. This circumstance was the existence of Gondwanaland, a former huge land mass comprising South Africa, southern South America, Peninsular India, Australia, and Antarctica. This great continent overlay the southern polar region in the late Paleooic, 300 million years ago, and there is evidence of extensive glaciation at that time throughout the southern Hemisphere and in India. In mid-Mesozoic time, about 170 million years ago, Gondwanaland began to split apart. One section drifted far to the north and impinged on the Asian continent; this is now Peninsular India. Another part, East Antarctica, remained over the South Pole. There is a striking similarity between the sedimentary rocks of different parts of Gondwanaland, and they have another thing in common - they are generally unfavorable as source and reservoir rocks for petroleum. Most are of non-marine origin, low in organic content except for coal and oil shales, and the sandstones and glacial deposits are poor reservoirs, with low porosity and permeability. Exploration programs have been carried out in most parts of the former Gondwanaland, except Antarctica. Results have generally been negative, though there has been limited success in central Australia, and good success in certain places in the downwarped margins of the continents that formed after the splitup of Gondwana. The hydrocarbon prospects of large parts of the desert regions of the Southern Hemisphere are adversely affected by their Gondwana heritage.

The history of the petroleum industry shows that oil fields have generally been discovered in humid climates having abundant creeks, springs, and rivers before they have been found in deserts. In humid regions the water table is near the surface, and oil floats to the level of the ground where it is easily recognized by its iridescent film on pools of water. On the other hand, in deserts oil seeps are rarely seen because the ground water level is far below the surface and running water is rare. There are oil springs and gas seeps at the surface on the Iranian side of the Persian Gulf, where water is fairly plentiful, but on the Arabian Peninsula, which has even greater oil resources, manifestations of petroleum are few.

We will now consider the oil and gas resources, known and potential, of some of the large deserts of the world, but will make only casual reference to the major producing regions of the Persian Gulf, North Africa, and Soviet Middle Asia. Our chief interest is in the potential for discovery of large petroleum resources in sedimentary basins not now productive.

The enormous desert area that extends across northern Africa for more than 5,000 km from the Atlantic coast to the Red Sea already yields prolific oil and gas production in central Algeria and Libya, and there are small fields in the Western Desert of Egypt, but there is no commercial production yet in the more remote interior between 27° and 14° north latitude. This interior desert region is geologically varied, with broad mountainous to rugged uplifts of granitic basement accompanied by volcanic rocks in the Hoggar and Tibesti, separated by low-lying sedimentary basins. Individual basins and parts of basins have thick sections of prospective rocks, including strata of marine origin.

In mid-Cretaceous time, about 100 million years ago, during a major phase of opening of the Atlantic Ocean by the separation of Africa from South and North America, a great fracture system developed in North Africa. Large rift basins developed between bordering faults. Good examples are the Sirte basin of Libya, and the Benue trough of eastern Nigeria, which was depressed as the Cameroun Mountains rose to the east, accompanied by volcanic activity. The fracture system extended far into the interior, and deep rift basins developed in Chad, Niger, and along the Niger River near Gao. These basins were invaded by the sea during late Cretaceous time, and a thick sequence of marine sediments was deposited, intercalated with non-marine strata.

These interior basins were little known until recently, but with improvement of seismic techniques their subsurface structure has been successfully mapped, and a series of widely spaced deep wells have been drilled in Chad and Niger since 1974. Several oil discoveries have been made in two separate basins in Chad, one of them north of Lake Chad and the other about 600 km south-southeast. One of the wells, Miandoum 1, in the southern basin, tested oil at the rate of 1,660 barrels per day. The program of exploratory drilling being carried out by a group of three large oil companies is continuing. The remote interior location requires that the international price for oil be high as operations would otherwise be uneconomic.

In Sudan recent geophysical work has disclosed the existence of interior basins between basement uplifts south of Khartoum, and a preliminary evaluation of these will be made.

Elsewhere in the North African desert there are various basins containing rocks of early Paleozoic age - the Taoudeni basin of Mauritania and northern Mali, the Djado basin of northern Niger, the Murzuk and Kufra basins of southwestern and southeastern Libya, and the Reggane and Tunduf basins of southwestern Algeria. Their sedimentary rocks are of the same age as those productive in Algeria, but except for the Reggane and Tinduf basins, marine source rocks are thin. Although only a few wells have so far been drilled, the results have been discouraging, and the amounts of oil found have been small. Only the Reggane and Tinduf basins appear to have much promise. The Taoudeni basin is of vast extent, and only two wells have been drilled. There are extensive intrusions of igneous rocks, and unfavorable factor.

On the west coast of Africa, the Asiun basin of the former Spanish Sahara is a coastal embayment with Jurassic and Cretaceous sediments, largely marine, but reservoir rocks are poorly developed, and the search for oil there, which was active in the decade of the 1960's, has been suspended.

The Western Desert of Egypt has so far yielded only small oil discoveries, and is unlikely to provide major fields like those to the west in the Sirte basin of Libya.

Turning to other desert parts of Africa, many unsuccessful wells have been drilled in Somalia and southeastern Ethiopia, where sedimentary formations of the same age as those of the Persian Gulf basin were deposited in a sea that extended southwest from Arabia. A non-commercial gas discovery was made on the Somali coast from a Cretaceous formation.

In southern Africa, the Kalahari Desert of Namibia and Botswana covers several shallow basins, but they are filled by sediments characteristic of the old Gondwana continent which overlie pre-Cambrian limestones. In the Republic of South Africa these strata have been drilled in numerous wells without success, and further prospects are poor.

Surface oil seeps in various parts of the Red Sea have encouraged drilling, but the only successes have been a gas discovery made in 1976 off the coast of Sudan, a gas well off the coast of Eritrea, and another in the northern part of the sea off the Saudi Arab coast. The Red Sea is characterized by immense thicknesses of salt, high pressures, and high temperatures; the high temperatures probably converted most of the organic matter in the sediments into gas rather than oil. There are large salt domes in the coastal plain of western Yemen, but wells drilled in this area have been dry.

In South Yemen there is a basin with numerous salt domes in the Hadramaut. The sedimentary rocks are of the same age as those of the prolific fields of Al Hasa province of Saudi Arabia. The Hadramaut basin has not been fully evaluated.

The Makran coast of southeastern Iran and southwestern Pakistan has a thick section of sedimentary rocks, but it is generally deficient in good source and reservoir potential.

Except for small oil fields in the Potwar basin near Lahore, only gas has been found in Pakistan. Sufficient exploration has been carried out to justify the conclusion that there is only a small likelihood that large oil fields will be found along the Indus valley. A sedimentary section thins eastward from the Indus and wedges out beneath the Thar Desert of western India, but prospects there are not encouraging.

Gas is produced in northern Afghanistan, but elsewhere in that country there has been little exploration. The Helmand depression in the southwestern part of Afghanistan has petroleum prospects, and these are scheduled to be evaluated.

In the Iranian Plateau there are several separate basins. Oil has been found near Qum and gas in a field to the southeast, but in other parts of central Iran exploration so far has not been successful, though an equivalent of the Asmari limestone, highly productive in the Zagros foothills of Khuzistan, is present. The Dasht-e-Lut is an enormous desert area with shallow basement rocks, but the Dasht-e-Kavir covers parts of sedimentary basins that appear to have still untested prospects.

In northeastern Iran a major gas field, Khangiran, in the Sarakhs basin near Meshed is being connected by pipeline to industrial centers. In the adjacent Turkmen S.S.R., in the Soviet Union, there are large gas fields, notably Shatlyk, and recently another large gas field, Dauletabad, has been outlined on the Soviet side of the frontier with Iran. These fields are in the Kara Kum basin, and production is chiefly from Jurassic limestones. Elsewhere in Soviet Middle Asia, a chain of large gas fields in the Kyzyl-Kum Desert along the northeast side of the Amu Darya River of Uzbekistan are the chief source of gas transported by pipeline to central Russia and to industrial centers in the Urals.

In northern and northwestern China there are several sedimentary basins of immense size (1, 2). Because they are near the center of the Asian continent they were beyond the reach of seas overlapping the continental border, and they are filled with fluviatile and lacustrine sediments. The prolific nature of some of the fields in these interior basins is an exception to the general worldwide rule that large oil reserves are found only where a considerable thickness of marine source rocks is present.

In the far northwest of China, in Sinkiang, there are three basins, Dzungaria, Tarim, and Turfan. In the Dzungaria basin, the Karamai field has a reserve of 750,000,000 barrels of oil, a true giant sized accumulation of oil. The Tarim and Turfan basins have been little developed. To the east are several basins of smaller size; being nearer to consuming centers of the country than the Sinkiang basins, they have been more fully explored and developed. In the Yumen district of Kansu are several fields with total oil reserves of about one billion barrels, and in the Tsaidam basin to the southwest in Tsinghai Province there are also large fields which have not been fully developed. Still farther east small fields have been found in the Ordos basin of Shansi and Kansu. No doubt many additional fields will be found in the immense interior basins of China.

The desert interior of Australia contains various sedimentary basins. Many of them are characterized by rocks of early Paleozoic and pre-Paleozoic age which have been altered by solution and recrystallization, resulting in low porosity and permeability. Oil and gas have been found in formations of this type in the Amadeus basin of central Australia. The Cooper basin of northeastern South Australia has a number of gas fields, which are connected by pipeline to consuming areas. Production here is from the Permian, in which the sandstone reservoirs have not been severely altered. In general, there is a deficiency in source rocks in central Australia, the strata being chiefly of non-marine origin. Extensive exploration programs were carried out during the 1960's with mediocre success. Attention in Australia is now primarily focused on the continental shelves on the borders of the continent, where large gas and oil fields are present.

In South America the only oil production on the desert west coast of Peru and northern Chile has been in the Talara district of northern Peru, where there are thick marine strata of Tertiary age. Farther south, basement rocks extend nearly to the coast line and undisturbed marine sedimentary formations are thin to absent. The greatest potential for oil is offshore, where a sufficient thickness of sedimentary formations is present. There has been successful offshore drilling in the Talara district, but only two exploratory wells, both dry, have been drilled farther south off the Peruvian coast, and the potential of this zone has not been fully evaluated.

## REFERENCES

(1) Meverhoff, A.A., Developments in Mainland China, 1949-1968: Amer. Assoc. of Petroleum Geologists, Bull., vol. 54, p. 1567-1580.

(2) Connell, H.R., 1974, China's Petroleum Industry - an Enigma: Amer. Assoc. of Petroleum Geologists, Bull., vol. 58, p. 2157-2172.

# Minerals in Deserts

George I. Smith

U.S. Geological Survey

Menlo Park, California

## ABSTRACT

Almost any kind of mineral deposit can occur in desert areas, and the lack of vegetation and soil cover makes finding them easier. Some kinds of deposits, though, are more likely to occur in deserts than elsewhere. Some of these result from processes genetically related to the present desert climate that improved lower grade deposits of ore. One such process, termed "secondary enrichment", is most effective in areas with deep water tables, and many low-grade copper, silver, and uranium deposits have been converted into mineable ore by the downward migration and redeposition of soluble metals. In a desert terrane, "placer" processes are effective whenever running water flowing over steep slopes erodes outcropoing ore bodies and transports and concentrates the heavier ore minerals at lower levels, thus converting low-grade or hard-to-mine bedrock deposits into economically workable concentrations. Other kinds of deposits are better preserved in deserts because the lower rainfall at the surface, and the lower volume of flow and the greater depths to groundwater, result in less destruction of soluble ores; deposits of salines and phosphates are the most notable ores affected by these factors. Still other ore deposits are created as a consequence of the arid climate, mostly because the high evaporation rates operating on standing bodies of water produce brines that can lead directly to concentrations of salts and indirectly to secondary minerals, such as zeolites, that are produced by reaction of silicate minerals with saline waters.

## INTRODUCTION

Some of the largest and most valuable deposits of minerals in the world are in deserts. For many deposits, this association is simply a coincidence; for others, however, there are geologic reasons that genetically relate the mineral deposits and the present arid climate. This paper is a brief outline of the geologic relation between arid climates and certain mineral deposits. Thus, it can be used by those involved in planning and developing desert areas to understand better the types of mineral deposits that are especially likely to be found and how they relate to the environment.

Deserts are areas that have several advantages over more humid regions for the discovery and development of mineral resources. Almost any type of deposit can occur in desert regions, and the greatest single advantage to those prospecting for them is the lack of deep chemical weathering and extensive soil and vegetation cover. The bedrock of mountain areas is more easily observed, mapped, and assessed, and the valley floors are more easily transversed or surveyed by geophysical techniques. I suspect that a much larger percentage of the outcropping deposits of ore in the world have been discovered in the desert regions than in the temperate and tropical regions.

Mineral deposits that were formed by geologic processes more than about 10 or 20 million years ago tend to be in belts that bear little or no genetic relation to today's climatic or geologic belts. Geologically, these ages are young, and unfortunately most mineral deposits are older. Some, however, were created during the last 20 million years, and they will commonly be in the same belts as are geologically young mountains, active volcanos, and present earthquakes. These belts are shown, generallized, in figure 1a. The deserts of the world are shown in figure 1b. The most promising areas for the discovery of <u>geologically young</u> mineral deposits in deserts, therefore, are where these zones overlap. The overlap of these areas is greater than random chance alone would suggest because many of the deserts in the world are caused by the rain shadow on the downwind side of high mountain ranges, and such ranges are one of the indications of geologically recent activity.

Some mineral deposits, however, are <u>improved</u>, <u>preserved</u>, or <u>created</u> by geologic processes that occurred as a consequence of an arid climate. Such deposits are clearly among the most likely to exist in deserts, and discovery and development of them could become a major item in an area's economic future. However, disadvantages also stem from the arid character of an area. The extraction and processing of many resources require large quantities of water, and the cost--or impossibility--of getting it makes some deposits uneconomic. The energy needed in mining and transporting mined products to markets is also a large part of the ultimate cost of the product. Arid and sparsely populated regions that do not have established electrical, rail and highway networks commonly cannot compete with more favorably placed deposits of the same resource.

## <u>DEPOSITS THAT ARE IMPROVED IN DESERT ENVIRONMENTS</u>

Deposits of metallic and non-metallic minerals are concentrated or converted into more easily worked forms by two processes that are common or intrinsic to deserts. The more important of these is the process of <u>secondary enrichment</u> of low-grade ore bodies that converts them to mineable grade. The process is one of leaching the ore minerals by downward moving, naturally acidified water, and redepositing their components in zones just above or below the water table (<u>see</u> Anderson, 1955). Some of these remobilized components are redeposited and concentrated in zones just above the water table as oxides, carbonates, or silicates; others are greatly enriched in zones just below the water table as sulfides (fig. 2). The basic requirements for

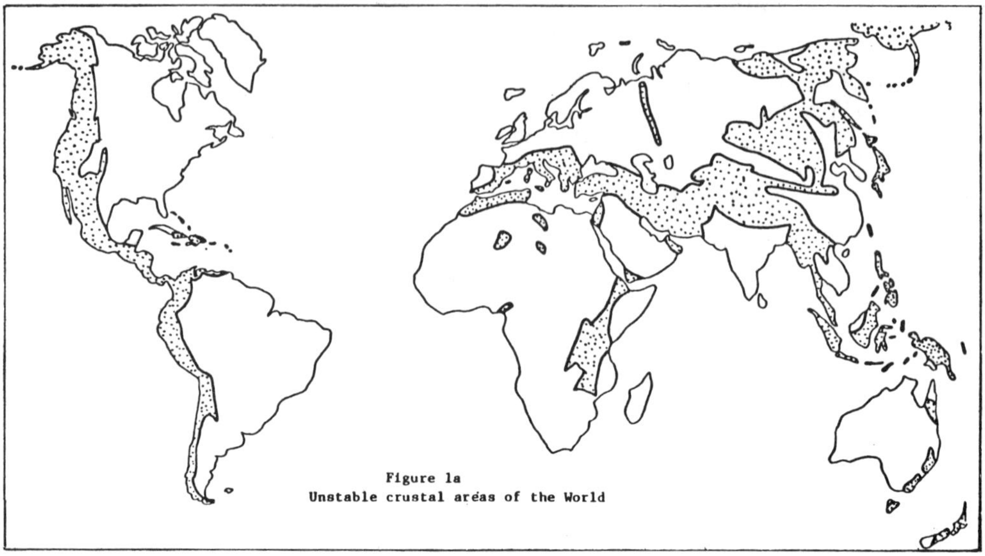

Figure 1a
Unstable crustal areas of the World

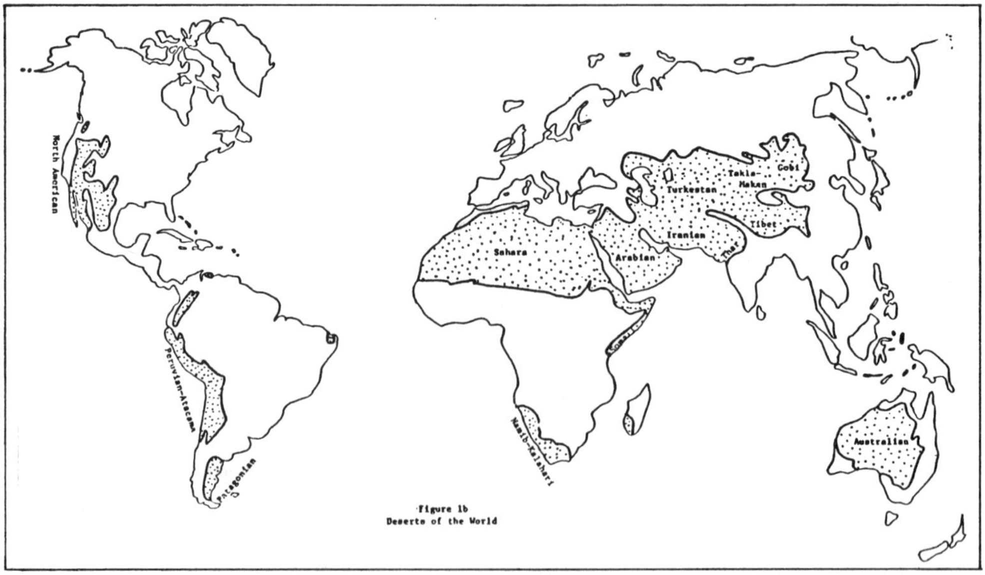


Figure 1b
Deserts of the World

secondary enrichment processes to work are (1) a water table that is substantially below the surface, (2) an ore body that extends above the water table, and (3) sufficient volumes of acid water--at sometime in the past--to leach the ore minerals and transport them downward to the zone of redeposition.

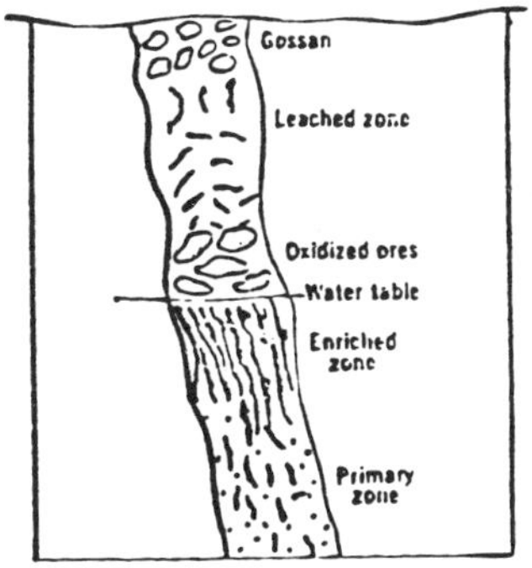


Figure 2. Diagrammatic relation between various parts of an ore body that of an ore body that has been improved by secondary enrichment. From Bateman (1950, p. 245).

Desert environments are favorable for this process because the water table is much more likely to be at moderate to great depth than in humid regions. The greater the depth to the water table, the larger the volume of low-grade ore that is leached and redeposited in a narrow zone. For example, at Bisbee, Arizona (copper, zinc, lead, silver, and gold), more than 400 m of overlying rock was leached by ground water; at Tintic, Utah (lead, silver, zinc, copper, and gold), nearly 700 m of rock was leached. Areas having the hydrology of the Amazon basin, for example, could not expect enrichment to these depths because the water table is virtually at the surface.

The large-scale changes in the climate that occurred during the Pleistocene ice ages (approximately the last 3 million years) caused major fluctuations in the amount of water available for leaching and in the position of the ground water table. Many deserts, especially in the temperate latitudes, now contain numerous ore bodies that owe their value to those climatic fluctuations and this process.

The mineral deposits most markedly affected by this process are those composed of minerals that react with ground water most easily; deposits containing copper, silver, and uranium, have been beneficially affected by this process the most frequently, although some lead, vanadium, zinc, iron, and nickel deposits owe part of their value to it. Of the 15 major copper deposits in the Western Hemisphere listed by Ridge (1972, p. 672) as having been strongly and favorably affected by ground water, 13 are in deserts; all of the 3 silver deposits are in deserts; both of the zinc deposits are in deserts; and one of the iron deposits is in a desert. Nearly half of all the copper-producing deposits in the Western Hemisphere listed by Ridge (p. 660) are in deserts. Uranium deposits are not tabulated individually in Ridge's book, but Walker (1963) concludes that nearly all of the uranium bearing veins in the United States have been altered by limited movements of ground water, and a large percentage of the commerically exploited uranium deposits in this country are in its arid to semi-arid regions.

The second enrichment process that occurs in desert environments is concentration by placer processes. By no means are placer processes limited to desert environment--the more common examples are in fact in humid, tropical, or shoreline environments--but they are locally important in deserts as well. The requirements for placer processes to be beneficial are to have outcropping ore bodies composed of minerals that are (1) not easily dissolved in water or rapidly weathered, (2) durable during transport by running water, and (3) heavy. Gold is the most frequently mined mineral from placer deposits, but minerals containing tin, tungsten, chromium, titanium, iron, zircon, and elemental platinum and copper are concentrated by placer processes.

Desert environments provide the basic requirement for placer concentration during brief periods of intensive rainfall. In some instances, the present climate produces the required conditions, in others, only past climates were adequate.

A form of placer deposit resulting from wind has been reported from some deserts. Weathered gold veins that crop out at the surface are eroded by wind, with the lighter minerals blown away and the heavier gold being left as a concentrate.

## DEPOSITS THAT ARE PRESERVED IN DESERT ENVIRONMENTS

Some types of mineral deposits consist of minerals that dissolve in water, some slowly and some rapidly. Deserts are much more favorable sites for the preservation of such deposits, and in extremely arid regions they even crop out.

The most obvious example of ores having these qualities are the saline deposits--beds composed of soluble minerals such as salt and gypsum. Some saline deposits are products formed by the evaporation of seawater from large coastal basins, others by the evaporation of water from inland lakes. Deposits of both occur in himid as well as desert regions, but preservation in the humid (as well as some arid) regions requires that the ore body be enclosed in virtually impermeable rock to avoid being dissolved.

Discovery of saline deposits--especially those small portions that contain the most valuable salts--is relatively easy in deserts as compared to other environments. In extremely arid regions, outcrops of halite (salt) occur, and discovery of the purest deposits or the valuable zones containing potash salts is a much simpler and rapid matter than in any other environment. In semi-arid areas that have a near-surface ground water table or too much rainfall to allow soluble salts to survive and crop out, small closed topographic depressions caused by incomplete solution of shallowly buried salts sometimes reveal the presence of deposits. Solution of saline deposits underlying humid areas is usually more complete, and surface expressions caused by it are less obvious or non-existent.

Besides saline deposits, certain other mineral commodities tend to be better preserved in deserts because of the dry climate. Probably the most important of these are deposits of phosphate and uranium. Most phosphate deposits are composed of the minerals in the apatite group that are slightly soluble in water, especially rain water saturated with atmospheric carbon dioxide, and climates that are characterized by moderate to large quantities of rain tend to destroy surface occurrences. Six-valent uranium minerals produced by secondary enrichment processes are also easily dissolved by rainwater. In desert environments, on the other hand, both are preserved and occur in outcrops, and prospecting for them is more simple and inexpensive.

Preservation of the near-surface evidence of where mineral deposits have been partially destroyed can be nearly as useful as outcrops of the deposits themselves. The geologic processes that lead to secondary enrichment of sulfides in arid regions, with their deep water tables, ultimately lead to the destruction of all near-surface expressions of the original sulfide deposits. In many deposits in arid regions, however, leaching has been incomplete and distinctively colored "gossans" (composed of the iron oxides that remained behind when the other sulfide minerals were leached) guide mineral exploration in a manner that is impossible in areas having deep soils and vegetation cover. Also, moderately soluble and commonly brightly colored secondary minerals (such as malachite and azurite) survive in a desert climate and may be followed by the prospector for many miles in alluvium back to the outcropping source. Many important discoveries of rich copper maines in the southwestern United States and northern Mexico were made in this way.

## DEPOSITS THAT ARE CREATED IN DESERT ENVIRONMENTS

Several types of mineral deposits form in deserts as a consequence of the low rainfall and high evaporation rate. This paper is concerned with areas that are deserts under our present climatic regime. Therefore, the mineral deposits that are of interest here are those that formed in geologically recent times when the climate of the area was similar. Desert conditions conductive to the formation of appropriate mineral deposits may have also existed in a given area in much older geologic periods as well, but that would generally be a coincidence.

The most important type of deposit in this category is continental saline deposits--salt layers, in some instances accompanied by the original residual brines, formed by evaporation of water from inland lakes and seas. Deposits of this type come in all sizes. Small deposits occur in many parts of the world; large deposits containing a billion dollars or more worth of industrial chemicals occur in several areas. Three such deposits occur in the arid parts of the western United States (trona deposits in the Green River basin of Wyoming, borax deposits in the Kramer basin of California, and mixed salts in Searles Lake of California). Two of these are genetically related to the present aridity of the area under which they lie; the one in Wyoming occurs in an arid region only by coincidence. Other arid parts of the world seem likely to be underlain by comparable deposits.

Conditions that led to the accumulation of such deposits were (1) the existence of a large closed basin, usually a consequence of crustal faulting or warping but in some instances of wind erosion, (2) large volumes of water from high mountains within the drainage area or from adjacent less arid regions, and (3) a source of unusual components that can remain disolved in the inflowing water until it reaches the basin where desiccation occurs. Most examples of the third requirement are found in areas where older saline deposits were being dissolved or where mineralized springs associated with geologically recent tectonic and volcanic activity drain into the accumulation area. Smaller volumes of water containing higher concentrations of dissolved solids would produce the same end result, but this combination is geologically less likely.

In many instances, the valuable components of these deposits are in the brines rather than the solids. Brines are, in effect large low-grade ore bodies which, because they are inexpensive to extract and process (commonly using solar energy to achieve concentration), can make economic extraction of the dissolved solids economic realities.

Besides salts and brines, the alteration of silicate minerals that are immersed in saline waters and brines produced by the desert environment leads to commercial deposits of minerals. Numerous deposits of naturally occurring zeolites are apparently the result of this process; they are becoming used more frequently to replace synthetic zeolites as catalysts, selective sorbents, and desiccants. Low-grade deposits of fluorite, a major source of fluorine, are known where volcanic ash layers were apparently submerged in fluoride-rich saline lake waters of arid regions, thus allowing chemical alteration to occur. Deposits of nearly pure potassium feldspar, used by glass and ceramic industries, have also been attributed to reaction of volcanic ash with brines.

An extreme instance of an ore deposit being "created" by an arid environment is the Achasai lead-zinc deposit in Turkestan. This deposit is a very low grade ore body that is minable only because a very large amount of fresh water drains into the mine. The water must be pumped to the surface where it becomes, in the desert, a sufficiently valuable "coproduct" to make operation of the mine economically feasible.

In the last few years, extensive speculation among geologists has revolved around the role of the large tidal flats that exist in arid regions--sabkhas--in the formation of metallic ore deposits. The sabkhas under consideration are environments that existed tens or hundreds of millions of years ago, and the mineralization processes involved were slow and required a large part of the intervening time, so those mineral deposits have no genetic relation to our present arid belts. Nevertheless, many of our present arid belts lie in areas that have been arid for millions of years, and possibly incipient and low-grade mineralization has occurred in zones that were sabkhas and now lie buried beneath the surface of the present desert.

## SUMMARY

Desert areas are exceptionally favorable areas for the existence of valuable deposits of minerals. Some have been improved by processes genetically related to the present arid climate, others have been better preserved in it, and still others are created as a direct consequence of it. All of these types of deposits, as well as deposits of other types, are more easily discovered in desert areas because of the lack of extensive vegetation, deep chemical weathering, and soil cover.

## REFERENCES AND SELECTED BIBLIOGRAPHY

Amiran, D.H.K., and Wilson, A.W., (1973), *Coastal Deserts, Their Natural and Human Environments*, Arizona University Press

Anderson, C.A., (1955), Oxidation of Copper Sulfides and Secondary Sulfide Enrichment, in Alan M. Bateman, ed., *Fiftieth Anniversary Volume*, The Economic Publishing Co., Lancaster Press Inc.

Bateman, A.M., (1942), *Economic Mineral Deposits*, J. Wiley & Sons, New York, N.Y.

Brobst, D.A., and Pratt, W.P., (editors)(1973), *United States Mineral Resources* U.S. Geological Survey Professional Paper 820

Lefond, S.J., et al., (editors) (1975), *Industrial Minerals and Rocks*, American Institute of Mining, and Petroleum Engineers, Inc. New York, N.Y.

McGinnies, W.G., et al., (1968), *Deserts of the World*, Arizona University Press, Tucson, Arizona

Ritcher-Bernberg, G., (editor), (1972), *Geology of Saline Deposits*, Unesco, Paris

Ridge, J.D., (1972), *Annotated Bibliographies of Mineral Deposits in the Western Hemisphere*, Geological Society of America Memoir 131

Walker, G.W., (1963), *Supergene Alteration of Uranium-bearing Veins in the Conterminous United States*, U.S. Geological Survey Professional Paper 455-E

# RECOVERY OF METALS AND MINERALS IN DESERT REGIONS

Roshan B. Bhappu
Mountain States Research and Development
Tucson, Arizona 85731

## ABSTRACT

Like any other part of the earth, nature has endowed the arid and semi-arid regions with valuable mineral resources. However, the difficulty of obtaining an adequate water supply at reasonable cost has hampered the development of mineral resources in these water-short areas. This paper is concerned with the dry and low-water consuming mineral processing and extractive technology currently available for recovering these mineral resources. The paper also covers potential extractive processes and techniques that could be used in the future and provides guidelines for education, research and development input necessary for extracting these mineral resources in the developing countries located in water-short areas.

## INTRODUCTION

Man, by his ability to utilize the mineral resources of the earth, has made tremendous progress in a relatively short geologic time. At the dawn of civilization, he used crude stones, flint, chert, and bone implements which were the only useful raw materials available. Chaldeans, Babylonians, and Egyptians made extensive use of clay for pottery, bricks, and tiles. Neolithic man was acquainted with native metals such as gold, copper, and tin, mostly from placer sources. Herodotus (484?-425 B.C.), Theophratus (372-287 B.C.), Strabo (A.D. 19), Avicenna (980-1037), and Agricola (1494-1555) in his famous De Re Metallica, have all described the occurrences, methods of extraction, and uses of metals, gemstones, earths, and salts. At the termination of the Dark Ages, the major metals in use were iron, copper, lead, tin, gold, silver, and mercury.

With the advent of the Industrial Revolution three centuries ago, man learned to produce iron from ores using coal and mechanical power. Since that time the use of metals increased rapidly and each progressive step in the history of man brought about increasing demand for old metals and the discovery and development of newer ones. During World Wars I and II, more metals and minerals were used than in all preceding history.

Today more than one hundred minerals are involved in international trade, some being necessary and vital for our essential industries, while others contribute to unessential industries or luxuries. Regardless of the end use, it is evident, that with increasing population and with the industrialization of developing countries, the demand for our mineral resources will continue to increase. This trend is clearly indicated in Table I which shows the development over the last century of two major base metals, iron and copper.

Like any other part of the earth, nature has endowed the arid and semi-arid areas with valuable mineral resources. However, the difficulty of obtaining an adequate supply of water at reasonable cost is one of the major problems seriously hindering the economic development of these resources in such water-short regions. This problem is particularly acute for extraction of mineral values from naturally occurring ores using the well-established mineral processing and extraction processes which require considerable amounts of water. For this reason, the use of dry concentration and extraction techniques requiring no or limited amounts of water should be considered for the development of mineral resources in those areas.

The minerals in Table II, unlike renewable resources of the soil, are "one-crop" nonrenewable resources. They are vanishing assets, most of which, once mined, are gone forever.

These concepts bring up a very crucial question about our civilization. Does the earth contain sufficient quantities of our depleting mineral resources to support and expand our complex civilization? The answer to this question is not simple, since estimating the capabilities of the earth to supply mineral resources involves a large number of uncertainties. In the first place, many estimates of the available reserves and locations of different mineral deposits are not reliable. Secondly, whether a particular type and grade of mineral material constitutes or can become an ore (a mineral deposit that can be exploited at a profit), depends on numerous economic factors that vary in time. With a breakthrough in mining or extractive technology or an increase in price, worthless rock can be converted to an ore overnight.

At this point we are confronted with another crucial question. Are there any limits to the grades of ore that can be processed? From a theoretical as well as a technological viewpoint there is no lower limit as was illustrated beautifully by the enrichment of $U_{235}$ from uranium metal at the Argonne Laboratory during World War II. In this case, the content of $U_{235}$ in uranium metal amounted to only a few parts per million and the extraction involved a very complex chemical processing scheme which obviously was very expensive and uneconomical.

Today, we are economically extracting copper from ores containing around 0.5 percent copper, and there is no doubt that if there is a critical need for the metal in the distant future, we will continue to extract copper from lower and lower grade ores containing from 0.1 percent to 0.01 percent copper. However, such an achievement could be possible only if a suitable metallurgical process and the required energy for processing are available. Keeping this in mind any material in the earth's crust can be considered a potential source if sufficient energy is readily available. Thus from a technical standpoint, we can safely state that our resource potential will be proportional to the available energy reserves. Since all the technical reports concerning the availability of energy to man indicate an enormous reserve to satisfy the highly industrialized world for many years, it is evident that we will continue to obtain the required metals from lower and lower grade ores and ultimately from mediocre metal-containing-rocks such as shale and granite.

Table I. Development of World Production of Base Metals

| YEAR | WORLD POPULATION | IRON PRODUCTION | | | COPPER PRODUCTION | | |
|---|---|---|---|---|---|---|---|
| | | TONS | PER CAPITA (KGS) | INDEX | TONS | PER CAPITA (KGS) | INDEX |
| 1800 | 850,000,000 | 830,000 | 1 | 100 | 9,000 | 0.01 | 100 |
| 1850 | 1,070,000,000 | 4,700,000 | 4.4 | 440 | 50,000 | 0.05 | 467 |
| 1900 | 1,440,000,000 | 39,810,000 | 27.6 | 2,764 | 600,000 | 0.42 | 4,170 |
| 1950 | 2,625,000,000 | 134,000,000 | 51.4 | 5,140 | 2,930,000 | 1.1 | 11,100 |
| 1970 | 3,700,000,000 | 444,000,000 | 120 | 12,000 | 6,300,000 | 1.7 | 17,000 |
| 2000 (projected) | 6,500,000,000 | 1,400,000,000 | 215 | 21,500 | 20,000,000 | 3.0 | 30,770 |

Table II. Abundance and Availability of Minerals

The major metals and minerals used in the industry are grouped as follows:

| METAL OR MINERAL GROUP | EXAMPLES |
|---|---|
| MINERAL FUELS | COAL, OIL, GAS |
| NUCLEAR FUELS | URANIUM, THORIUM |
| IRON | HIGH GRADE AND TACONITE ORES |
| IRON-ALLOY METALS | MANGANESE, CHROME, NICKEL, TUNGSTEN, MOLYBDENUM, VANADIUM, COBALT |
| NONFERROUS METALS | COPPER, ZINC, LEAD, TIN, ALUMINUM, MERCURY, ANTIMONY, TANTALUM, TITANIUM |
| PRECIOUS METALS | GOLD, PLATINUM, SILVER, PALLADIUM |
| METALLURGICAL MINERALS | FLUORSPAR, BARITE, GRAPHITE, BAUXITE, REFRACTORIES |
| CHEMICAL MINERALS | SULFUR, BORAX, SALINES |
| FERTILIZER MINERALS | PHOSPHATE, POTASH, NITRATES |
| STRUCTURAL MINERALS | STONES, CEMENTS, GYPSUM, BITUMENS |
| CERAMIC MINERALS | CLAY, FELDSPAR, SILICA |
| ABRASIVE MINERALS | DIAMOND, CORUNDUM, EMERY, GARNET |

## MINERAL RESOURCES IN ARID REGIONS

Because of the random distribution of mineral resources throughout the world, it is likely that as many mineral deposits including saline deposits of recent origin are locked in arid regions as there are in the wet areas. Such deposits are easy to locate and to identify due to lack of vegetative cover, and they have been mined extensively. They are currently the principal economic element in large parts of the arid regions, and these resources may be expected to assume even greater significance in the future.

The vast deposits of copper in the Southwestern USA, Chile, Peru, Yugoslavia, Turkey, and Iran; tin in Bolivia; phosphates in Egypt, Turkey, Israel and Jordan; lead-zinc in Australia, Iran, Turkey, Arab Republic; iron ore in Australia, Peru, Brazil, and Turkey; chromite in Turkey, Yugoslavia, and Pakistan; gold-silver in Western USA, Northern Mexico, Australia; and uranium deposits in Southwestern USA and Western Australia are classic examples of valuable minerals located in arid regions of the world.

Some of the most important non-metallic mineral resources such as pumice, perlite, scoria, beryllium, mica, lithium minerals, phosphates and clays are also found in arid regions of the world and these are being exploited at increasing rates to meet the growing demands.

In the current exploration activities, geologists and geophysicists are using modern airborne equipment for reconnaissance exploration equally effectively over jungle and deserts. In recent years, the increase in success factor in the search for hidden mineral deposits can be directly attributed to the dynamic evolution of exploration geophysics as a spin-off resulting from significant contributions in the field of space geosciences; especially, due to the development of super-sensitive detecting, measuring, and recording equipment which allow the pickup of weaker and weaker signals against backgrounds of geologic and instrumental noise. Such techniques are specifically effective in desert regions covered with sand dunes and alluvium.

After a mineral deposit has been located, explored, and its ore reserves and grades determined, the next phase is the recovery of valuable metals or minerals by mining. Over the years, mining has evolved into a very scientific and exact engineering discipline and, currently, a large number of surface and underground mining methods are available for mining any type of mineral deposit even in the arid and semi-arid regions as long as the economics are favorable. Thus, today there are open pit mines capable of mining 100,000 to 300,000 tons per day of ore and waste in the Southwestern US porphyry copper operations, and there are underground mines capable of mining gold ores as deep as 10,000 to 12,000 feet in South Africa. Moreover, new mining techniques such as in situ or solution mining and ocean mining are now being developed to extract mineral resources heretofore not exploited.

## EXTRACTIVE METALLURGY IN ARID REGIONS

Metallurgy may be defined as the art and science of extracting metals and minerals economically from their ores and other metal-bearing products and adapting these for human utilization. The field of metallurgy includes all the concentration processes and techniques involved in separating the valuable metals or minerals from the worthless host rock (gangue) as well as the smelting, refining, and working of the extracted minerals. In achieving this goal, metallurgists utilize all the basic laws of physics, chemistry, and even biology to the concentration, extraction, purification, alloying and working of metals to meet the demand of our minerals-based society. The following Table III lists the various processes involved in extractive metallurgy and shows their interrelations.

The metal industry of the world is exceeded in importance only by agriculture. Currently, more than 6,000 separate metals and alloys are used commercially and new alloys are being developed constantly.

One only has to look around in order to appreciate the significant contribution that metals and minerals have made to our existence by providing the necessities of life, health, security, economy and well-being.

### Mineral Processing

Mineral concentration, mineral processing or ore dressing is the art and science of treating natural ores to separate valuable minerals (concentrates) from worthless gangue (tailings) on the basis of physical or chemical properties of the minerals. The concentrate is next subjected to a chemical treatment to break up the metal-bearing compounds to liberate the metallic elements. Such chemical processes are quite expensive as a rule, and the cost is proportional to the bulk of material treated. Thus, it is economically more advantageous to treat a small amount of enriched concentrate than a large volume of low grade ore.

Separation of valuable minerals on the basis of physical properties may depend on the size, shape, and color of the particles; on their magnetic susceptibility; and on their electric or heat conductivity. The above separations can be carried out in water or in air under dry conditions. In general, wet processing shows a more effective separation than dry concentration.

Table III. Process Involved in Extractive Metallurgy

RUN-OF-MINE ORE → MINERAL PROCESSING

↓

CRUSHING—GRINDING
HAND PICKING—GRAVITY
MAGNETIC—ELECTROSTATIC
ELECTRONIC—FLOTATION

↙ CONCENTRATES OR UNPROCESSED ORES

↘ TAILINGS (TO WASTE)

| PYROMETALLURGY | HYDROMETALLURGY | ELECTROMETALLURGY |
|---|---|---|
| ROASTING | LEACHING | ELECTROWINNING |
| CALCINING | PRECIPITATION | ELECTROREFINING |
| SMELTING | AMALGAMATION | FUSED SALT ELECTROLYSIS |
| REFINING | CYANIDATION | |
| | ION EXCHANGE | |
| | SOLVENT EXTRACTION | |

In mineral processing, the amount of new water required depends on the method employed and on the extent of water reclamation utilized for recovering process water for recycling. The following table lists the water-to-ore requirements for the known concentration methods:

Table IV. Water Requirements for Various Concentration Methods

| PROCESS | WATER:ORE | WATER RECLAIMED (%) |
|---|---|---|
| SCRUBBING AND WASHING | 4-20:1 | up to 50 % |
| GRAVITY CONCENTRATION | 4-15:1 | up to 75 % |
| FLOTATION | 2-4:1 | up to 80 % |
| HYDROMETALLURGICAL | 2-1:1 | up to 85 % |
| HEAVY-MEDIA-SEPARATION | 2-1:1 | up to 90 % |
| DRY CONCENTRATION | 0:1 | - |

As can be seen, the majority of the wet concentration methods require a rate of 2:1 to as much as 20:1 for water to ore treated. However, it is also clear that with appropriate water reclamation systems it is possible to recover 75 to 90 percent of the process water for recycling. Thus, even in water-short areas, wet processing cannot be ruled out for the treatment of a given ore. It may be that the mine itself will provide the required water for processing.

Assuming that there is very little water available for processing, there are several dry processing methods and techniques available for recovering metallic mineral resources in water-short areas. These processes include:

1. HAND PICKING
2. DRY CONCENTRATION
3. DIFFERENTIAL DISINTEGRATION
4. MAGNETIC SEPARATION
5. ELECTROSTATIC SEPARATION
6. ELECTRONIC SEPARATION
7. HEAVY LIQUID SEPARATION
8. THERMAL CONCENTRATION
9. RADIO-ACTIVITY SORTING
10. FLUORESCENCE SORTING

The above mentioned methods clearly indicate that the majority of dry concentration procedures are based on physical properties of the mineral such as separating heavy galena (PbS) from quartz ($SiO_2$) by dry pneumatic tabling, or separation of iron bearing minerals by magnetic separation. To a lesser extent, techniques are also available for separation, based on the chemical properties of the minerals such as separation of sulfur or stibnite from their ores by melting, and recovering mercury by retorting.

While some dry concentration methods such as gravity, thermal concentration, and magnetic separation have been practical for years, other techniques, such as electrostatic separation, sorting by electronic selection and heavy-liquid separation have resulted from very recent technological advancements. Through the application of such newer procedures, a large number of metallic, non-metallic and rare-metal minerals can now be recovered with a minimum use of water at or near the mine.

## Hand Picking

Though considered menial and unscientific, hand picking may still be considered an effective mineral concentration method in water-short areas. This is specially true if the valuable mineral pieces are different enough in color or physical nature to be identified, large enough to be picked, the mineral value sufficiently high and the cost of labor sufficiently low to result in favorable economics.

Some of the mineral resources currently concentrated by hand picking include diamonds, gems and semi-precious stones, gold-bearing host rock, coal, chromite, manganese, molybdenite, mica, clays, barite, fluorite, and tin. In recent years, several innovations in hand picking technology have been developed such as modified conveyor belts, hand sorting tables and many labor-saving and ease-of-operation devices.

## Dry Gravity Concentration

The specific gravities of a representative group of minerals arranged by classes and decreasing gravities are presented in Table V. This table shows a wide range of specific gravities from near 1 to 19. It is interesting to note that most of the valuable minerals of base and rare metals show the higher specific gravities (with specific gravity range from 2.6 to 3.3). Accordingly, the separation of these valuable minerals from their ores is relatively easy, such as separation of galena from quartz. On the other hand, if the two or more such heavy minerals are present in the ore (such as cassiterite and wolframite) the separation of one from the other would be difficult if their specific gravities are close.

Currently, a large number of dry (pneumatic) concentration devices are available for economic recovery of valuable minerals in the arid areas. These devices use air to effect differential movement of mineral particles of varying specific gravities, shapes and sizes. These include pneumatic jigs and tables, fluidized bed separators using fine particles of a heavy substance in air having an apparent

Table V. Specific Gravities of Minerals

| MINERALS | COMPOSITION | SPECIFIC GRAVITY |
|---|---|---|
| GOLD | Au | 19.3 |
| URANINITE | $UO_2$ | 10.7 |
| COPPER | Cu | 8.8 |
| GALENA | PbS | 7.5 |
| CASSITERITE | $SnO_2$ | 7.0 |
| WULFENITE | $PbMoO_4$ | 6.8 |
| SCHEELITE | $CaWO_3$ | 6.0 |
| CHALCOCITE | $Cu_2S$ | 5.7 |
| HEMATITE | $Fe_2O_3$ | 5.3 |
| MAGNETITE | $Fe_3O_4$ | 5.2 |
| ILMENITE | $FeTiO_3$ | 4.8 |
| MOLYBDENITE | $MoS_2$ | 4.7 |
| PYROLUSITE | $MnO_2$ | 4.7 |
| CHROMITE | $FeCrO_4$ | 4.6 |
| BARITE | $BaSO_4$ | 4.5 |
| FLUORITE | $CaF_2$ | 3.8 |
| DIAMOND | C | 3.5 |
| APATITE | $Ca_5(F,Cl)(PO_4)_3$ | 3.2 |
| MAGNESITE | $MgCO_3$ | 3.2 |
| BERYL | $Be_3Al_2(SiO_3)_6$ | 2.8 |
| TALC | $H_2Mg_3(SiO_3)_4$ | 2.7 |
| QUARTZ | $SiO_2$ | 2.6 |
| KAOLINITE | $H_4Al_2Si_2O_9$ | 2.6 |
| GRAPHITE | C | 2.2 |
| SULFUR | S | 2.1 |

density intermediate with the mineral species to be separated (dry HMS), and air classifiers and cyclones utilizing centrifugal action.

Typical minerals concentrated by dry gravity separation devices include coal, sulfide and oxide minerals of copper, lead and zinc, pyrite and pyrrhotite as sources of sulfur, barite, fluorite, monazite, wolframite, cassiterite, uranothorianite, manganese minerals, phosphates, gold, and tantalite-columbite.

These methods have been successfully used for dry concentration of various minerals such as clays, micas, cement rock, barite, iron ores, carnotite, asbestos, spodumene, fluorite, coal and many other minerals. The techniques involved in differential disintegration are either based on differences in hardness of minerals (scrubbing, scuffing, and differential grinding) or on internal stresses set up by heat (decrepitation).

The equipment involved for differential dry grinding, scrubbing and scuffing include attrition mills, impact mills, jet or fluid energy mills, pan or chilean mills, rolls and roller mills, and tumbling mills. In recent years, the development of the Aerofall mill (autogenous or semi-autogenous mills) and the Majac mill (jet pulverizer) systems have been used increasingly for differential grinding and concentration of selected minerals including iron ores, mica, barite, graphite, talc and titanium, and uranium minerals. Figure I shows a schematic flowsheet for dry processing of mica using the above techniques.

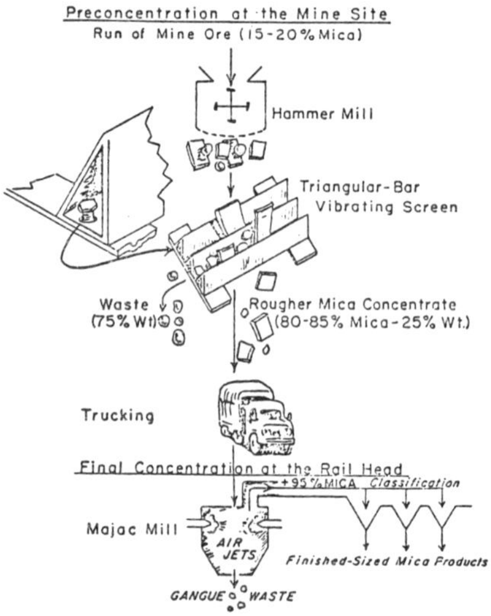


Figure I. Proposed flow-sheet for dry processing of mica.

## Magnetic Separation

Magnetic separation of minerals in a dry state uses differences in their magnetic attractability. Table VII shows the relative magnetic attractability of minerals. This table clearly shows that besides iron ores (hematite and magnetite), magnetic separation can be effectively used for dry concentration of titanium, zirconium, chromium, and manganese minerals. Currently, a large number and types of very sophisticated magnetic separators are available.

## Electrostatic Separation

Electrostatic separation, on the other hand, is based on differences in electric charging by friction (tribo effect), electrification by heating (pyroelectric effect), dielectric constants, and conductivities of different minerals. In addition, separation may also be based on ionization and electric wind attending corona discharge in gases (corona method). In recent years, considerable research and development in this field has taken place in USSR and based on their efforts, a large number of metallic and industrial minerals can be effectively separated in a dry state. Classical examples of electric methods of concentration include iron ores and ferruginous products by effectively removing non-conducting gangue minerals such as apatite, biotite and hypersthene; diamonds; manganese ores; coal fines; and selective separation of many rare metal ores and products (titanium-zirconium, monazite-xenotime, tantalumniobium and cassiterite-scheelite).

Table VII. Relative Magnetic Attractability of Minerals

| | SUBSTANCE | RELATIVE ATTRACTABILITY |
|---|---|---|
| | IRON (TAKEN AS STANDARD) | 100.00 |
| STRONGLY MAGNETIC | MAGNETITE | 40.18 |
| | ILMENITE | 24.70 |
| | PYRRHOTITE | 6.69 |
| | HEMATITE | 1.32 |
| | ZIRCON | 1.01 |
| | LIMONITE | 0.84 |
| WEAKLY MAGNETIC | CORUNDUM | 0.83 |
| | PYROLUSITE | 0.71 |
| | MANGANITE | 0.52 |
| | CALAMINE | 0.51 |
| | GARNET | 0.40 |
| | QUARTZ | 0.37 |
| | RUTILE | 0.37 |
| | CERUSSITE | 0.30 |
| | ARGENTITE | 0.27 |
| | PYRITE | 0.23 |
| | SPHALERITE | 0.23 |
| | MOLYBDENITE | 0.23 |
| | APATITE | 0.21 |
| | TETRAHEDRITE | 0.21 |
| | TALC | 0.15 |
| NON-MAGNETIC | ARSENOPYRITE | 0.15 |
| | MAGNESITE | 0.15 |
| | CHALCOPYRITE | 0.14 |
| | FLUORITE | 0.11 |
| | CELESTITE | 0.10 |
| | CINNABAR | 0.10 |
| | CHALCOCITE | 0.09 |
| | CUPRITE | 0.09 |
| | SMITHSONITE | 0.07 |
| | STIBNITE | 0.05 |
| | GALENA | 0.04 |
| | CALCITE | 0.03 |

## Electronic Separation

Electronic separation of minerals utilizes sensitive instrumentation to detect difference in radiation emission, reflectivity, color and luster of minerals. The elements necessary for effective mechanical electronic sorting devices are: (a) presentation of mineral pieces for examination by sensers; (b) a sensing device for selecting the pieces to be removed; (c) an electronic system to act upon the information provided by the sensing device; and (d) a means of removing the selected pieces. Figure II shows the major elements of the Sortex mineral separator.

Presently, a large number of commercial electronic separators are available for effective separation of various minerals. These include Lapointe Picker (radioactivity sensed by beta sensitive Geiger tube) for uranium ore processing, Goodman Sorter (X-ray detector) for molybdenite and tin ores, Sortex Separator (based on

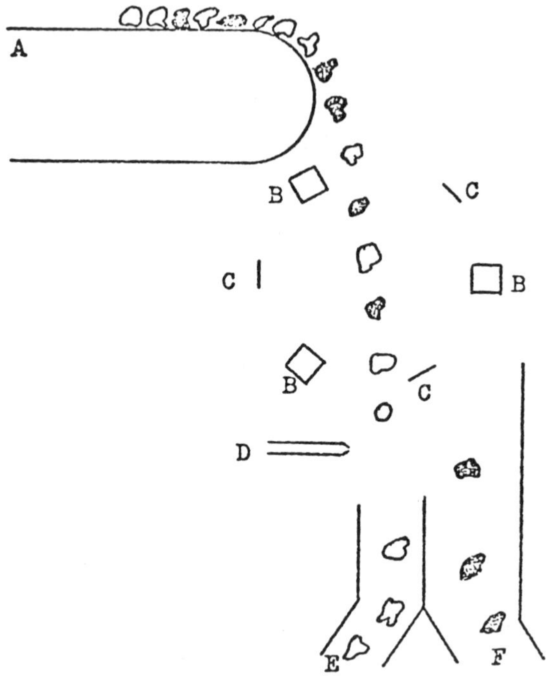


A Conveyor
B Photocells
C Backgrounds
D Air jet
E Ore
F Waste

Figure II. Sortex mineral separator.

reflection or transmittance of light) for separating calcite from dolomite, Gromax Electronic Sorter (based on light reflectance, color and shade differences) for barite, talc, and titanium ores, Ore Sorters (based on electrical conductance and reflected or transmitted light) for processing iron ore.

The most popular application of electronic sorting is in the recovery of diamonds based on the fact that diamonds are more transparent to certain X-rays than the accompanying gangue, and on the property of diamonds to transmit and scatter light, where the gangue will not.

It should be noted that this concentration method is still in its infancy and that there are possibilities of using many other sensing properties and devices. These include electromagnetic spectrum, infrared range, chemical staining of selected minerals, fluorescence, neutron activation.

## Heavy Liquid Separation

Heavy liquid separation utilizes organic liquids of selected specific gravities to obtain separation between the valuable and the gangue minerals. Thus by using tetrabromo-ethane of specific gravity 2.94, the typical gangue minerals (silica and silicates) with specific gravity of 2.6 to 2.8 can be separated from valuable minerals (oxides and sulfides) having specific gravities above 2.94. In this case, the gangue minerals sink in the liquid thereby concentrating the valuable minerals. Since no water is required in this separation process, this technique can be classified as a dry concentration process.

In the past, this method was not widely used because of the lack of inexpensive industrial heavy liquids of sufficiently high specific gravities. Today, heavy organic liquids formulated around bromine are available in tonnage quantities at economically favorable prices. In recent years, considerable research and development was carried out in Israel on this versatile process. Besides tetrabromoethane (2.94 sp.gr.), the other commercially available liquids include: methylene iodide (3.31 sp.gr), bromoform (2.89 sp.gr.),and methylene bromide (2.49 sp.gr.). The industrial applications of heavy liquid separation for selective concentration of valuable minerals include: coal, clays, cassiterite, andalusite, beryl, potash, chromite, magnesite, phosphate, diamonds and iron minerals.

## Thermal Concentration

Thermal concentration is one of the most widely used methods of processing ores to recover valuable minerals, compounds and metals. In this process minerals are separated by the application of heat. In many metal extraction practices this would be a preliminary step to an actual reduction (pyrometallurgical) process.

Table VI shows the application of thermal concentration techniques to recover various minerals, compounds, and metals.

It should be noted that most of the arid and semi-arid areas of the world contain natural fuels (oil, gas, and coal) which can be readily used for thermal concentration of ores and pyrometallurgical recovery of metals. A classical example would be the extraction of nickel from oxide and silicate ores by reduction with fuel oil to obtain ferronickel which can be separated by magnetic means. Similarly, reduction of hematite to magnetite by reduction with fuel oil followed by magnetic separation of magnetite can be readily achieved.

## Metal Extraction

Once the wanted minerals are concentrated, the next steps in metal extraction systems are: (a) to decompose the mineral to free the metallic component as a crude metal and (b) to refine this impure product into a marketable or commercial grade commodity. Since this phase of extraction involves chemical properties of the mineral rather than physical (as is generally the case in concentration processes), it is often referred to as "chemical metallurgy" whose three important subdivisions are pyrometallurgy, hydrometallurgy, and electrometallurgy.

Ever since the first metal was extracted from an ore by the action of wood fire in ancient times, the pyro or fire metallurgy procedures have played a significant role in the extraction of various metals. A large number of major metals such as iron, nickel, tin, copper, zinc, lead, gold, and silver as well as many minor and rarer metals such as molybdenum, tungsten, and titanium are won from their ores or concentrates by pyrometallurgical methods.

Table VI. Application of Thermal Concentration Techniques

| TECHNIQUES | APPLICATIONS |
|---|---|
| 1. DRYING | Shipment of minerals or pretreatment of ores |
| 2. DECREPITATION | Kyanite, spodumene, vermiculite and perlite |
| 3. CALCINATION | Limestone, alumina, rhodochrosite |
| 4. ROASTING | (a) Sulfides $\rightarrow$ Metal oxides + $SO_2$<br>(b) Reduction $\rightarrow Fe_2O_3 \rightarrow Fe_3O_4$<br>(c) Sulphation $CuSO_4$, $NiSO_4$, $ZnSO_4$<br>(d) Chlorination $\rightarrow PbCl_2$, $ZnCl_2$, LiCl, $MnCl_2$ |
| 5. MELTING | Liquation of stibnite, bismutite |
| 6. FUMING OR VOLATILIZATION | Recovery of Zn, Pb, As, Bi, Sb, Sn |
| 7. DISTILLATION | $Hg_2S \rightarrow Hg^o + SO_2$ |

Smelting is a major pyrometallurgical process in which the high grade ore or concentrate is mixed with suitable fluxing compounds and reducing agents such as carbon or fuel oil and reacted in appropriate furnaces to produce an impure (crude) metal and a worthless slag. The slag, being lighter than molten metal, floats on top of it and is readily removed. The crude metal is next refined by either heating in the presence of purifying reagents or by electrorefining to produce pure marketable products.

For those which do not readily respond to physical concentration of valuable minerals from gangue, suitable chemical or hydrometallurgical processing is applied for extracting the metal or salable product. In such procedures, the ores or concentrates are digested with selective solvents to dissolve the valuable metals or minerals and then the metal bearing solutions are treated further by concentration, precipitation or electrolysis to recover the metals or compounds. In recent years, the advent of selective leaching reagents (lixiviants), specific extraction techniques (solvent extraction and ion exchange), and the use of autoclaves (using high pressures and temperatures) have allowed metallurgists to recover metals and compounds more effectively and economically from refractory and lower grade resources. Thanks to these techniques, currently copper, uranium, nickel, gold, and silver are extracted economically from their marginal and low grade ores.

Of course, in arid and semi-arid regions, due to lack of water, hydrometallurgical processes for the recovery of metals cannot be readily applicable. Nevertheless, under certain favorable conditions there is a place for hydrometallurgical extraction in desert areas, especially since it is possible to recover and recycle 75 to 80 percent of the process water. Also, it is entirely possible that in certain cases, the mine itself will provide the required water for processing.

It should also be noted that, in general, process water need not be pure for it is entirely feasible to use brackish water for processing. Especially, physical concentration of minerals in a wet state is not affected by the quality of water. Furthermore, certain flotation processes involving selective recovery of sulfide minerals does not require pure water and separations can be readily achieved in brackish water or sea water. Even in hydrometallurgical extractions, presence of chloride ions under acid leach conditions may increase the dissolution rate of metals and minerals. However, the most important detrimental concern above the use of brackish or sea water for processing is its corrosive nature, abatement of which requires the use of expensive materials of construction and constant maintenance of pumps, pipelines, etc.

Since the art or science of extractive metallurgy deals in the separation of valuable metals from gangue materials either in the solid, liquid or gaseous phase, it can be readily appreciated that mineral processing engineers and scientists have a special competence to understand and to remedy pollution problems associated with extraction. The same basic principles that govern metallurgical extractions also control pollution abatement technology and most of the common processing machinery can be applicable to pollution problems. Accordingly, it is up to the mineral engineers and metallurgists to pave the way for improvements and innovations in the pollution abatement technology and to share their knowledge with environmental engineers in providing the badly needed mineral resources and metals with a minimum degradation of the government.

## FUTURE OF EXTRACTIVE METALLURGY

In looking at the future of extractive technology, it is evident that the new sociopolitical forces will play a dominant role in its development. It has and it will compel mineral engineers to seek new processes and techniques that will produce the required metals and minerals with the minimum degradation of the environment and maximum conservation of our finite mineral resources. This goal may be achieved by (1) bigger and/or better processes, (2) new processes by virtue of engineering design, and (3) new processes based on novel physics, chemistry and biology.

Cost reductions either by increasing the size and output of an existing process or by modification of its performance, have in the past paid handsome dividends. Thus, modernization of the iron blast furnace, improvement in the Hall process (fused salt electrolysis of alumina) used in producing aluminum, and the computerized electrolytic copper tank-house are classical examples of such innovations. Similarly, improved methods of extraction provided by new engineering design such as basic oxygen steel making, the Imperial smelting blast furnace for lead and zinc reduction, cementation of copper in cones, and recent developments in continuous copper-smelting processes, should be beneficial.

The most important contribution to the future metal extraction technology, however, must be provided by wholly new processes based on novel physics, chemistry and biology. The basic information on the development of such processes can be found in existing literature but additional research and development will be required to put these ideas into practice. Recent developments include: (1) electrowinning of nickel developed by International Nickel Company, (2) autoclave processes for nickel, cobalt, and copper developed by Sherritt-Gordon Mines, Ltd., (3) and the solvent extraction-electrowinning process for copper recovery utilized by several copper producers in the southwestern United States and Zambia. There is no doubt that more effective metallurgical processes will be developed in the future based on pyro-, hydro-, electro-, and vapor- chemistry as well as on basic principles of physics, biology, and geology.

As far as the future extraction of metals and minerals in the desert regions is concerned, it is obvious that the existing technology, though particularly applicable to specific minerals, is not sufficient to enable large scale mining of all the available mineral resources in these water-short areas. Accordingly, the following steps must be taken:

1. New processes and techniques using no or very little water must be developed. Special emphasis should be placed on dry processing and concentration of finely-divided valuable minerals and metals.

2. Since fuel (oil or gas) may be readily available in desert regions, efforts should be made to use this energy in various pyrometallurgical techniques to recover metals and minerals. These include: thermal concentration, distillation of compounds, vaporization, and direct reduction to metals using fuel as a heat source as well as a reducing agent.

3. Use the available fuel for generating electricity which then can be used for dry processing of ores using magnetic, electrostatic and electronic separations. After concentration, use additional electricity and fuel as a reductant in direct smelting and refining in electric furnaces.

4. Also because of the availability of oil and gas, develop new chemical reagents based on these commodities for processing and extracting metals and minerals.

5. Develop new extractive processes or techniques based on solar energy which is readily available in these regions. Solar energy, if properly harnessed, would provide very high and very clean heat which may facilitate smelting and refining of metals and minerals from their ores.

6. If it is absolutely necessary to use water for processing and/or extracting specific metals or minerals, efforts should be made to improve water reclamation and recycling technology.

7. Developing improved materials of construction is necessary so that brackish or sea water can be used for economical processing and extracing of metals and minerals.

## EDUCATION, RESEARCH AND DEVELOPMENT

From the preceding discussions of the extraction of metals and minerals in the desert areas, it is quite evident that though currently available technology may provide partial means of extracting their resources, it will require a concentrated research and development effort to develop effective processes for maximum recovery of their resources. In this respect, the implications of science and technology for mineral processing and metal extraction were clearly indicated since most of the processes utilize the physical, physico-chemical or chemical properties of the minerals and reagents.

While some concentration and extraction techniques such as magnetic concentration and conventional smelting have been known by years, other procedures, such as electrostatic and electronic separations and electric smelting have resulted from very recent technological achievements. Through the application of such processes, a large number of metallic, non-metallic and rare metals can now be extracted economically.

Furthermore, from the above discourse on metal and mineral recovery in desert regions it is quite evident that there are some inherent problems associated with the concentration of minerals regardless of the method used. Of particular importance is the treatment of fine particles of any ore on a large enough scale for it to be economical. Another area of great importance is the harnessing of solar energy not only for producing the required electrical energy, but also for direct application in pyrometallurgical processes and techniques for recovery of metals and their compounds. Problems such as this can only be solved with a great deal of research and development and this will require specially trained scientists and engineers using up-to-date laboratory and pilot plant facilities.

Moreover, application of extractive technology in desert areas, especially those located in the developing countries, will require the necessary education and training facilities to produce the required scientists, engineers, plant operators, and skilled technicians for process development and plant operations.

In order to satisfy the growing need for metals and minerals on a world wide basis with part of the demand supplied by existing mineral resources in desert regions, more trained mineral scientists, engineers and technical personnel, as well as more international aid in research and development will be required. For this reason, it is imperative that the United Nations through its various agencies should increase its activities in the dissemination of such knowledge to developing countries in the arid and semi-arid regions, by organizing appropriate seminars, conferences, assignment of experts, initiating new projects, and undertaking large-scale development programs. In this connection, an international seminar on "Ore Concentration in Water-Short Areas" was held in February, 1966 at the United Nations Headquarters. It is high time that similar seminars on related topics of mineral and metal extraction be undertaken again in the near future.

## BIBLIOGRAPHY


(1) Agricola, G. (1556) "De Re Metallica", Hoover Edition (1950), Dover Publications, Inc., New York, New York

(2) Cloud, P.E. Jr. (ed.) (1969) "Resources and Man" by Committee on Resources and Man, National Academy of Sciences - National Research Council, W.H. Freeman and Co., San Francisco, California

(3) Lovering, T.S. (1969) "Resources and Man", 6 - Mineral Resources from the Land, W.H. Freeman and Co., San Francisco, California

(4) Skinner, B.J. (1969) "Earth Resources," Prentice-Hall Inc., Englewood Cliffs, New Jersey

(5) Newton, J. (1947) "An Introduction to Metallurgy," John Wiley and Sons, Inc., New York, New York

(6) Bray, J.L. (1947) "Non-Ferrous Production Metallurgy," John Wiley and Sons, Inc., New York, New York

(7) Kellogg, H.H. (1971) Extractive Metallurgy in the Years Ahead - New Process to Meet New Problems, "A.I.M.E. Centennial Volume," New York, New York

(8) Deju, R., Bhappu, R., Evans, G., and Baez, A., (1972) "The Environment and Its Resources", Gordon and Breach, New York, New York

(9) United Nations - "Proceedings of the UN Interregional Seminar on Ore Concentration in Water-Short Areas", (1968) United Nations Publication E.68.11.B.4, New York, New York

# THE PRESENT STATE OF GEOLOGICAL CARTOGRAPHY AND MINERAL EXPLORATION IN THE SAHARA

Robert Giraudon

## ABSTRACT

The Sahara, in spite of difficulty of access, has been covered by aerial photographs for the largest part of its surface and by topographical mapping at a scale often as large as 1:200,000. Geological mapping at this comparatively accurate scale extends only over some scattered patches, whereas the western and central parts have been surveyed at the scale of 1:500,000 and the eastern part at the scale of 1:1,000,000.

Fostered by an oil potential which was suspected more than 30 years ago, geophysical surveys are of exceptional density when compared with the rest of Africa. The methods used have been mostly airborne magnetism and radiospectrometry. Mineral exploration has focussed only on substances of high unit value in areas not too remote from a seaport. The lack of transport, water supply, and availability of a qualified labour force are the main problems faced by mining ventures. The principal mineral output of the Sahara is iron ore, uranium, phosphates, copper and foremost, petroleum of which the Sahara is a major producing area in the world, and which has boosted the economic growth of countries where oil and gas fields have been discovered at an unprecedented rate.

The Sahara is the largest desert in the world. It is 5,500 kilometres long from the Atlantic to the Red Sea (including the Nubian Desert in the Sudan) with an average width of 1,500 kilometres. Its maximum width is 2,000 kilometres and the total surface area covers 8,000,000 square kilometres, which approximately equals that of the United States.

This desert spreads over eleven countries: Algeria, Chad, Egypt, Libya, Mali, Morocco, Mauritania, Niger, Sudan, Tunisia and Western Sahara (formerly under Spanish administration). Among these countries Mauritania and Western Sahara have their entire territory included in the Sahara. Libya and Egypt are also almost completely in the desert, except for a coastal fringe in Libya and the rich Nile Valley and Mediterranean coast of Egypt. Algeria occupies the largest share.

It is often thought that geological information on the Sahara is scarcer than for any other area of Africa except for the areas which have oil potential, such as Algeria and Libya. This belief is incorrect: the basement complex and the rim of some sedimentary basins are much better known in the Sahara than they are in equatorial forest areas for instance. In addition, one has to remember that no oil exploration would have been possible without preliminary general geological information having been gathered. In spite of particularly hard working conditions - difficulty of access, complete lack of available local resources, not to mention the well known scarcity of water, and a very trying climate - the Sahara has always attracted geologists. One possible explanation is that outcrops of rock are displayed over immense stretches of land without being "spoiled" by any soil or vegetation, and thus give an opportunity to read the geological facts in a direct way, which is seldom possible elsewhere. Another is perhaps the beauty of the desert awakening in the limpid air of the early morning hours, or the warm and changing colours of sand and rocks before the sunset, or the exceptional feeling of freedom which envelops the geoscientist who, like the nomads, can head in any direction of the limitless horizon and erect his tent anywhere on a land without restriction.

The latest developments in techniques, such as aerial photographs, satellite imagery, remote sensing, etc. have broadened the geologist's investigation capability making his task altogether more accurate and fruitful. However, these developments require more and more investment, and today the limit on the investigation is therefore more a matter of financing than of technical capacity. Available funds were naturally directed to the most promising areas and, thus, the density and quality of geological information varies considerably from one Saharan country to another.

## ORGANIZATION OF GEOLOGICAL SURVEYING AND MINERAL EXPLORATION IN SAHARAN COUNTRIES

All of the countries in the Sahara were at one time or another colonies of European powers, once the influence of Turkey came to an end.

This fact is of certain importance because in many of these countries, the bases for the present geological survey departments were set up by the colonial administrations which carried out the first geological reconnaissances. Except for Egypt, all of these countries acquired independence only after the Second World War, some as early as 1951 (Libya) others as late as 1962 (Algeria). Initially, the local geological survey departments depended on centralized organizations which were the Overseas Geological Survey (United Kingdom) for the former British colonies, La Direction des Mines de la France d'Outremer for the former French territories, and Instituto Geologico y Minero de Espana for the former Spanish colonies. These organizations, or their successors,(1) are still valuable sources of documentation for maps and mineral exploration performed before the independence of each country, and sometimes also subsequently. Each of the new countries has since set up its own organizations. Except for Algeria ("Universite d'Alger") none of them has an institute specializing in desert investigation but, on the other hand, such a specialization is implied for many of the countries which have almost the entirety of their territory in the Sahara. These organizations are given in Annex I attached hereto.

---

(1) For France the "Bureau de Recherches Geologiques et Minieres", Orleans.

## BASIC TOPOGRAPHIC CARTOGRAPHY

Although the purpose of this note is restricted to geological surveying and mineral exploration, it may be pertinent to review what kind of topographic documents are now available as a base for geological mapping.

In brief, it can be said that almost all of the Sahara has been covered by aerial photographs. Former colonies or protectorates, including Egypt, were photographed at the scale of about 1:60,000 with many sections at a larger scale by the British, or at 1:50,000 with some areas at a larger scale by the French. The only exception seems to be the Western Sahara, formerly under Spanish administration, for which, as far as we know, no aerial photographic cover on a directly usable scale is available. Of course, most of these photographs are old, always in black and white, and not as good as those which are produced now with advanced technology. They are, nevertheless, easily readable and of sufficient accuracy to support the work of the photogeologists. In addition, the fact that rocks are denuded of any cover over vast areas makes it possible to have, in photogeological maps, an accuracy which is unobtainable elsewhere, even though more elaborate photographic documents may be available.

With the support of satellite imagery, which adds a synthetic view of the lineament pattern, and a four channel scanning, which lends itself to computer processing, it is now possible to achieve an impressive output of geological mapping with coarsely spaced traverses and some carefully selected spots for ground checking, provided that the responsible geologists are well qualified and familiar with the geology of the area surveyed.

Nevertheless, in spite of these exceptional factors, much remains to be done in terms of detailed geological mapping, because geologists have to overcome numerous obstacles. A primary obstacle is a certain lack of large-scale topographic maps. However, if one considers the extent of the Sahara desert, its difficulty of penetration and scarcity of population, it is rather surprising that the documents available have achieved such quality and density. Of course, with some exceptions, most of these topographic maps have been derived from the ortho-restitution of aerial photographs. A tentative list of these is attached as Annex II of this note.

A second handicap is the fact that the geological surveys set up after independence do not have as yet sufficient numbers of qualified geologists. This is not only a matter of training, a problem which could be solved after some years with adequate assistance such as scholarships etc. The main problem is that the operational cost of a geological survey - fully equipped to face the challenge of surveying very vast and inaccessible areas - is out of reach in terms of budget for countries with low income per capita and a very scattered population.

A third bottleneck for systematic surveying is the priority granted to investments which give the promise of more immediate returns. In desert countries where infrastructure is generally poor, and where the weight of investment per capita is particularly heavy, such preference is logical. In the field of mineral investigation, this means that priority is given to areas which are the most attractive, either in terms of mineral potential or of transport facilities, water supply, availability of qualified labour etc. As a consequence, detailed geological mapping has a tendency to develop around certain privileged points rather than on a systematic basis.

Last but not least, the fourth bottleneck is that a large part of the detailed surveys and logs of drill holes carried out by private companies are generally retained by the latter as confidential documents.

The geological maps generally available are therefore the result of surveys compiled by the colonial powers prior to independence, by bilateral or multilateral technical assistance programmes and, for the more advanced countries, by the recently established national organizations. A tentative list of these documents, country by country, is contained in Annex III.

In conclusion, it can be said that almost all of western and central Sahara is covered by geological maps at the scale of 1:500,000 whilst the eastern part is mapped at the 1:1,000,000 scale.

Many areas have been surveyed at a larger scale, in particular in Mauritania, Morocco, Algeria, Niger, Egypt and the Sudan. Furthermore, systematic mapping is more or less in progress in Egypt, the Sudan, Tunisia and Algeria at scales from 1:100,000 to 1:250,000.

## MINERAL AND OIL EXPLORATION

Mineral exploration has been very active in those parts of the Sahara which are not too far from the Atlantic or Mediterranean coasts. Oil exploration does not experience this geographic restriction, and it can be said that all of the sedimentary basins of the Sahara have been investigated at one time or another by oil companies, even in such remote parts as eastern Niger (Djado Basin) or Chad (Erdi Basin).

As far as oil is concerned, this means that almost all of the areas explored have been surveyed by seismic and airborne magnetism plus, in some cases such as Algeria, by gravity.

Airborne geophysical surveys for prospection of base or radio-active metals are distributed in more irregular and scattered patches. Algeria is certainly the country which has the most impressive cover since, in addition to magnetic airborne surveys carried out by the oil companies and the radio-spectrometry performed by CEA(1) before independence, the Canadian International Development Agency has recently financed, at the request of the Algerian Government, an airborne magnetic survey combined with radio-spectrometry of the entire country (850,000 km flight lines).

The central part of Mauritania, between Akjoujt and Adrar (edge of the Taoudeni Basin) was flown over for magnetism and radio-spectrometry in the 1960's. Previous airborne magnetic surveys had covered the Akjoujt area, Tasiast, Inchiri and the Zouerat area in the late 1950's. The first test in Africa of input electro-magnetism (Barringer method) was conducted over the copper deposit of Akjoujt around 1965. Another interesting study, and one of the first tests of this method, was the airborne High Sensitivity Magnetism Survey carried out over the iron ore deposit of Zouerat in the late 1960's. In Mali, the massif called Adrar des Iforas, which is a "horst" symmetric to Niger's Air range has recently been flown over with a 2 kilometre spacing for magnetism and gamma ray spectrometry. In Niger, airborne radio-spectrometry has been carried out in the Agades area. In Egypt airborne radio-spectrometry and magnetism have been carried out in the Eastern Desert (south-east of Aswan) and in other places.

In Egypt also, the facilities offered by satellite remote sensing have been applied beyond the simple compilation of the imagery: the Geological Survey

---

(1) French Atomic Energy Commission

Department as well as oil companies have taken advantage of the possibility of computerized enhancement and processing of the taped data.

Therefore, mainly due to its oil potential, the Sahara is perhaps the part of Africa which has the largest proportion of its surface covered by geophysical airborne surveys.

## MINERAL AND OIL RESOURCES OF THE SAHARA

Numerous ore indications, deposits and, of course, oil and gas fields have been discovered, almost all of them in the last few decades.

- Iron ore: Several occurrences are known, some of them among the outstanding accumulations known in the world.

The most important in terms of economy is the deposit of Zouerat in Mauritania, the first description of which was made by a French Army officer about 50 years ago. This deposit was considered for exploitation only in the 1950's when the demand for steel plants quickly increased after the Second World War and technology reached the level which enabled it to meet the requirements for exploitation of a mine located in a completely desertic area 600 km. from the Atlantic coast, (actually 300 km. in direct line of flight across the western Sahara).

The mine was in operation in 1964 and increased its output to 10,000,000 tons of ore with 64 - 66% iron content in 1974. The proved reserves were initially estimated at 150,000,000 tons, but have been extended somewhat since. Transport accounts for more than one half of the production cost. The mine was nationalized at the end of 1974 and output then dropped to 8,000,000 tons annually. New investments should make it possible to increase it beyond 11,000,000 tons, but recent troubles in this area have not as yet made it possible to reach this target.

A search has been made to find another source of iron ore to be shipped by the 600 km railway constructed between Zouerat and the outport of Nouadhibou after the Zouerat deposit has been exhausted. Four and a half billion tons of low grade magnetic iron ore - a possible reserve - have been found in the same area. Studies have been carried out to upgrade this ore by dry magnetic separation. Investment required would be in the range of several hundred million dollars.

Whereas the iron ore of Zouerat belongs to the Precambrian IBQ type (iron banded quartzite), the deposit of Gara Djebilet, 135 km south-east of Tindouf, in Algeria and close to the Mauritanian border, is of the sedimentary oolitic type. Its age is Devonian, and it occurs on the rim of the Tindouf sedimentary basin.

The possible reserves are estimated at three billion tons, but opinions differ regarding the grade of iron: According to W. A. Hance (3) this tonnage applies to an iron-grade of 55%. We will be more conservative and suggest 200 or 300 million tons at 57%, the remainder being between 45% and 57%. The ore is penalized by a relatively high content of phosphorus. A pilot plant for concentration was set up before the independence of Algeria, but did not upgrade the ore significantly (only from 57% to 59%). The deposit is 400 km. from Tarfaya, in Morocco, and about 1,300 km. from the Algerian Mediterranean coast. The construction of a railroad was envisaged at one time, but the deterioration of the relations between Algeria and Morocco has led to its postponement.

A 70 million ton iron ore deposit has been described in the central part of the western Sahara, but so far has not been considered of economic interest.

A low grade iron ore deposit has been studied in Libya near the border between Tripolitana and Fezzan in the Shatti Valley area. Possible reserves are estimated to be about two billion tons with 40 - 50% iron content of which 720,000,000 tons are between 45% and 50%. The deposit is 550 km south of Tripoli.

Since 1955 iron ore has been mined in Egypt, 48 km east of Aswan and more recently in the Bahariya Oasis, which has been linked to the Helwan steel complex by an 837 km railroad line.

A small quantity of iron ore is also known to exist in the Red Sea Hills, Sudan.

- Uranium and radioactive minerals: In the Sahara, by far the most important area for radioactive minerals is located near Agades in Niger. The first deposit was discovered at Arlit, 250 km north north-west of Agades. A consortium, SOMAIR, was formed in 1968 and production has increased steadily to reach an output of 1,500 tons of concentrate per year. The Arlit deposit is of sedimentary type like those in Colorado, and mineralisation is located in the channels of an eroded surface at the top of a monotonous sequence of continental sandstone, capped by a transgression. The ore is mined by open pit after removal of 35-40 meters of overburden. It is then concentrated in situ, an operation which required construction of a sulphuric acid plant, a 9 MW power plant and the discovery of a sufficient supply of water to provide two cubic meters per ton of crude ore treated. Water comes from pumping the mine pits and an aquifer uncovered by the mining operations. The mine requires the import of 40,000 tons of material, mainly sulphur and oil, and also equipment. The resulting high operating costs under desert conditions are aggravated by the need for 1,600 km of trucking preceded by 400 km of rail transport from the ocean port of Cotonou (Dahomey). The supporting town, which houses 2,000 persons, has been built from scratch and is supplied only by imported goods.

Uranium ore deposits have also been found at Akokan, 20 km to the southwest, and have been studied by a CEA and Japanese-Nigerian group. Another discovery has been made at Imouraren near Agades. Underground mining will be required at the Akokan deposit, the mining of which would require an investment of more than $100 million.

Total reserves of uranium in Niger are now estimated to be one-fourth of those of the United States.

In Algeria, a deposit whose possible reserves might amount to 14,000 tons of uranium metal has been identified in the Hoggar massif. In Egypt a possibly important deposit of rare earths has been reported in the eastern desert. In Mauritania a deposit of rare earths is known at Zellaga, 300 km south south-east of Nouakchott. It was mined for about two years around 1967 and 1968. The deposit also contains fluorite. Occurrences of uranium have been identified in the "Dorsale Regueibat" where they are linked with quarternary lake limestones known as "Hamada".

- Copper: A deposit was discovered around 1947 at Akjoujt, 270 km east of Nouakchott, in Mauritania. Reserves have been estimated at between 400,000 and 500,000 tons of metallic copper. The ore grade is 1.79% to 2.7%. Mining was delayed by the difficulties encountered in concentrating the gossan ore, one of the richest parts of the deposit, consisting of copper oxides in a carbonated matrix. A solution was found with the Torco process developed by Charter Limited, which became the main shareholder of the new company called SOMIMA. An output of

50,000 tons of 80% concentrate was planned, but production from 1971 to 1973 was disappointingly low (maximum 20,000 tons of concentrate a year) because of strikes and environmental problems. In 1975 the shareholders of SOMIMA agreed to sell the mine to the Mauritanian Government.

Some other indications of copper ore are known to exist in Mauritania in the southern part of the country (M'bout area and the rim of the "Massif de l'Affole") and, of lesser interest, in the northern part (east of Chegga). Further study of the former is envisaged.

Copper indications have also been reported in Egypt and in Niger where they are linked to uranium, but do not appear to be of economic interest.

- Phosphates: This is the fourth mineral of economic importance in the Sahara. The largest deposit, by far, is located at Bou Craa, in the Western Sahara, where it was discovered in 1963. According to W.A. Hance (3) "Measured reserves total 1.7 billion tons which ranks it second only to Morocco but the total may prove to be much larger. An investment of $300 million permitted the shipment of 3.3 million tons per year of 75%-80% concentrate. Operations began in 1973 and plans call for increasing the level to 10 million tons. The rock is moved by a 100 km conveyor belt to the outport of El Aaioun where loading facilities have been constructed for 100,000 ton carriers."

Phosphates have also been discovered in Egypt at Isna in the Nile Valley, in the Western desert and between Khaoga and Dakhla Oases where a one billion ton deposit is to be linked by a 500 km railway line to the Red Sea port of Quseir. The projected production will amount to 5 to 10 million tons a year of 72%-74% concentrate.

- Coal: Low grade ore was formerly mined at Kenadza, in Algeria, near Colomb-Bechar. The mine was totally uneconomic and was closed after the discovery of hydrocarbons in the country.

- Other minerals: Niger has a small production of tin - about 140 tons a year of cassiterite concentrate - which is produced in the Air, 250 km north of Agades, and taken to Nigeria for smelting. Tungsten has been reported in the Hoggar (Algeria) and in Tibesti (Chad) but the remoteness of the deposits from the sea and the environmental conditions have impeded any attempt of exploitation so far. Gold has been reported in small quantities in the Red Sea Hills (Sudan). Mauritania produces a small quantity of gypsum sand from dunes near Nouakchott; the goal is a yearly output of 15,000 tons to be exported to Senegal for use in cement plants. Gypsum has been reported in numerous places and the Sahara also contains large reserves of salts (mainly ClNa) scattered in shallow deposits which are the remainders of dried up lakes ("Sebkhras").

- Petroleum: In this review we have purposely left to the last the hydrocarbon resources which, in terms of economic value, are by far the most important production from the Saharan subsoil. The reason is that a detailed description of oil exploration, discoveries, production and the subsequent political and economic impacts would have greatly exceeded the scope of this note. Many years ago geologists had predicted that, because of the enormous size of the geological structures, if oil were found in the Sahara it would be found in large quantities. That has been proven true. The Sahara is now a major area of the world as far as hydrocarbons are concerned. The first significant find occurred in Algeria in 1952. It was followed by many others, first in this same country, then in Libya (as from 1957) whose production rapidly surpassed Algeria's in Egypt (Suez area) and in Tunisia (El Borma). Evidence of oil has been proved to exist in Chad (Kanem) and is expected to be found in Eastern Niger. All of the Saharan countries have been explored for hydrocarbons no matter how remote from the sea or how difficult the environment. Of course the inner landlocked countries were investigated last but the returns from petroleum wells, particularly since prices started to skyrocket in 1973, make exploration rewarding even under conditions which were considered as a deterrent previously.

The Sahara has billions of tons of oil reserves and trillions of cubic meters of gas reserves (Algeria alone accounts for 4 to 7 trillions of the latter, depending on the sources of information, i.e. more than 1/8 of the world reserves). It attained a total yearly production of 174,000,000 tons of oil in 1974, according to W.A. Hance (3) or 137,000,000 tons in 1975, plus 8.9 million tons of LNG (liquefied natural gas) according to a United Nations publication (7). Revenues of some Saharan countries like Libya have increased more than 36 times their level at the time of independence and have permitted infrastructure investments which would have been completely out of reach otherwise.

## CONCLUSION

These impressive achievements should not obliterate the fact that the Sahara remains nevertheless an immense desert with all of the deterrent factors that such a condition implies. It is true that mineral and oil production have given to certain countries an unprecedented economic boost inconceivable only 30 years ago. But the future of the Sahara still depends on water supply which is the only way to have a population settled there and to develop normal and complete activity. All of the present mineral output is exported and exploration is carried out only when the returns from the exported products are sufficient to pay for the investment made. This means that immense stretches of the Sahara desert are not being considered for systematic exploration of many minerals because, even though they would be discovered in large quantities, their transport or their beneficiation *in situ* would not allow them to meet the prices of the international market.

It has to be kept in mind also that operations in the Sahara are very demanding in terms of human effort, equipment and, consequently, of investments. In this respect we would like to quote once more W.A. Hance about the difficulties faced by the oil companies in Algeria: "Numerous problems accompanied development. The Sahara environment is a difficult one: crews can work at full time only 8 months of the year; they must be changed frequently, and heat and sand expose machinery to tremendous wear and tear. These factors resulted in the high cost of drilling......" The same could have been written about the difficulties encountered by the company which mined iron ore at Zouerat in Mauritania and the subsequent high costs of operation.

The Sahara has known a prodigious development over the last 30 years. Oil and, to a certain extent, mineral resources account for the largest part of this development but in spite of rather good knowledge of its geological features, in particular when compared to the rest of the continent, and in spite of an intensive ongoing search for oil, a large percentage of its surface still remains to be explored in detail for minerals. Its potential can be expected to be in proportion to its dimensions. We believe, however, that extensive activity will flourish in this part of the world only if and when water is made available in large quantities.

Then the Sahara would no longer be a desert but a patchwork of oases and everything will become feasible. The recent petroleum discoveries may have been only the starting point of this new era.

## ANNEX I

### A List of National Organizations Dealing with Geological Cartography and Mineral Exploration in the Sahara

| | |
|---|---|
| Algeria | "Direction des Mines", Algiers.<br>SONAREM, Algiers ("Societe Nationale de Recherches et d'Exploitation des Ressources Minieres").<br>University of Algiers ("Institut de Recherches Sahariennes"). |
| Chad | "Direction des Mines et de la Geologie", N'Djamena. |
| Egypt | Geological Survey and Mineral Resources Department, Ministry of Industry, Cairo.<br>Desert Institute, Cairo.<br>General Desert Development Authority, Cairo.<br>University of Alexandria.<br>University of Cairo. |
| Libya | Geological Mining Services, Mines Department, Ministry of National Economy, Tripoli. |
| Mali | "Direction Nationale de la Geologie et des Mines", Bamako.<br>SONAREM, Bamako ("Societe Nationale de Recherches et d'Exploitation des Ressources Minieres"). |
| Morocco | "Direction des Mines et de la Geologie", Rabat.<br>BRPM, Rabat ("Bureau de Recherches et de Participation Minieres"). |
| Mauritania | "Direction des Mines", Nouakchott.<br>SNIM, Nouakchott ("Societe Nationale Industrielle et Miniere"). |
| Niger | "Direction des Mines", Niamey.<br>OFRMIG, Niamey ("Office des Recherches Minieres et Geologiques"). |
| Sudan | Geological Survey, Ministry of Mineral Resources, Khartoum.<br>University of Khartoum. |
| Tunisia | "Service Geologique de la Direction des Mines", Tunis.<br>"Office National des Mines", Tunis.<br>SOTEMI, Tunis ("Societe Tunisienne d'Expansion Miniere" - Affiliated with ONM). |
| Western Sahara | Formerly "Instituto Geologico y Minero de Espana". |

## ANNEX II

### List of Topographic Maps Covering the Sahara

**Algeria**

The desert area is covered at the scale of 1:200,000 by 179 sheets (south type) out of 203.

**Chad**

The desert area is covered by 1:200,000 maps compiled by I.G.N., France.

**Egypt**

In 1963 43 sheets at 1:250,000 were published out of 69 (U.S. P.502 type); 10 additional maps at the same scale had been compiled by the British Directorate of Overseas Surveys which, in addition, had published 98 sheets at 1:100,000; 17 other maps at the same scale but of the P.676 type are also available.

**Libya**

All of the country is covered by the "Carta demonstrativa della Lybia" at the scale of 1:400,000 compiled in Rome before the area was covered by aerial photographs. Another map at the scale of 1:500,000 has been compiled by the British DOS. Fezzan is covered by 47 sheets out of 121, compiled at the scale of 1:250,000 by the U.S. Army Map Service (Series P.501 and P.502). Tripolitana is covered by 160 sheets compiled at the scale of 1:200,000 by the Italians, out of which seven are in five colours - and 153 were published in colour or in black and white between 1927 and 1933. In Cyrenaica only 23 sheets of 1920 type (Rome) at the scale of 1:200,000 are available. This does not include the topographic documents which may have been established by the oil compnaies.

**Mali, Morocco (southern part), Mauritania, Niger**

These countries are almost completely covered by maps at the scale of 1:200,000 compiled by I.G.N., France.

**Western Sahara** (formerly under Spanish administration)

Only a 1:500,000 map is available.

**Sudan**

170 sheets of a map at the 1:250,000 scale (i.e. a complete set) and 140 sheets of a map at the 1:100,000 scale had been published.

**Tunisia**

The Saharan part of the country is completely covered by 1:200,000 scale maps.

ANNEX III

## List of Geological Maps of the Sahara

<u>Algeria</u>

A large part of the Saharan part of the country is covered by "La Carte Geologique du Sahara" at the scale of 1:500,000 ("Service de la Carte Geologique d'Algerie"). It has been completed by "La Carte Geologique du Hoggar" and by "La Carte de la Dorsale Regueibat et de la Region des Eglab", both of these at the same scale of 1:500,000 and available from BRGM (France). Some sections have been studied at a larger scale, e.g. "Les gneiss precambriens de l'Aleksod" mapped at the scale of 1:100,000 by J.M.L. Bertrand (CNRS -"Centre de Recherches pour les Zones Arides, Service Geologique" No. 19, 1974, France).

<u>Chad</u>

The Saharan part, which includes most of the well known Tibesti massif, has been surveyed at the scale of 1:500,000. An explanatory note was drafted by Ph. Wacrenier ("Direction des Mines et de la Geologie de l'AEF" - Brazzaville 1958, now available from BRGM, France and in Djamena).

<u>Egypt</u>

Egypt is covered by a geological map at the scale of 1:1,000,000 in six sheets (Overseas Geological Surveys, 1910, now part of the Institute for Geological Sciences, London). Seven sheets at the scale of 1:125,000 were known to be published in 1963. A systematic cover at the scale of 1:100,000 is in irregular progress.

<u>Libya</u>

There is a geological map of the country at the scale of 1:2,000,000 by L. C. Conant and G. H. Goudarzi, published in 1964 by USAID.

Some parts of the country have been surveyed on a larger scale:

- "Carte geologique de reconnaissance du Sahara Lybien" at the scale of 1:500,000 on the border with Algeria (8 sq. degrees ) -"Institut de Recherches Sahariennes de l'Universite d'Alger;

- "Carte geologique de Fort Polignac" at the scale of 1:500,000 by "Compagnie de Recherches et d'Exploration de Petrole au Sahara" - 1958;

- "Carte geologique du Massif du Fezzan" at the scale of 1:500,000 by C. R. Collomb (Compagnie des Petroles Total, 1961);

- "Carte geologique du Bassin du Djado" at the scale of 1:500,000 compiled by P. Jacquemont and others (Bureau de Recherches Petrolieres - 1959);

- Geology of the Garian area at the scale of 1:100,000 by the United Nations Technical Assistance Programme.

Mali — The Saharan part of Mali is covered by geological maps of reconnaissance at the scale of 1:500,000 published by the "Direction des Mines et la Geologie de l'AOF" and are now available from BRGM, France, or in Bamako.

Morocco — The southern part, like the rest of the country, is covered by a geological map in six sheets (sheets Marrakech and Ouarzazate) at 1:500,000. Three 1:200,000 geological maps straddle the Dra "Oued", which can be considered the limit between the Anti Atlas and the Sahara.

Mauritania — Mauritania has a geological map at the scale of 1:1,000,000 (Nouakchott-Mauritania) and, except for some sections along the border with Mali, the territory is covered by geological reconnaissance maps at the scale of 1:500,000 (BRGM, France and Direction des Mines et de la Geologie, Nouakchott). All of the basement complex, a large part of the rims of the Taoudeni and Tindouf basins, the entire Mauritanides range and the Tasiast in the northeast of the country have been studied in more detail with the support of aerial photographs: The eastern part of the Dorsale Regueibat by the United Nations Office of Technical Co-operation (a map at the scale of 1:500,000 is about to be published), the eastern part of the same Dorsale at the scale of 1:200,000 by BRGM, maps of Amsaga, Tasiast and the southern part of the Mauritanides at the 1:200,000 scale by BRGM, Akjoujt area at the 1:50,000 scale BRGM, Zouerat area and many locations along the Mauritanides range at the scale of 1:50,000 or larger by "Le laboratoire de Geologie Structurale de l'Universite de Marseille", France.

Niger — Besides a composite map at the scale of 1:1,000,000 of the border with Mali between the 0° and 9° East meridians, and the 11° and 19° North parallel, the country has been covered by 1:500,000 scale geological maps of reconnaissance, some of them having been reviewed with the support of aerial photographs and satellite imagery (Air). More detailed maps have been compiled by CEA (French Atomic Energy Commission), North of Agades, and by Texaco in the Bilma area, near the Chad border, but we do not know to what extent these documents are available.

Sudan — A 1:1,000,000 scale geological map of the Anglo-Egyptian Sudan has been published in Khartoum in 1952 under the supervision of Miss F. Delaney. There exists, also, a geological map of the Nuba mountains at the same scale. Systematic geological mapping is now being done at a scale of 1:250,000.

Tunisia — The southern part of the country is covered by a 1:500,000 scale geological map published in 1951 under the supervision of G. Castany. Basic mapping at a scale of 1:200,000 is now in progress (22 sheets were published in 1963 for the entire country).

Western Sahara — A geological map at the scale of 1:500,000 was published in 1958 by the Instituto Geologico y Minero de Espana, Madrid.

On a smaller scale two composite maps are of particular interest: the "Carte Geologique de l'Afrique Occidentale au 1:2,000,000" edited and published by BRGM in 1960 and the "Carte Geologique du Nord Ouest Africain au 1:2,000,000", Algiers, 1952, reviewed in 1961. An explanatory note of the first by J. Barrere and M. Slansky was published by BRGM in 1965.

## BIBLIOGRAPHY

(1) Furon, R. and Lombard, J. - 1964. Explanatory Note - Geological Map of Africa (1:5,000,000). Association des Services Geologiques Africains, UNESCO-ASGA.

(2) Hance, W. A. - 1975. Geography of Modern Africa. Second Edition. Columbia University Press, New York and London.

(3) Van Chi-Bonnardel, R. - 1973. The Atlas of Africa. Prepared and published by "Jeune Afrique" under the direction of --. The Free Press, a Division of Macmillan Publishing Co., Inc., New York, N.Y.

Authors unknown:

(4) - 1971 - Symposium on the Geology of Libya. Papers presented at the Symposium in Tripoli, April 14-18, 1969. Edited by Carlyle Gray. Faculty of Science, University of Libya, Libyan Arab Republic.

(5) - Date unknown - Geology and Mineral Resources of Libya -a Reconnaissance. Geological Survey Professional Paper 660. Prepared in cooperation with the Ministries of Industry and National Economy of the Government of Libya under the auspices of the Agency for International Development, U.S. Department of State, U.S. Department of the Interior.

(6) - 1963 - A Review of the Natural Resources of the African Continent. Published by UNESCO, Paris.

(7) - 1976 - World Energy Supplies 1950-1974. United Nations - Statistical Papers Serial J No. 19, New York, N.Y.

# Saline Deposits in Deserts

George I. Smith

U.S. Geological Survey, Menlo Park, Ca.

## ABSTRACT

Saline deposits are mostly products formed by the evaporation of water; evaporated seawater produces marine deposits, and lake water produces non-marine deposits. Marine deposits are our main source of gypsum, salt, anhydrite, and potash. Characteristically, they formed in large coastal basins in desert environments, but mostly during periods that are too ancient for those environments to have any areal relation to present deserts. When they coincidentally do underlie present deserts, though, they are more easily discovered and tend to be more completely preserved than deposits underlying humid regions. Nonmarine deposits supply most of our sodium carbonate, sulfate, borate, and nitrate, and some of the salt, gypsum, calcium chloride, potash, lithium, bromine, iodine, magnesium, and strontium. The deposits are mostly products of arid climates where large inland lakes evaporated to dryness or near-dryness, and many are geologically young enough for the deposits to underlie areas that are still deserts.

## INTRODUCTION

"Saline deposits" is a term used here to include solid salts, the brines that accompany some of them, and brines that have little or no relation to salts. The term "evaporites" is commonly used for the solid salt parts of such deposits. Saline deposits are the main source of a number of industrial and agricultural chemicals, and several countries have an annual production from them worth hundreds of millions of dollars. Some of the products are important because of the properties attributable to one element in them, others because of the physical properties of the product itself. What they have in common is that their ingredients were at one time dissolved in large bodies of water, and later concentrated or crystallized into deposits by one or more geological processes which can be inferred from their present settings.

Components extracted from saline deposits generally come from one of three geologically different types. Potassium compounds, salt, gypsum, and anhydrite mostly come from sediments formed by the evaporation of seawater during the relatively distant geologic past. Sodium carbonate, sulfate, borate, and nitrate, and some lithium, bromine, iodine, calcium, and strontium compounds come from sediments and near-surface brines formed by the evaporation of inland bodies of water, often during geologically more-recent times. Bromine, iodine, calcium chloride, and some magnesium compounds are extracted from deeply buried interstitial brines that apparently started as seawater trapped in the sediments but subsequently underwent extensive postburial change.

Brines and salts represent extremely diverse forms of "ore" which may vary in grade by one or two orders of magnitude and require quite different technologies for processing. Brines, in essence, are very low grade ore bodies that are exploitable only because they are very large and can be inexpensively processed. Many brines contain less than 5 percent of the commercial product, and some contain only a few tenths or hundredths of a percent. Solid salts, in contrast, are generally very high grade ore bodies that are more expensive to mine; operating potash mines normally extract ore containing more than 20 percent potassium mineral; salt, sodium carbonate, and sodium borate mines normally extract ore that is 75 to 95 percent mineral.

Predicting the commercial value of many saline deposits is difficult; most of the products have low sales prices relative to their weight, and their value is determined largely by how much is left after paying the cost of transporting them to markets. Deposits near rail and highway systems have an advantage over those that are more remote, deposits near seaports have advantages over all others in shipping to some domestic and most international markets.

## GEOLOGY OF SALINE DEPOSITS

All natural waters contain dissolved solids. Seawaters contain about the same components regardless of what part of the world they come from, but nonmarine waters in rivers and lakes reflect the composition of the rocks they flowed across and thus contain a variety of components in widely differing concentrations. Upon partial to total evaporation in closed or nearly closed basins, though, both sea and lake waters form bodies of concentrated brines and salts--saline deposits. They are discussed here according to whether they represent evaporation of (1) marine or (2) nonmarine waters, or (3) are brines formed by geological processes other than the evaporation of water.

### Marine Saline Deposits

Marine saline deposits almost always contain halite, gypsum, and anhydrite, but the most valuable also contain potash minerals. They typically form stratified rock units that extend hundreds of kilometers in both dimensions and attain thicknesses of a thousand meters or more. All marine saline deposits are thought to have been formed by evaporation of seawater in subsiding coastal basins or on broad shelf areas where the sea water circulation was restricted and the evaporation rate was high.

Marine saline deposits are related to desert areas in that they originally formed along arid coastal zones. Areas that were arid at the time of their deposition, however, are not necessarily arid today. Geologically young sediments underlying present deserts were probably also deposited in an arid region, but most marine saline deposits in the world are too old for this relation to hold. Nevertheless, deserts are favorable sites for the preservation and discovery of such deposits. Preservation of salts is favored in areas that have never been underlain by large quantities of migrating subsurface water which dissolves and removes the most soluble (and valuable) components; deserts are areas where this is not now, at least, happening. Discovery of salts is more likely in very arid areas where soluble salts can crop out and be identified, or lie close enough to the surface for their affect on the topography to be evident.

## Mineralogy of Marine Deposits

Almost all potash-bearing evaporite deposits have halite (NaCl), and anhydrite ($CaSO_4$) or gypsum ($CaSO_4 \cdot 2H_2O$) associated with them. In desert areas, gypsum commonly crops out and is easily identified in some areas by its paper-thin light and dark layers caused during deposition by seasonal variations in the amount of other minerals or organic material. Halite commonly forms units hundreds of meters thick, and commonly encloses any deposits of potash minerals. Because it is a highly soluble mineral, it rarely crops out at the surface except in very arid regions. Rocks that contain potash minerals generally have complex mineralogies; the major potash minerals in marine deposits are sylvite (KCl), carnallite ($KCl \cdot MgCl_2 \cdot 6H_2O$), kainite ($KCl \cdot MgSO_4 \cdot 3H_2O$), langbeinite ($K_2SO_4 \cdot 2MgSO_4$), and polyhalite ($K_2SO_4 \cdot 2CaSO_4 \cdot MgSO_4 \cdot 2H_2O$). Most commonly mined ores contain sylvite.

## Geologically Favorable areas

No large marine saline deposits are being formed today along the arid coasts of the world. Desert areas of today that are most likely to be underlain by marine saline deposits are those with a geologic history that included deposition of relatively young near-shore marine sediments. Areas that are both arid and geologically stable are the most favorable for the development of large coastal shelf areas, but unstable coastal regions characterized by broad persistent downwarping or faulting, like the Red Sea and the Persian Gulf areas, are most likely to have been characterized by subsiding marine basins in the geologically recent past.

The growth of the world's major ice sheets during the Pleistocene ice ages (the last 3 million years) caused repeated lowering of the ocean surface to elevations 50 to 100 m below its present level. These changes caused some of the nearly landlocked arms of the present ocean (like the Black Sea and the Red Sea) to become partially or totally cut off, and probably resulted in the deposition of salts for geologically brief periods. Smaller sea level changes caused by growing ice sheets began about 10 million years ago, and desert areas that bounded similar arms of the sea during the earlier part of this period may be underlain by saline deposits for the same reason.

## Methods of Prospecting

The preliminary evaluation of areas for marine saline deposits is essentially one of reconstructing their geologic history. This allows geologists to determine changes in the positions of the shoreline during past periods of marine and continental sedimentation, infer the prevailing climate at the time of deposition, seek areas with favorable geological structures, and study outcrops, cores, and landforms for signs of buried saline layers. If favorable environments are found, a more detailed study must follow. Prospecting for marine saline deposits is a highly developed phase of economic geology, and several deposits having great value have been located and developed in the last few decades. About 25 percent of the world's continental area is now known to be underlain by marine saline deposits of one or more geologic ages.

Inasmuch as marine saline deposits represent areas that were large and persistent marine basins along edges of the continental masses, geologic reconstruction of the positions of past coastlines during some part of geologic time helps narrow the target in exploration. Deposits such as sandstone, shale, and limestone represent progressively more offshore environments, and the regional distribution of such rocks and the facies changes in them help reconstruct the direction toward land. Deposits of reddish-brown conglomerate, sandstone, or shale generally represent rocks that were formed from sediments deposited subaerially in deserts. Sediments having these colors (as well as others) were mostly deposited by aeolian, fluvial, or alluvial processes, and they generally have distinctive sorting and bedding characteristics that make them easily identified as nonmarine sediments. The line along which these marine and nonmarine rock types meet was the coast line and is the area most favorable for the existence of buried saline deposits.

The normal sequence of crystallization that occurs during evaporation of seawater is (1) limestone, (2) dolomite, (3) anhydrite or gypsum, (4) halite, and (5) potash salts. The potash salts crystallize only when the seawater has been evaporated to a point where it has 1.5 percent of its original volume. Where a basin filled with seawater was evaporated to dryness, these rock units are found superimposed on each other--limestone at the bottom and potash salts on the top. But reasonable depths of seawater can only produce a thin layer of these sediments. and mineable deposits are much thicker and therefore almost had to be the result of the following variation of this simple process: During the period needed for currents to carry seawater across a very large basin, a major amount of evaporation occurs; this results in differences in the salinity of the surface waters in various parts of the basin, and in the development of zones within the basin that are characterized for long periods of time by the precipitation of only one of the different minerals of the sequence. Deposits containing mineable thicknesses of potash, the most valuable salt recovered from marine deposits, are commonly attributed to such a mechanism, and many very thick deposits of salt are thought to be the result of it. Prospecting for such deposits, therefore, requires the identification of areas that had large enough marine basins for this process to operate.

In basins containing potash deposits--as well as in many that contain other salts but no potash--the other saline rock types are found in concentrically zoned environments within the saline basin. The regional pattern of these rock types can thus be used as an indication of the zone where the conditions most favorable for the accumulation of potash existed. A recognition of these rocks, as well as a knowledge of their probable relation to each other and to the potash-bearing areas, is thus essential to a program of geologic exploration.

In addition to concentrically zoned rock types, there is commonly a similar zonation in the amount of bromine that occurs as an impurity in halite. The deposits of halite formed from brines that were nearly concentrated enough to produce potash salts have the highest amounts. The direction in which the bromine content of halite increases, and the age of the halite beds containing the highest content, help narrow the target for intensive exploration.

## Nonmarine Saline Deposits

Nonmarine saline deposits are mostly the result of the evaporation and desiccation of inland lakes and seas. Most basins that contain lakes, but have no external drainage, are products of faulting or warping of the earth's crust or damming by volcanic activity. To persist, they require crustal movements that will lower the floor of the basin faster than sedimentation can raise it, or elevate barriers across possible outlets faster than erosion can destroy them. Inasmuch as sedimentation and erosion rates in humid areas are normally faster than crustal movements, closed basins are mostly found in arid or semiarid areas where both rates are lower.

Nonmarine saline deposits are normally smaller than marine deposits. As a consequence of this, plus the fragile nature of all water-soluble minerals, most of the world's nonmarine saline deposits are geologically young and therefore lie in areas that are still deserts.

Mineralogy of Nonmarine Deposits

The components that are dissolved in natural waters are largely determined by the components of the rocks in the drainage area. Inasmuch as the types and proportions of rocks in these areas are highly variable, the compositions of the waters--and the salts that crystallize from them--are variable. As shown in figure 1, most natural waters tend to be dominated by one or two positively charged elements, most commonly calcium (Ca), sodium (Na), or potassium (K) but locally magnesium (Mg). Figure 2 shows, however, that when those waters enter closed basins and evaporate, concentrate, and deposit salts, most brines become dominated by sodium (plus potassium). The same waters also contain negatively charged ions or radicals, with most inflowing waters (fig. 1) being dominated by carbonate ($CO_3$) and bicarbonate ($HCO_3$), whereas the closed basin brines (fig. 2) may have almost any balance of these components as well as concentrations of borate or (rarely) nitrate.

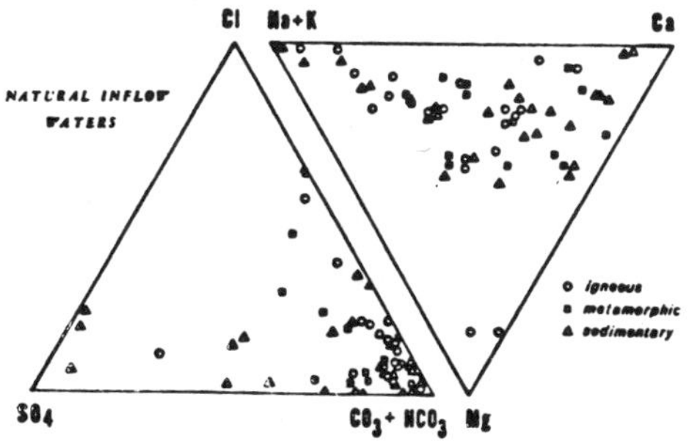


Figure 1. Diagrams showing the ratios of components dissolved in selected natural flowing waters. The dominant rock type in the drainage area is also indicated. From Hardie and Eugster (1970)

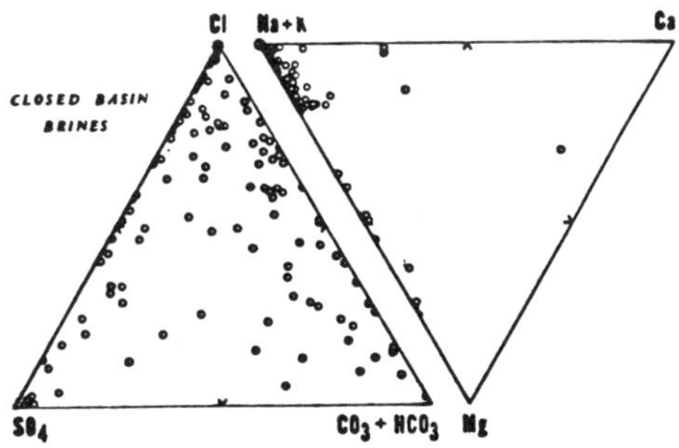


Figure 2. Diagrams showing the ratios of components dissolved in closed basin brines. From Hardie and Eugster (1970)

Upon crystallization of these components as nonmarine salts, hundreds of minerals can result. Among the more common or important of them are the following:

| | | | |
|---|---|---|---|
| Halite | ($NaCl$) | Mirabilite | ($Na_2SO_4 \cdot 10H_2O$) |
| Gypsum | ($Na_2SO_4 \cdot 2H_2O$) | Bloedite | ($Na_2SO_4 \cdot MgSO_4$) |
| Glauberite | ($Na_2SO_4 \cdot CaSO_4$) | Hanksite | ($9Na_2SO_4 \cdot 2Na_2CO_3 \cdot KCl$) |
| Thenardite | ($Na_2SO_4$) | Burkeite | ($Na_2CO_3 \cdot 2Na_2SO_4$) |
| Natron | ($Na_2CO_3 \cdot 10H_2O$) | Borax | ($Na_2B_4O_7 \cdot 10H_2O$) |
| Nahcolite | ($NaHCO_3$) | Kernite | ($Na_2B_4O_7 \cdot 4H_2O$) |

| | | | |
|---|---|---|---|
| Trona | ($Na_2CO_3 \cdot NaHCO_3 \cdot 2H_2O$) | Ulexite | ($NaCaB_5O_9 \cdot 8H_2O$) |
| Gaylussite | ($Na_2CO_3 \cdot CaCO_3 \cdot 5H_2O$) | Colemanite | ($CaB_6O_{11} \cdot 5H_2O$) |
| Soda-niter | ($NaNO_3$) | Inyoite | ($Ca_2B_6O_{11} \cdot 13H_2O$) |

Deposits of saline minerals that contain calcium (Ca) or magnesium (Mg) are mostly formed in lakes or seas that have nearly neutral waters; deposits that contain carbonate ($CO_3$) or borate ($B_4O_7$) formed in alkaline lakes; deposits containing the other elements and radicals can be formed by either type of water.

Many nonmarine saline deposits are altered after burial. Minerals such as hanksite, gaylussite, kernite, and colemanite are mostly products of such alteration.

### Geologically Favorable Areas

Geological conditions favorable for the accumulation of such deposits were those that provided: (1) a closed basin as a consequence of crustal faulting or warping (or of wind erosion), (2) large volumes of water from nearby high mountains or from adjacent less arid regions, and (3) valuable components in the inflowing water that remained in solution until desiccation occurred. Most modern examples of streams that contain such valuable components are found in areas where older outcropping salts are being dissolved or where mineralized springs associated with geologically recent volcanic activity drain into the accumulation area. Modern volcanic activity, along with closed topographic basins and adjacent mountainous regions, occur chiefly in areas characterized by presently active geological processes. Most known deposits lie in the arid belts that are in the geologically active zones along the west coast of North and South America and the zone between Turkey and southern China.

### Methods of Prospecting

During parts of the geologically young past (the last 100,000 to 500,000 years), lakes of substantial depth occupied many of the world's closed desert basins. Those areas are also likely to be deserts today, and what follows is a discussion of some of the clues that should be sought for the existence of buried saline deposits. Fortunately, many of the climatically-controlled lakes existed recently enough for some of the sediments that were deposited on the basin floors to be preserved as outcrops. These outcrops, by their character, can provide firm evidence to geologists that large lakes did exist, and allow geologic criteria to be applied, indicating the probabilities of valuable evaporite deposits being concealed in the basin center.

The questions that need to be asked when evaluating a desert area for saline deposits are: (1) Did the area at some time have suitable topographic and climatic conditions for the creation of large lakes or inland seas? (2) Were there sources of evaporite components within its drainage area? (3) Did favorable conditions for salt concentration exist? (4) Did the lakes exist long enough for large volumes of valuable components to accumulate? (5) Were those periods followed by arid episodes that caused most of the concentrated components to crystallize as salts?

Topographic and Climatic Requirements: Questions regarding the topographic requirements are easily answered when seeking geologically young deposits because those settings were nearly the same as now. A knowledge of the past climatic variations--their magnitudes and ages--is necessary for each area to assess the favorability in terms of climate. Some precision is needed to evaluate whether the lakes did or did not completely fill; overflowing causes the valuable components to be lost to other basins or to the sea.

Sources of Evaporite Components: The components in the salts and brines of valuable lake deposits may be derived from one or more of at least three sources: the atmosphere, rocks, and thermal springs. Much of the carbonate in saline deposits comes from atmospheric $CO_2$ that dissolved in the river and lake waters. Some of the sodium chloride and other seawater components are introduced into saline lake basins by air masses moving inland from the sea. Some of the sodium, potassium, and magnesium, and the balance of the carbonate is derived from the solution of those components from the rocks that crop out within the drainage area. Halite, gypsum, and other components in any exposed evaporite deposits within the drainage area are also dissolved, and they contribute chloride and sulfate to the total. The most valuable components, however, are commonly derived from thermal springs in the drainage basin. In many areas, however, such springs are no longer active or only discharge underground, and indirect evidence of past activity must be sought. Drainage areas containing evidence of older saline deposits or of thermal springs are most favorable for the existence of younger desert salt lake deposits containing valuable components.

Conditions Favorable for Concentration: The conditions that prevail when dissolved components are concentrated into a saline deposit are basically simple: large lakes composed of mineral-bearing waters must form in the valley, and throughout much of their history they must not overflow, thus allowing the dissolved components to be trapped in the basin and the salinity of the lake waters to increase.

Fortunately, a number of geologic criteria indicate the former existence of saline lake waters. One of the best criteria is the distribution of fossils. In sediments formed during periods characterized by saline lake waters, fossils of freshwater animals are restricted to a zone near the inlet--the main source of fresh water; in sediments deposited in fresh lakes, those fossils are generally in sediments exposed throughout the basin. A second criterion of salinity is when the exposed outcrops of lake-deposited clays and fine silts have efflorescences of soluble saline materials. The composition of the efflorescent minerals is likely to reflect the composition of any salines buried in the middle of a basin. Another result of the small amounts of soluble saline minerals dispersed throughout the lake sediments is the local development of karst topography (internally draining depressions and collapse pits). Additional features indicative of an appreciable sodium carbonate content of the lake waters come from the "common ion" effect of such waters on inflowing calcium bicarbonate waters; one observes an increasing percentage of calcium carbonate in progressively younger sediments, a "chemical delta" at the inlet, and an abundance of finely laminated lake sediments caused by seasonal stratification of the saline lake.

Conditions Favorable for Accumulation of Large Volumes of Salts: Inflow into a closed depression has to be large enough over a long enough time to introduce substantial quantities of dissolved materials into the basin.

Theoretically, this quantity could be provided by a small volume of water if it came from an area containing a very rich source of dissolved solids, but it is more likely to be provided by a large inflow which drains an area containing several sources of salines. The most favorable settings for these conditions are near the edges of deserts created by the rain shadow that lies on the downwind side of major mountain ranges. Such ranges extract most of the moisture in passing air masses, and some of it drains into basins lying near that edge of the desert. This provides the large amounts of water favorable to the accumulation of saline deposits, and large mountain ranges commonly have thermal springs discharging along their bases as a result of faults that occur along the bounding zones of uplift.

Climatic Changes Adequate to Crystallize Salts: The final step in accumulation of solid salts occurs when the large saline lakes evaporate and ultimately desiccate. These periods tend to have been the same over large areas because they mostly reflect a change in regional climate. The ages of major climatic fluctuations in the past are sufficiently well known in some parts of the world to know the ages of the most favorable horizons. The desiccation event causes a temporary interruption in the deposition of deep lake sediments and produces the deposition of saline layers in the central part of a valley. In the areas that were formerly covered by water, such episodes are recorded by the deposition of alluvium or by erosion; where those processes operated slowly, soils and desert pavements formed. These events became part of record and now are clues to buried saline layers that should be sought during the initial stages of prospecting.

### Subsurface Exploration

Once a desert area is considered favorable, the presence or absence of nonmarine salines can only be determined by a program of drilling. Drill cores of buried sediments that immediately surround the saline bodies provide clues not obtainable from exposed and weathered sections. In sediments deposited in basins that were accumulating salts, soluble minerals are preserved, organic materials are retained, and some of the original pore water may still be present. The detailed nature of these components is commonly indicative of the presence or absence of associated salines in the center of the basin. These mud layers also tend to have a uniform mineralogy and organic content over large areas because they were formed in lakes that had uniform water chemistries. The tendency for uniform mineralogy over large areas is of major importance in the search for salines. If a drilling program is started in the right basin but the wrong place, there is a good chance that the mud layers far off to the sides of the saline body will have the same mineralogy as the muds more closely associated with it. If the mineralogy of the mud layers is diagnostic and correctly interpreted, this knowledge may provide the necessary justification to continue a program that would otherwise be abandoned.

## BRINES UNRELATED TO SALINE DEPOSITS

Brines that have no obvious relation to bodies of solid salts are obtained from occurrences of pore waters in petroleum-bearing formations, from pore waters in rocks peripheral to marine salines, and from thermal springs or wells. Much of the world supply of bromine, iodine, and calcium chloride, and some magnesium come from them. The locations of these deposits bear no predictable relation to arid regions and are not discussed further.

SUMMARY

Saline deposits are both created and preserved in desert environments. Most marine deposits are too old for the arid climate belts in which they formed to have any relation to present arid belts, but they are now better preserved and can crop out in deserts. Nonmarine deposits that still exist were almost all deposited within the last 65 million years, and a very large number are less than 1 million years old. The more recent deposits mostly lie in areas that are deserts today. Favorable geologic environments can be identified easily, but discovery of buried deposits generally requires drilling.

## REFERENCES AND SELECTED BIBLIOGRAPHY

Braitsch, O., (1971), *Salt Deposits and Their Origin and Composition*, Springer-Verlag, New York

Briggs, L.I., (1958), Evaporite Facies, *Journal of Sedimentary Petrology*, 28, 1, 45-46

Buzzalini, A.D., et al., eds., (1969), Evaporites and Petroleum, *American Association Petroleum Geologists Bulletin*, 53, 4, 773-1032

Hardie, L.A., and Eugster, H.P., (1970), The Evolution of Closed Basin Brines, in *Mineralogical Society of America Special Paper 3*, Washington, D.C.

Livingstone, D.A., (1963), Chemical Composition of Rivers and Lakes, in *Data of Geochemistry, 6th ed.*, U.S. Geological Survey Professional Paper, 440-G

McGinnies, W.G., et al., (1968), *Deserts of the World*, Arizona University Press, Tucson, Arizona

Richter-Bernberg, G., ed., (1972), *Geology of Saline Deposits*, Unesco, Paris

Smith, G.I., (1966), Geology of Searles Lake - A Guide to Prospecting for Buried Continental Salines, in *Second Symposium on Salt, Cleveland, Ohio*, Northern Ohio Geological Society, J.L. Rau, ed.

Smith, G.I., et al., (1973) Evaporites and Brines, in *United States Mineral Resources*, U.S. Geological Survey Professional Paper 820

Stewart, F.H., (1963) Marine Evaporites, in *Data of Geological Chemistry*, 6th ed., U.S. Geological Survey Professional Paper 440-Y

White, D.E., et al., (1963), Chemical Composition of Subsurface Waters, in *Data of Geochemistry, 6th Edition*, U.S. Geological Survey Professional Paper 440-F.

# NITRATE RESOURCES IN DESERTS

George E. Ericksen
U.S. Geological Survey
Reston, Virginia 22092

## INTRODUCTION

Accumulations of nitrate-bearing saline materials occur in deserts throughout the world, but only the unique caliche deposits of the Atacama Desert of northern Chile have significantly large resources of nitrate and have been exploited on a large scale. The nitrate deposits of Chile were the main source of the world's nitrate for nearly 100 years prior to 1930. They have since declined in importance, having been displaced by synthetic nitrogenous compounds, which have shown a phenomenal growth, particularly for fertilizers, since World War II (fig. 1). In 1940, world consumption of fixed nitrogen products, used largely as fertilizer and for munitions, was about a million metric tons of contained nitrogen; about 25 percent was supplied by the Chilean deposits. In 1970, world production was nearly 39 million metric tons of contained nitrogen (about 80 percent used as fertilizer), of which Chile contributed only 100,000 metric tons, or 0.26 percent of the total. On the basis of present growth rates, projected word consumption will be about 165 million metric tons in the year 2000 (fig. 1). The Chilean contribution in 2000 will probably not be significantly greater than it was in 1970.

The resources of Chilean nitrate are small in terms of present-day requirements for nitrogeneous chemicals. For example, the total production of sodium nitrate from the Chilean deposits (including small amounts of potassium nitrate) has been about 140,000,000 metric tons for the period from 1830, when nitrate from this area was first exported to Europe, until 1976. This is equivalent to about 23 million metric tons of contained nitrogen, about half that in synthetic nitrogeneous chemicals consumed in the year 1976 alone (fig. 1). Reserves of $NaNO_3$ in commercial-grade Chilean nitrate ore (>7 percent $NaNO_3$) are estimated to be about 200,000,000 metric tons, equivalent to 33 million metric tons of contained nitrogen, still less than world annual consumption in 1976. Probable resources of lower grade caliche (1-7 percent $NaNO_3$) are several times greater than those of the currently commercial-grade ore, but probably equivalent to not more than 2 or 3 years of current world consumption of fixed nitrogen.

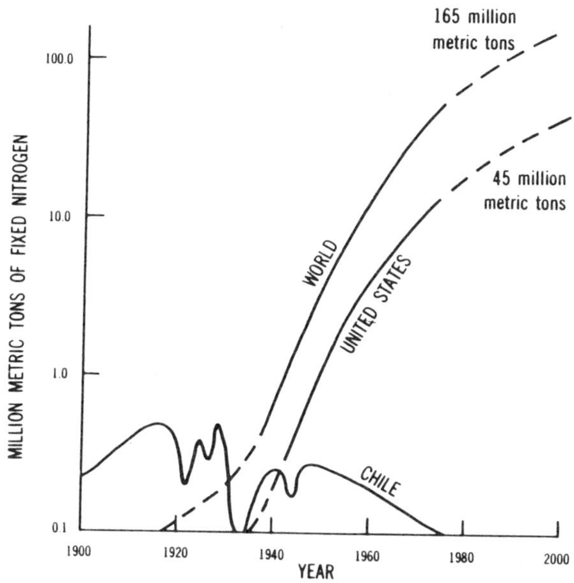


Figure 1. Production of natural nitrate in Chile and of synthetic nitrogenous chemicals in the United States and in the world as a whole, in terms of contained nitrogen, for the period 1900-1970 and projected to 2000. From U.S. Bureau of Mines, Mineral facts and problems (1970); Minerals yearbook (1933-1970) and Ericksen (unpublished data, 1974).

Other than the Chilean nitrate deposits, natural nitrate deposits, of which only part occur in deserts, have yielded only small amounts of nitrate. Only the deposits in India, which are largely in non-desert areas, have had a significant long-term production. These deposits consist of efflorescences of nitrate-rich saline material in and near ancient village sites. Elsewhere, cave and cavernous earth deposits were formerly exploited for nitrate that was used chiefly for explosives. Deposits of sea bird guano in coastal deserts, which have been exploited as a source of phosphate fertilizer, also are rich in nitrate and ammonium salts. Such deposits are widespread along the coasts of Chile and Peru.

## CHILEAN NITRATE DEPOSITS

The Chilean nitrate deposits are among the most unusual of all mineral deposits, and their origin has been the subject of lively discussion for more than 100 years. They accumulated over a long period of time - perhaps several million years - in the nearly rainless Atacama Desert, where paucity of rainfall allowed large amounts of highly soluble saline minerals, including nitrate, to accumulate in the soil (Ericksen, 1975). Dozens of published reports propose a wide variety of sources for the nitrate and associated saline minerals in the deposits and for the mode of their accumulation. Evidence now at hand (Ericksen, 1975) indicates that the saline materials accumulated as atmospheric precipitates - dry fallout and soluble saline materials in fog condensate and rainwater. Sea spray from the nearby Pacific Ocean and emissions associated with intense volcanism in the nearby Andes during late Tertiary and Quaternary time probably have been major sources of the saline materials now found in the deposits.

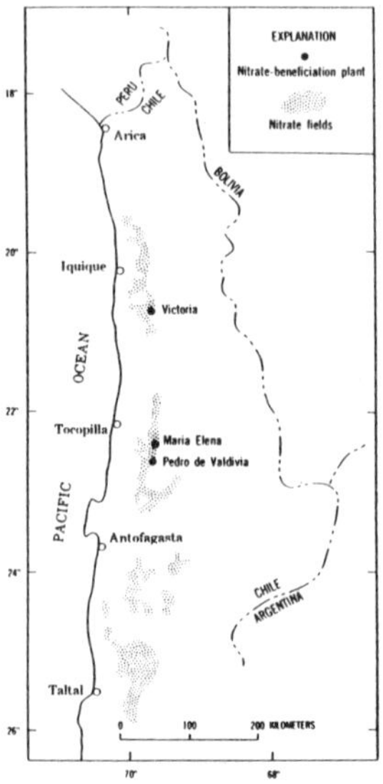


Figure 2. Index map of northern Chile showing location of the nitrate fields and of nitrate beneficiation plants in operation in 1977.

The deposits extend from about lat. 19°30'S. to 26°S., a distance of nearly 700 km (fig. 2), chiefly along a belt less than 50 km wide, between long. 69°3'W. and 70°W. The region is characterized by a broad Central Valley, bordered on the west by the relatively low Coastal Range, where few peaks are more than 2000 m in altitude, and on the east by the Andes Mountains, where peaks extend to altitudes of 6000 m or more. Nitrate deposits are most widespread along the eastern side of the Coastal Range, a region of low relief characterized by rounded hills and ridges and broad, shallow debris-filled valleys. The deposits are present in all topographic positions, from tops of hills and ridges to the centers of the broad valleys. However, many of the richest deposits, which were worked extensively during the 19th century, are on the lower slopes of hills, marginal to a salt pan or playa. Nitrate and other saline minerals form veins or impregnations in such diverse rock types as granitic intrusions, andesite and rhyolite flows and tuffs, limestone, sandstone, and shale. They also form a cementing material in unconsolidated sediments such as slope debris, alluvium, windblown silt and sand, rhyolitic ash, stream gravels of probably Tertiary age, and lacustrine sand and silt. Chemical and mineralogical composition of the saline material is unrelated to the type of host rock.

Two major types of nitrate ore, locally called caliche, can be recognized, one (alluvial-type caliche) in which the saline minerals are chiefly a cement in unconsolidated sediments and the other (bedrock caliche) in which they are impregnations, veins, and irregular masses in bedrock. The alluvial-type caliche generally rests on poorly cemented or uncemented rock debris similar to that in the caliche itself, whereas the bedrock type gives way downward to less fractured or more imprevious rock relatively free of saline minerals. The typical alluvial-type nitrate deposit is considered to be made up of several layers (fig. 3): chuca and costra (overburden), caliche, and coba and conjelo (low-grade or barren material beneath the caliche). Alluvial caliche, by far the most widespread and abundant type, is generally 1-3 m thick, but high-grade material as

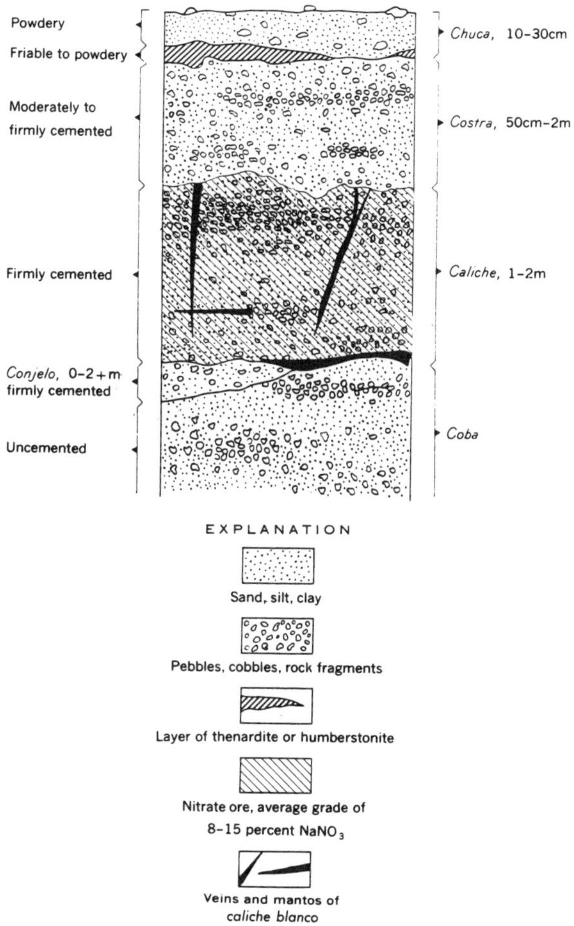


Figure 3. Diagram showing layering in alluvial-type nitrate deposit in northern Chile.

thin as 50 cm has been mined locally. Veins and mantos (horizontal to gently inclined layers) of high-purity white nitrate-rich saline material (called caliche blanco by the miners) are widespread in the alluvial caliche of some fields or areas but relatively rare in others. They commonly are 10 to 50 cm thick. Mantos generally occur at the base of the caliche layer and tend to be most prevalent in areas of thick caliche. Bedrock caliche consists of saline impregnations in porous rocks and of veins and mantos of caliche blanco along fractures in otherwise massive or poorly bedded rocks such as ash-flow tuffs, and along bedding planes and in cross-cutting fractures in layered sedimentary rocks. Such veins and mantos, which generally range from a few centimeters to several tens of centimeters in thickness, formed along incipient cracks and narrow fissures that were gradually forced open by the crystallizing saline minerals. Mantos of caliche blanco in bedrock are generally at depths of 3-6 m, but at places are at depths of as much as 12 m. In a given nitrate field, the layer tends to be at a relatively constant depth, and, like mantos of caliche blanco in alluvial-type caliche, tends to be parallel or subparallel to the land surface.

The nitrate deposits show great local and regional variations in chemical composition, so that estimation of the composition of an average or typical nitrate ore is difficult. Perhaps the best estimates are those based on large tonnages of ore treated in the nitrate-beneficiation plants, as, for example, ore treated at the Maria Elena and Pedro de Valdivia plants (fig. 2). Monthly average grades of such ores (table 1) indicate the variations in bulk composition of the water-soluble components in material containing a minimum of about 5.5 percent $NaNO_3$ (4.0 percent $NO_3$) (Grossling and Ericksen, 1971). Daily averages show greater variation even though the ore is mined selectively to maintain a grade of about 8 percent $NaNO_3$. High-grade nitrate ores, mined selectively in other fields during the 19th century, contained several times as much nitrate and lesser amounts of sulfate and chloride. Such ore commonly averaged 25 percent or more $NaNO_3$.

Table 1. Ranges in Monthly Averages (1) in Chemical Composition of Nitrate Ore Treated in Oficinas Maria Elena and Pedro de Valdivia, the Two Largest Nitrate-Beneficiation Plants in Chile.

| | Range in Composition (weight percent) |
|---|---|
| $NO_3$ | 4.05 - 7.56 |
| Cl | 2.44 - 9.23 |
| $SO_4$ | 4.04 - 20.28 |
| $IO_3$ | .021 - 0.096 |
| $ClO_4$ | .007 - 0.067 |
| $B_5O_9$ | .021 - 0.82 |
| Na | 3.92 - 10.83 |
| K | .25 - 1.90 |
| Ca | .73 - 3.54 |
| Mg | .06 - 1.30 |
| Total soluble saline material in ore | 21.3 - 47.8 |

(1) For the period 1932-1967 when the two plants treated about 1 million metric tons of nitrate ore per month.

The deposits are unique in having relatively high concentrations of iodine, which occurs as the iodate ion ($IO_3^-$) and perchlorate ($ClO_4^-$). They have been a major source of the world's iodine since the mid-19th century. Perchlorate, not yet identified in mineral form, is known to exist in nature only in these deposits.

The saline minerals that have been identified in the Chilean nitrate deposits (table 2) include borates, iodates, chromates, and sulfates in addition to nitrates and a single chloride mineral, halite. Halite and soda niter are by far the most abundant, being the dominant minerals in nearly all nitrate ore. Although the total sulfate may equal or exceed nitrate, as indicated by the ranges in chemical composition cited above, it is contained in several minerals, and any single sulfate mineral in rarely more abundant than soda niter. The most abundant sulfate minerals are glauberite, bloedite, gypsum, anhydrite, darapskite, and humberstonite. Thenardite, which is an abundant mineral in other desert saline deposits, is rare in the nitrate ore because it is unstable in solutions from which soda niter precipitates (Ericksen and Mrose, 1970).

Table 2. Saline Minerals of the Chilean Nitrate Deposits.

| | |
|---|---|
| Halite, $NaCl$ | Ginorite, $Ca_2B_{14}0_{23}.8H_20$ |
| Soda niter, $NaNO_3$ | Hydroboracite, $CaMgB_60_{11}.6H_20$ |
| Niter, $KNO_3$ | Kaliborite, $HKMg_2B_{12}0_{16}(OH)_{10}.4H_20$ |
| Darapskite, $Na_3(SO_4)(NO_3).H_20$ | Thenardite, $Na_2SO_4$ |
| Humberstonite, $K_3Na_7Mg_2(SO_4)_6(NO_3)_2.6H_20$ | Glauberite, $Na_2Ca(SO_4)_2$ |
| Lautarite, $Ca(10_3)_2$ | Bloedite, $Na_2Mg(SO_4)_2.4H_20$ |
| Bruggenite, $Ca(10_3)_2.H_20$ | Kieserite, $MgS0_4.H_20$ |
| Dietzeite, $Ca_2(10_3)_2(Cr0_4)$ | Epsomite, $MgSO_4.7H_20$ |
| Lopezite, $K_2Cr_20_7$ | Gypsum, $CaSO_4.2H_20$ |
| Ulexite, $NaCaB_50_9.8H_20$ | Anhydrite, $CaSO_4$ |
| Probertite, $NaCaB_50_9.5H_20$ | Bassanite, $2CaS0_4.H_20$ |

## NITRATE DEPOSITS IN THE MOJAVE DESERT, SOUTHWESTERN UNITED STATES

Nitrate deposits in the Mojave Desert of the southwestern United States were studied intensively by the U.S. Geological Survey during and shortly after World War I (Noble and others, 1922; Noble, 1931; Mansfield and Boardman, 1932). Three types of deposits can be recognized: (1) caliche (2) cave and cavernous cliff; and (3) clay playa. Many nitrate mining claims were staked and explored at this time, but no nitrate was produced. There is no record of earlier production.

Caliche-type deposits are restricted to the driest part of the Mojave Desert - chiefly areas near Death Valley in southeastern California - where they are associated with poorly consolidated Tertiary lacustrine clay deposits that tend to weather to a terrain characterized by smooth rounded hills, essentially barren of vegetation, locally called clay hills or mud hills. Thin discontinuous layers of nitrate-bearing caliche occur on the tops and sides of these hills. The caliche layer, which is generally less than 15 cm thick, consists of compact saline-cemented clay at the bedrock surface and is overlain by a saline soil as much as 30 cm thick. The saline material in the caliche layer consists chiefly of halite, and the content of sodium nitrate generally ranges from traces to a few percent. Chemical analyses also indicate the presence of gypsum and smaller amounts of other sulfate minerals (Noble and others, 1922). The soil above the caliche is less compact, containing abundant soft white saline material consisting chiefly of sulfate minerals, and is capped by a compact, relatively impervious surface-soil layer several centimeters thick, which inhibits percolation of rainwater (Noble and others, 1922).

Cave and cavernous cliff deposits are much more widespread than the clay-hill type deposit in the Mojave Desert and have been reported by Mansfield and Boardman (1932) from many localities in southeastern California, Nevada, Arizona, Utah, and New Mexico. Most prevalent are encrustations of nitrate-rich saline material in cliffs of cavernous, fractured, and brecciated volcanic rocks, where overhanging ledges protect the salts from rain. In some places, efflorescences have spalled from the cliff face to form accumulations of saline and rock material several tens of centimeters thick at protected bases of cliffs. Elsewhere, nitrate-rich saline material occurs as efflorescences on the roofs and walls of caves and as accumulations of efflorescences and spalled material on cave floor. These deposits consist chiefly of potassium nitrate rather than sodium nitrate and contain varying amounts of halite and sulfate minerals. Not uncommonly, they consist of relatively pure potassium nitrate.

Noble (1931) reported nitrate in amounts of as much as 1 percent in a 15-cm thick layer of puffy soil just beneath the surface of Leach Lake, a dry clay playa in a basin of interior drainage about 30 km south of Death Valley, California. He reported still smaller amounts of nitrate in one of the dry clay playas of the Coolgardie Lakes near Barstow, California. Moist saline clay at depths of 30-100 cm below the playa surface was found to contain small amounts of nitrate, estimated in the field to be less than 1 percent $NaNO_3$ (Noble, 1931).

The origin of the nitrate deposits in the Mojave Desert has been discussed by Gale (1912) and Noble (1931). The clay hill deposits are similar to the Chilean nitrate deposits and probably have a similar origin. Cave and cavernous cliff deposits are considered to have formed by action of bacteria on animal wastes and plants. Noble (1931) considered that the nitrate in clay playas results from present-day accumulation and that the nitrate was formed either by nitrifying bacteria in the playa or was washed into the playa from the surrounding area.

## Nitrate Deposits in Other Deserts

The literature on nitrate deposits in other deserts, although scanty, is adequate to give an idea of the distribution and character of the deposits in some areas. Cave and cavernous cliff deposits similar to those in the Mojave Desert probably occur in most desert areas; clayhill caliche deposits, such as those in southeastern California may be present at a few places, but extensive deposits of nitrate-rich caliche such as those in northern Chile evidently are absent.

Asia. - India was a chief source of potassium nitrate for gunpowder for several centuries before being displaced by Chilean nitrate in the 19th century. However, most of the Indian production came from nondesert areas, of which Bihar Province in eastern India was most important, and relatively little came from arid and semiarid regions of western India. Indian production was about 73,000 metric tons of potassium nitrate per year at the end of the 18th century, which is probably near the all-time maximum production, but had declined to about 18,000 metric tons by the end of the 19th century (Brown, 1936). During the period 1929-1939, annual production of potassium nitrate in India was 5,500 to 12,600 metric tons (Dunn, 1942). During this period, deposits in the Punjab region of northwestern India, in part within and marginal to the Thar Desert, were major sources of the nitrate. Estimated annual production from this area in the late 1930's and early 1940's was about 5,000 metric tons (E.G. Newman, unpublished data, 1964).

The Indian nitrate deposits consist of efflorescences on soils in and near old villages. Dunn (1942, p. 70-71) gave the following description of nitrate accumulations in Bihar Province, which is applicable to deposits elsewhere in India:

> "The nitrates necessary for the formation of saltpetre have accumulated in the course of centuries in the soil of this highly populated area. The North Bihar villages, built on the sites of previous tumbled villages of mud huts, become raised above the level of the surrounding country. The floors of these huts are invariably made of mud and cow-dung, forming a nitrogeneous deposit to which is added other animal refuse and the ashes of innumerable fires, which contain potash. The decaying refuse undergoes nitrification, and the product drains from the raised

> village sites into the surrounding fields at a lower level. In the dry season the salts are brought to the surface as a mixed efflorescence of sodium chloride, sodium sulphate, sodium carbonate and nitrates of potash and magnesia."

Africa. - Nitrates have been reported from several regions in Africa. Little-known nitrate deposits, said to be analogous to the Chilean deposits, occur at several localities in the Sahara Desert of southern Algeria (Launay, 1903). The distribution, grade, and nature of the deposits have not been reported. Bruggen (1938) stated that efflorescences of nitrate-rich saline materials form on the surface of a sandy crust at the oasis of Touat, Algeria, after heavy rains. Nitrate deposits, which apparently are analogous to the clay-hill deposits of California, have been reported in the Esnch Shale near Luxor in Nile River valley 760 km south of Cairo (Bruggen, 1938; Launay, 1903). According to Bruggen (1938), outcrops of the shale contain as much as 18 percent $NaNO_3$. The material is now being utilized by local farmers as a natural fertilizer and soil conditioner (A.A. Hussein, Egyptian Geological Survey and Mining Authority, oral communication, May 1977). This nitrate is probably a surface accumulation and not a primary constituent of the shale. Bruggen (1938) mentioned a similar occurrence near Mariental, South West Africa, where the weathered rock was reported to contain as much as 80 percent $NaNO_3$.

Australia. - Nitrate occurrences in arid parts of Australia have been briefly described by Mawson (1930). Cave deposits are found in the Macdonnell Ranges, about 100 miles west of Alice Springs in central Australia, in an area of dolomitic limestone where former caves have been largely obliterated by collapse and by infilling with cave earth. Nitrate-rich saline materials occur as efflorescences on walls of the few remaining caves and in cave earth beneath siliceous and ferruginous breccia that formed by collapse of near-surface caves. Rock and soil impregnated by saline materials were found to contain highly variable amounts of potassium nitrate, ranging from less than 1 percent to nearly 19 percent $KNO_3$. Mawson suggested that the nitrate formed from animal waste products and was preserved in the caves and in the soils of collapsed caves by the layer of relatively imprevious surface material. Bat guano rich in ammonium salts occur in limestone caves of the Flinders Range in the arid northern part of South Australia. Near Mt. Warren Hastings, farther to the north, the floor of a very large limestone cave is covered with a thick deposit of bat guano and other debris containing abundant ammonium sulfate and chloride salts. At Elder Rock, also in South Australia, a collapsed cavern area contains ammonium and nitrate-bearing saline material to depths of as much as 20 m below the surface.

Soils and ground water in some arid areas of Australia have notable concentrations of nitrate. For example, Mawson (1930) reported ground water in a large area in Western Australia - the area bounded by lat. 24° and 30°S. and long. 116° and 124°E. - to have concentrations of nitrate ranging from a trace to 0.036 percent; 39 analyses of water show an average of about 0.01 percent $NaNO_3$. Mawson also reported ground water of the Tarcoola district, South Australia, as having unusual concentrations of nitrate. Simpson (1952) reported "lake salts" from near Mt. Ridley in Western Australia as containing unusual concentrations of potassium nitrate. He cited 6 analyses of lake salts that show abundant NaCl and 1.8 to 19.4 percent $KNO_3$. The nature and extend of these salts are not discussed. Perhaps this is a playa-type deposit in which nitrate-bearing salts are presently accumulating.

## REFERENCES CITED

Brown, J.C., 1936, India's mineral wealth, a guide to the occurrences and economics of the useful minerals of the Indian empire: London, Oxford Univ. Press, 335 p.

Bruggen, Juan, 1938, El salitre: Bol. Minero, Santiago de Chile, v. 50., p. 737-754.

Dunn., J.A., 1942, The economic geology and mineral resources of Bihar Province: India Geol. Survey Mem., v. 78, 238 p.

Ericksen, G.E., 1975, Origin of the Chilean nitrate deposits: Geol. Soc. America Abstracts with Programs, v. 7, no. 7, p. 1068.

Ericksen, G.E., and Mrose, M.E., 1970, Mineralogical studies of the nitrate deposits of Chile. II. Darapskite, $Na_3(NO_3)(SO_4).H_2O$: Amer. Mineralogist, vo. 55, p. 1500-1517.

Gale, H.S., 1912, Nitrate deposits: U.S. Geol. Survey Bull. 523, 36 p.

Grossling, B.F., and Ericksen, G.E., 1971, Computer studies of the composition of Chilean nitrate ores: Data reduction, basic statistics, and correlation analysis: U.S. Geol. Survey open-file report, 85 p.

Launay, L., 1903, Les richesses minerales de l'Afrique: Paris, Librairie Polytechnique, 395 p.

Mansfield, G.R., and Boardman, Leona, 1932, Nitrate deposits of the United States: U.S. Geol. Survey Bull. 838, 107 p.

Mawson, Douglas, 1930, The occurrence of potassium nitrate near Goyder's Pass, McDonnell Ranges, central Australia: Mineralog. Mg., v. 22, p. 231-237.

Noble, L.F., 1931, Nitrate deposits in southeastern California with notes on deposits in southeastern Arizona and southwestern New Mexico: U.S. Geol. Survey Bull. 820, 108 p.

Noble, L.F., Mansfield, G.R., and others, 1922, Nitrate deposits in the Amargosa region, southeastern California: U.S. Geol. Survey Bull. 724, 99 p.

Simpson, E.S., 1952, Minerals of Western Australia: Perth, Gov't. Printer, v. 3, 714 p.

U.S. Bureau of Mines, 1933-72, Minerals yearbook 1932-1970: Washington, D.C., U.S. Gov't Printing Office (annual publication).

U.S. Bureau of Mines, 1970, Mineral facts and problems, 1970: U.S. Bur. Mines Bull. 650, 1291 p.

# DESERT SALINE DEPOSITS

Lonnie Hiebert

Saline chemical deposits are a common feature of deserts, to the extent that most people think of brackish water or salt formations synonymously with deserts. High rates of evaporation and many closed drainage basins lead to the frequent desert "playas," and many of these result in some sort of saline formation. Most such basins do not have "economic" salt deposits, but in almost every desert of the world there are some potentially economic saline formations. Only a few exceptions occur. Other saline deposits probably originally formed in arid or desert regions, even though many of them may not currently exist in deserts. Geological changes frequently have covered to isolate evaporite deposits in zones different than their original climatic conditions.

The formation of a saline chemical deposit has the following requirements: a depression and closed drainage basin; a source of the chemicals, such as the ocean, or dissolved solids in streams or rivers; isolation, such that the inflow of water is less than evaporation; and time--usually a very long period.

## TYPES OF SALINE DEPOSITS

There are three main types of desert saline deposits using classifications based upon the source of the chemicals. These are: 1. Sea water evaporites 2. Inland lake evaporites 3. Geothermal water evaporites.

### Sea Water Evaporites

For deserts near oceans the major source is sea water. Basins can occur that are either isolated or semi-isolated from the ocean. A combination of such basins and good evaporating conditions create a potential for saline salt deposition. The rate of deposition and the salts' thickness will depend on a number of factors: evaporation rate (net), inflow of fresh seawater, and depth-to-area ratio of the basin. As evaporation occurs, the sequence of crystallization proceeds in a fixed manner. Calcium carbonate is the first to precipitate, followed by gypsum (calcium sulfate) when the concentration is about 3-1/2 times the original, or when about 70 percent of the water has evaporated. Anhydrite, an anhydrous form of calcium sulfate starts forming when about 80 percent of the original water is evaporated. Then halite (sodium chloride) follows when only 8 percent of the original water remains. The potassium and magnesium minerals follow placing these most valuable salts usually in the central area of the deposit, or at the lowest portion of the basin.

If there is partial or periodic replenishment of the sea water into the basin, the effect on the deposit is a change in the layering of the salts. Also, the potassium and magnesium may be concentrated at the side away from the source of inflow. In the case where there is still current or recent inflow, these minerals may still be present in the concentrated residual brine.

## Inland Lake Evaporites

A common source of saline chemicals is the leaching of minerals, or from previously deposited salt beds. Examples of inland lakes that have been formed by such leaching are the Great Salt Lakes, Utah, Dead Sea, Lake Assale, Ethiopia, and the Salar de Atacama in Chile. Since there is more variation in this type of salt source, these inland lakes have a greater variety of salt compositions. Also since many still have water or brine inflow, some of the economic minerals are still in the brine fraction. Some of these are "bitter" lakes where sodium sulfate predominates. Many have a very thick salt deposit with the voids filled with brine. These brines may contain a valuable mineral content of potassium, magnesium, bromine, lithium, or other salts.

## Geothermal Water Evaporites

Hot spring, rising geothermal waters, or possibly volcanic flow such as carnotites account for many saline deposits. The minerals in these deposits are usually different from those of other saline deposits, often being high in soda ash, but many include borax and other less common chemicals.

An example of this type of deposit is the immense sodium borate deposit at Boron, California, in the Mojave Desert. It is rather unique in that only minor traces of other salines are present in the evaporite beds of sodium borate. One of the theories of its origin is that geothermal waters or a volcanic vent that was rich in borates discharged into a slightly alkaline lake, and the residual concentrated brines flowed for their final evaporation to Boron. Calcium borate deposits are found in another former evaporate basin about 20 miles away at Four Corners. No other remnants have been found. Somewhat similar deposits have been found in Turkey and Argentina, but borate deposits are still comparatively rare, and whenever the origin is known, it is always associated with volcanic activity. A sodium carbonate operation in Uganda is based on dissolution of magmatic carbonatites, and Searles Lake, California is known to have been derived from hot springs.

## COMMON SALINE CHEMICALS

Saline chemicals, as discussed in this paper, are primarily made up of the natural occurring salts of the alkali metals: lithium, sodium, potassium; the alkaline earth metals: magnesium, calcium, and on rare occasion, barium and strontium. The acidic or anion portion of these salts includes commonly chlorides, bromides, sulfates, carbonates, borates and more rarely nitrates, iodides and others.

These metals and anions are found in numerous combinations in saline deposits. Although calcium and magnesium carbonates, i.e. limestone and dolomite could be considered part of the above definition, they are insoluble and also found in many land formations and will not be considered in this paper.

The most common of the salines is sodium chloride, being found in most saline deposits, and in many cases as the major component. Other abundant salines are potassium and magnesium chlorides, sodium and magnesium sulfates and sodium carbonates. Many double salts of the above form distinct mineral or crystalline forms. Rarer chemicals, but sometimes found in concentrated occurrences in saline deposits, are: sodium borate, lithium sulfate, sodium and potassium nitrates, calcium chloride, iron chlorite, sodium iodates and bromates, etc. Some of these are normally found only in the brine fractions of the saline lakes.

## EXPLOITATION

### Historical

Early use or exploitation of mineral resources in arid regions was practiced exclusively with surface crystalline deposits. Such deposits with surface exposure generally found only primitive use, but as man's knowledge of their value became more sophisticated, so did the methods of recovery.

The first exploitation of saline minerals was confined to salt (sodium chloride). Trade resulted, and areas with salt deposits traded with areas deficient in salt. An example of such trade still exists in Africa. In the Danakil depression of Ethiopia, slabs of salt are harvested from surface deposits by hand labor, loaded on camel caravans and sent westward to parts of the Sahara for trade in areas deficient in salt. Such salt exploitation became more complex until even underground formations were utilized. Recently, an ancient mining activity was discovered by the Russians in one of their underground potash mines. Ancient mining tools were found when they cut into an old mined-out zone. This early mining had been for salt.

Marco Polo is credited with bringing borax to Europe from China or Tibet. Its use as a flux in goldsmithing created a demand that supported this trade for centuries. A colorful bit of similar history is the use of the 20 mule teams for hauling borax out of Death Valley. Again, a surface deposit of ulexite, a double salt of calcium and sodium borate, occurred as "cotton ball borax". This was scraped by hand, placed in vats with water and sodium carbonate (soda ash) and boiled. Upon cooling borax crystals formed. These were loaded into wagons pulled by teams of twenty mules across the desert to the nearest railroad over one hundred miles away.

In the nineteenth century the discovery of nitrate caliche in the Atacama desert of Chile, combined with the demand for nitrates in gunpowder, created a large extractive industry. Due to the extremely arid conditions these highly soluble nitrate minerals and other saline salts formed in a clayey-rock formation by what appears to be limited flow of brine into the formation and capillary action. Originally hand dug holes located areas of high nitrate concentration, and these ores were selectively mined, the nitrate leached, and crystallized from the solution. At one time there were between 100 and 150 "oficinas" or mine-plant complexes producing sodium and potassium nitrate. Since World War I, the advent of synthetic nitrate production, and old and inefficient operations, this number has declined to only three effective "oficinas" at present.

Potassium salts recovery is far and away the largest tonnage of the commercial saline mineral operations (although soda ash is gaining rapidly). However, the history of mining potash (KCl) only started early in the twentieth century. Prior to that time potash was recovered by leaching wood ashes in large iron pots, and thus the name. Once massive underground potash formations were found by deep drilling, Germany dominated the potash industry until after World War II, when part of the Rhine deposits were transferred to France. In about 1930 mining for potash was started in New Mexico near Carlsbad, and, in 1923, extraction of potassium and other salines commenced at Searles Lake, California. Since that time potash deposits have been located and mining has been started on every continent, only some of which are in present-day deserts. Potash demand is mainly for fertilizer, and its consumption is directly proportional to a growing population and the world food supply.

## Exploitation Planning

Before exploitation or development, saline deposits, like any other mineral deposit, must have a thorough prior exploration to determine their grade and reserves. Market surveys must be completed, transportation methods and costs determined, and then process feasibility studies made. If the preliminary studies indicate that an ore or brine deposit is of the required size and grade, a market survey should be performed to establish the size of the market. Process feasibility studies then can provide an estimate of the processing method, yield, and cost. The life of the deposit is next estimated, and a decision made on the optimum capacity of the mine and processing plant, keeping in mind that larger plants usually operate more economically. Finally, detailed studies would be made in which all of the processing alternates should be utilized. The various questions to be answered are:

- Method of mining or harvesting
  - Open pit mining
  - Underground mining
  - Dredging
  - Brine pumping
  - Solution mining
- Plant site
  - At or near mine
  - Topography and soil conditions
  - At a remote location
    - Railhead
    - Usage point or in conjunction with an existing plant
- Transportation
  - Rail, truck, pipeline
- Utilities
  - Water
    - Available on site, in sufficient volume
    - Brackish or potable
  - Electricity

Available from existing source
Generate own

Steam turbine
Gas turbine
Diesel engine

Fuel

Types of fuel available

Geothermal
Can solar energy be used
Fossil fuels

Processing

Current state of the art

Is there a known process
Are there alternates
Does this mineral or brine require new technology
What byproducts can be produced

Selection or design of preferred process

Maximum recovery
Lowest energy consumption
Least capital cost
Lowest operating costs

Manpower

Adequate supply in local area
Is new housing needed
New townsite
Training needed

Environmental effects

Mining effects

Mine wastes
Effects on vegetation

Wastes

Liquid, solid, gases and dusts

Local laws and requirements

Civic requirements

Economics

Capital source
Return on investment
Effect on local economy and that of the country

In reviewing the above outline it is obvious that many of the questions are interrelated. By using past examples of success and failure as a guide much time can be saved in designing the operation. A few examples of imaginative thinking that were successful in developing desert saline processes are:

Searles Lake operations (KCl, $K_2SO_4$, $Na_2CO_3$, $Na_2SO_4$, $LiCO_3$, Br, Borax)
- Complex fractional crystallization of brines

Great Salt Lake chemical ($K_2SO_4$, $Na_2SO_4$, NaCl)
- New solar pond-flotation technology for "sea water-type" brine

Solar evaporation of salt in Baja California (NaCl)
- Large scale
- Taking advantage of location
  - Solar energy
  - Sea transportation

Dead Sea (KCl, MgO)
- Large solar pond-crystallization process

Laguna del Rey, Mexico ($Na_2SO_4$, MgO)
- Highly efficient evaporative operation

Texas Gulf Sulfur Potash operations, Moab, Utah
- Solution mining combined with solar ponds under unusual terrain conditions

## POTENTIAL FUTURE DESERT OPERATIONS

It was not until very recent times with modern science and technology that large scale exploitation of the saline deposits has started. Even so there are many known deposits that have not been evaluated or exploited throughout the deserts of the world. The potential for developing needed resources is therefore promising. Especially for some of the underdeveloped areas. Local resources can be developed for internal needs. The needs for imports of these essential salines will be reduced or eliminated. And in many cases an export market may also be created.

Each saline deposit has a potential for exploitation. Imaginative thinking can tailor the methods of reclamation and processing to local conditions and needs. The rapidly increasing price of the saline minerals and the strategic importance of many of them to developing countries should allow many more to be operated in the future.

## RELATION BETWEEN PHOSPHATE DEPOSITS AND DESERTS

Richard P. Sheldon


### ABSTRACT

Many phosphate deposits, including some of the largest in the world, occur in deserts, and most of the great deserts of the world contain phosphate deposits.

Table I.

| Desert | Area of phosphate deposit | |
|---|---|---|
| Australian Desert | Queensland | |
| North Africa, Sahara Desert | Spanish Sahara<br>Morocco | Tunisia<br>Senegal |
| Arabia & Middle East, Arabian Desert | Turkey<br>Syria<br>Jordan | Israel<br>Egypt<br>Saudi Arabia |
| Indian Desert | Rajasthan | |
| USSR, Turkistan Desert | Karatau | |
| South America, Atacama Desert | Peru | |
| North American Desert | Mexico, California<br>Idaho, Utah | |

Is this due to the vagaries of geologic history or is it due to a genetic relationship? The question is important because if it is the latter, the desert areas should be studied to assess their expected phosphate resources.

Geologic analysis shows that marine phosphates are economically the most important type of deposit, and indeed are genetically related to adjacent coastal deserts. Modern marine deposits occur in the warm trade wind belt where cold deep ocean waters upwell into shallow water along the coast, causing the deposition of the phosphate offshore and the formation of a cold oceanic air mass adjacent to a warm continental air mass, which in turn causes a coastal desert. Thus, modern coastal desert areas should continue to be explored for undiscovered modern and geologically young genetically associated phosphate deposits both offshore and onshore.

Warm interior deserts have no genetically associated modern phosphate deposit, even though unrelated ancient phosphate deposits may occur in them. Thus interior deserts may have been coastal deserts in the geologic past and should be explored for genetically associated phosphate deposits. This can be accomplished by application of the modern theory of plate tectonics in order to trace the movements of the deserts backward in time to determine whether or not they were in more favorable positions for the occurrence of phosphate in the past.

Thus, the observation that the world's deserts seem to contain an abnormal amount of the world's phosphate deposits is not entirely due to geologic chance but to geologic reason. This relationship should be considered and exploited in any strategy of desert development and management.

## INTRODUCTION

A cursory look at the distribution of the world's large economic phosphate deposits reveals two things: first, they are mostly marine in origin; and second, they tend to occur more often than not in the desert areas of the world. The very large and rich phosphate deposits of Morocco come immediately to mind, as well as almost equally impressive deposits in nearby Spanish Sahara, the Karatau Formation in the Turkestan desert of the USSR; and the recently discovered large deposits in western Queensland, Australia. A large portion of the world's known resources occur in these three deposits but the list does not stop there. Significant deposits exist in the deserts of Peru, Mexico, U.S.A., India, the Middle East and other North African countries. These occurrences are listed in Table 1.

Is this observation due to the vagaries of geologic chance or to an underlying genetic relationship? If it is the latter, it is clear that where geologic studies have not been carried out, an opportunity exists for additional mineral development of desert lands.

Many of these desert lands are in developing countries where agricultural lands have been farmed for centuries so that their soils have been systematically depleted of their supply of available phosphate. They have become agriculturally marginal to the point that agricultural production cannot generate the capital necessary to finance the import of sufficient phosphate to help regenerate the land. Where capital is not available from other internal sources, it seems important to develop any local phosphate resources, thereby avoiding drawing on scarce foreign currency.

### Geologic Environments of Phosphate Deposits

Earth science studies have revealed that the earth is in constant motion. The "solid" earth moves due probably to geothermally driven convection currents deep within the earth and to the earth's rotation. These forces combine to drive the dozen or so gigantic plates that make up the earth's surface, a process referred to by geologists as plate tectonics. The oceans and atmosphere of the earth are also in constant and much more obvious motion due to solar driven thermal convection currents and the earth's rotation.

These dynamic earth processes result in the formation and movement and commonly the destruction of mineral deposits. Phosphate is no exception to this rule. It is concentrated at several phases of its "geochemical cycle," that is its cycle of movement due to the restless Earth. (Figure 1)

Figure 1 shows diagrammatically the various phases in which phosphorus is concentrated. The basic cycle consists of the formation and tectonic uplift of the land, its physical and chemical erosion, transportation of the erosion products to the sea, and sedimentation to become again a part of the landmass.

The first phase of the cycle at which phosphorus is concentrated is in the mineral apatite in certain types of igneous rocks, largely carbonatites, shown by the dike in the cross section. Phosphorus can be further concentrated when phosphate bearing rocks are weathered by the leaching away of some rock constituents, usually calcite, with the resultant formation of a phosphate-rich residuum. Much of the phosphorous, however, is dissolved and carried to the ocean by rivers. The major phase of concentration occurs on the shallow ocean floor by precipitation of apatite in areas of upwelling water. Minor phases of concentration occur by the formation of bat and bird guano and lagoonal sedimentation.

In mining terms, only three of these phases of concentration contribute economic deposits. The majority of phosphate, three quarters of the world's production, comes from ancient marine deposits; about one quarter is produced from igneous apatite or its leached residuum; and a small amount comes from phosphate derived from bird guano, the so-called insular phosphate.

## Marine Phosphate Deposits

Phosphate of the marine deposits accumulates on the floor of the sea in areas of upwelling ocean currents. These currents occur where the wind blows the surface water offshore and deeper phosphorous rich water wells up along the coast to take its place. The water is warmed and depressurized causing the formation of apatite. The upwelling current sets off one of nature's most interesting chain reactions. Some of the phosphate is deposited on the sea floor, but much of it, along with other nutrients, reaches the surface of the ocean in the zone of sunlight and literally fertilizes it. Minute floating plants or phytoplankton grow prolifically. Minute animals, or the zooplankton, feed off the plants, little fish off the zooplankton, bigger fish off the little fish, and so forth, resulting in one of the most productive areas of life on earth. The sediments deposited in this environment, both on and offshore are quite distinctive.

These areas of upwelling occur mainly but not entirely on the west coasts of continents in the trade wind belt (figure 2). Phosphate-rich sediments deposited in these areas are part of tomorrow's supply of phosphate, but the state of present technology does not favor their exploitation now, both because they are low grade and because they occur in water too deep for economic mining. Although they are not very well explored, there is no question that they constitute a very large resource.

Onshore from these areas of upwelling and contrasting with their incredible fertility, are desert areas with almost equally incredible sterility. These include deserts such as the southwest African Namib desert and the Chilean Atacama desert, famous for its nitrate deposits.

Brongersma-Sanders in 1948, pointed out the relationship between upwelling ocean water and coastal deserts and Kazakov in 1937, pointed out the relation between upwelling ocean water and phosphate deposition. Even though both hypotheses have

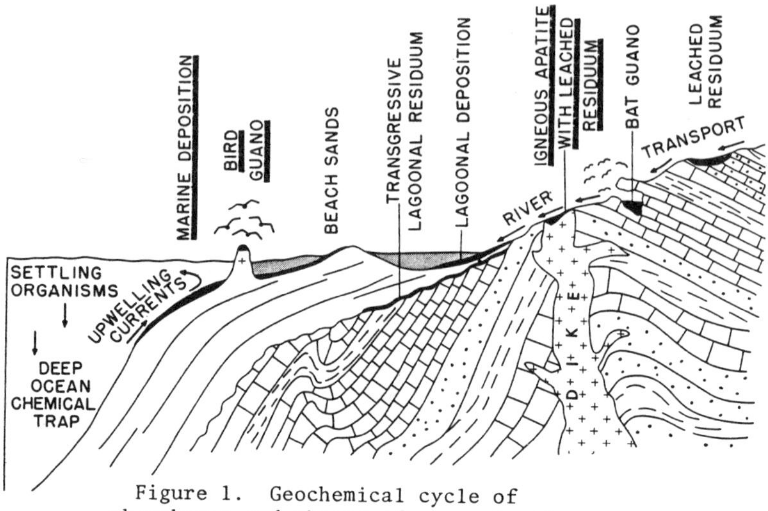


Figure 1. Geochemical cycle of phosphorus and phases of concentration

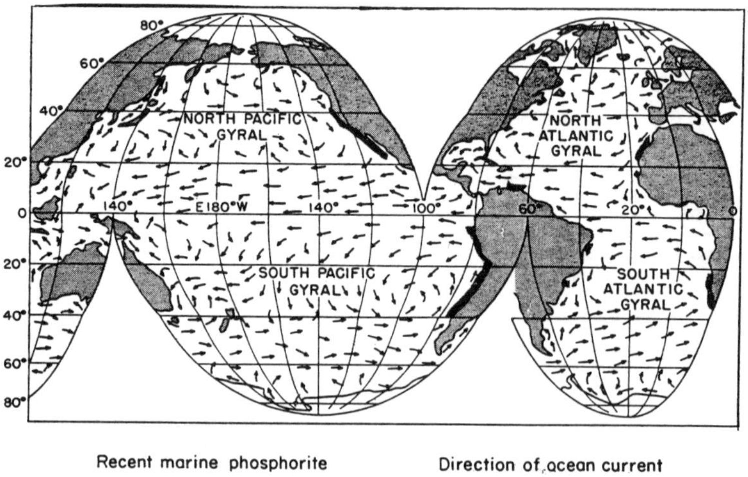


Figure 2. Distribution of recent marine phosphorite showing major ocean currents.

been improved upon since and even though much is still to be learned, Kazakov's and Brongersma-Sanders' basic concepts have withstood the rigors of scientific testing. Their hypotheses clearly lead to the conclusion that there is a genetic relationship between phosphate deposits and deserts at the present time.

How about in the geologic past? In areas of modern upwelling, deposits of geologically young marine phosphate, lagoonal salt, and desert sand occur, indicating that in the relatively recent geologic past similar upwelling ocean currents and coastal deserts occurred. The latitudinal distribution of these recent and young phosphate deposits are shown in figure 3. Have all such geologically young deposits been found in modern coastal deserts? Probably not, if the normal exploration history of other minerals is any guide. Exploration for such phosphate deposits has only been going on for about 15 years, so that the obvious deposits have probably been found, but obscure deposits probably still remain to be discovered.

When the older phosphate deposits are analyzed, no such simple latitudinal distribution is shown (figure 4A). However when plate tectonic movement after the formation of each phosphate deposit is taken into account (Sheldon, 1964) and the paleolatitudes of the deposits plotted (figure 4B), it can be seen that they had at the time of their formation about the same latitudinal distribution as the young phosphate deposits. In addition, when the paleogeography of these deposits is studied, it is found that they occur on the edges of ancient oceans and are commonly associated with salt deposits indicating arid conditions and with widespread Eolian sand bodies indicating desert sand dunes. Thus many ancient deposits were formed in trade wind belts where deep ocean currents welled up along coastal deserts.

Some of these ancient deposits occur now in modern deserts. In part, this is geologic chance as in the case of the Karatau phosphate deposits of the Turkestan Desert or the Queensland deposits of the Australian Desert. In the latter case, the phosphate deposits were formed near the equator about 500 million years ago. Subsequently Australia, locked in its plate, moved southward to the South Pole and then back north into its present position. Thus the occurrence of Australia's ancient deposits in the modern desert in the tradewind belt is only geologic chance, because the climate existing at the time of the formation of the phosphate was drastically changed as the Australian continent "sailed" into polar waters and then "sailed" back again to the trade wind belt. In Peru, changes were much less dramatic. In the Jurassic period of geologic time about 160 million years ago, phosphate was formed on the sea floor in the trade wind belt. Again in the Tertiary period about 30 million years ago, another phosphate deposit was formed in the same belt. On the sea floor, phosphate is forming now. Even though the South America plate was moving westward during this time as it split off and moved apart from Africa, it did not move a great deal either north or south. Thus Peru remained in the trade wind belt, and the Atacama desert with its three related phosphate deposits had its beginnings at least 160 million years ago. I coined the term "phosphogenic province" to apply to areas where phosphate deposits of different ages were formed successively in these upwelling marine environments that were stable over long periods of time. Another example is the Baja California phosphogenic province in western U.S. and Mexico, where in addition to the modern offshore deposits, are found deposits of about 30, 160, 250, 330, and 470 million years of age.

It is seen that the North and South American coastal deserts are quite ancient and make up phosphogenic provinces that contain a number of deposits of different age. The Australian coastal desert is relatively young and contains no economic phosphate, although a phosphate bed a few million years old has been reported from a well drilled offshore of western Australia. The northern and southern African

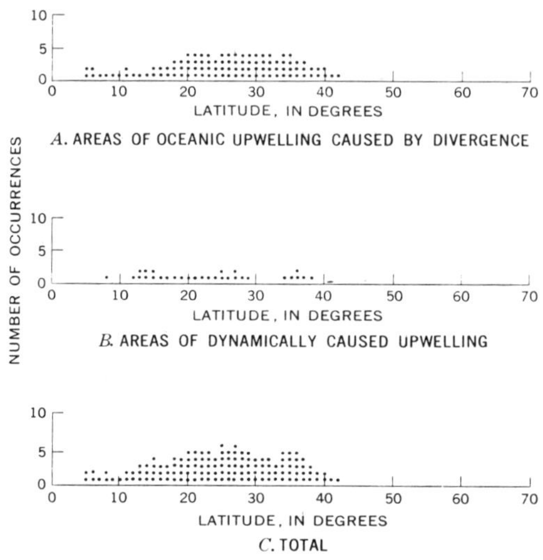


Figure 3. Distribution of young (Recent and late Tertiary) phosphorite.

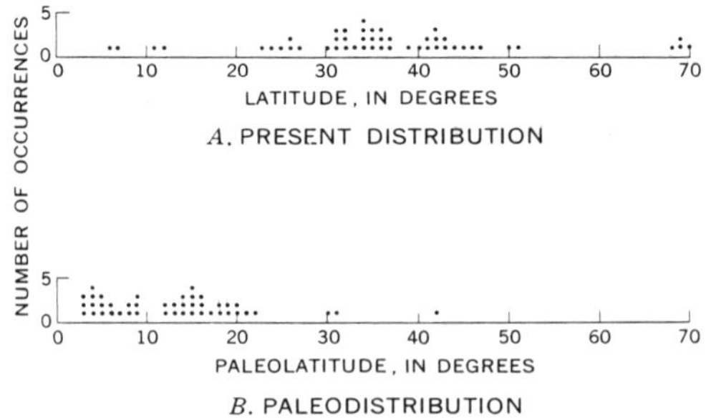


Figure 4. Present distribution and paleodistribution of ancient (early Tertiary and older) phosphorite.

Atlantic coastal deserts are young because the Atlantic Ocean is young, forming abut 100 million years ago when South America and Africa parted. The Atlantic Sahara and Namib phosphogenic provinces therefore are known to contain deposits no more than about 100 million years old.

Thus some coastal deserts are ancient and can be expected to contain multiple phosphate deposits and others are young and can be expected to contain few, if any, all depending on the geologic history of the area.

Interior deserts, which also occur mostly in the trade wind belts of the world may contain phosphate deposits which are within the phosphogenic province of the modern coastal desert. Such is the case, for example, with the 250, 330, and 470 million year old deposits of Idaho and Utah which occur in the Baja California phosphogenic province. On the other hand, some interior deserts may contain phosphate deposits that genetically are completely unrelated.

## CONCLUSIONS

This analysis reveals that there is a real correlation between phosphate deposits and deserts. It is not geologic chance that the world's desert areas are better endowed with phosphate than other areas. This relationship calls for recognition of several issues by national planners for desert regions. First, they should recognize that they can expect common problems attached to the development of phosphate mining and fertilizer production systems. These include the environmental impact of phosphate mining, land restoration problems of mined over sedimentary terrain, trace element pollution areas surrounding phosphate fertilizer plants and elemental phosphate electric furnace plants, mineral supply problems of sulfur or sulfuric acid and possible silica-rich rock and water supply for fertilizer and chemical plants. In addition, national planners should recognize that phosphate exploration opportunities in many desert regions have not been exhausted, so that geologic and geophysical studies and geodynamic analysis would be helpful in regional development.

## REFERENCES CITED

Brongersma-Sanders, M., 1948, The importance of upwelling water to vertebrate paleontology and oil geology: K. Nederlandsche Akad, van Wetensch, afd, Natuurk, Tweede Sect., v. 45, 112 p.

Kazakov, A. V., 1937, The phosphorite facies and the genesis of phosphorites, in Geological investigations of agricultural ores, USSR: Leningrad Sci. Inst. Fertilizers and Insecto-Fungicides Trans. (USSR), no. 142, p. 95-113.

McKelvey, V. E., 1967, Phosphate Deposits, Geological Survey Bulletin 1252-D.

Sheldon, R. P., 1964, Paleolatitudinal and Paleogeographic Distribution of Phosphorite, in USGS Prof. paper 501-C, p. C106-C113.

# PLANNING A MINERAL DEVELOPMENT PROJECT IN THE DESERT - SAINDAK PORPHYRY COPPER DEPOSIT, CHAGAI DISTRICT, BALUCHISTAN, PAKISTAN.

Mazhar Aslam[1] and S.A. Bilgrami[1]

## INTRODUCTION

The programmes and procedures followed in the evaluation of Saindak porphyry copper deposit are here described in detail mainly for the benefit of those in the developing countries who might undertake similar ventures. In many respects Saindak, with its location in an isolated desert, far removed from any supply centre, is typical of porphyry copper deposits. There is a paucity of documentation on how to go about proving a porphyry copper deposit and this paper might fill in this gap.

### Location and Accessibility

The Saindak porphyry copper deposit in the Chagai District is located on the extreme western border of Pakistan with Iran. The main deposit area called the 'Sulphide Valley' is bound by Latitudes 29° 14' - 29° 18' 30" N and Longitudes 61° 35' 40" and 61° 37' E. (Fig-I). The deposits can be reached from Quetta by a 670 km. partly paved road or by a standard guage railway which runs a weekly train service. There is a landing strip at Juzzak about 27 Km south of Saindak on which small twin engined aircraft can land. Saindak is connected with Quetta, Nokkundi and Karachi by RDC's private wireless.

### Climate

The area is extremely dry with an average annual rainfall of below 10 cm. The average summer and winter temperatures range around 35 - 50°C and 12 - 16°C respectively.(1-3)

---

[1]Resource & Development Corporation, Karachi

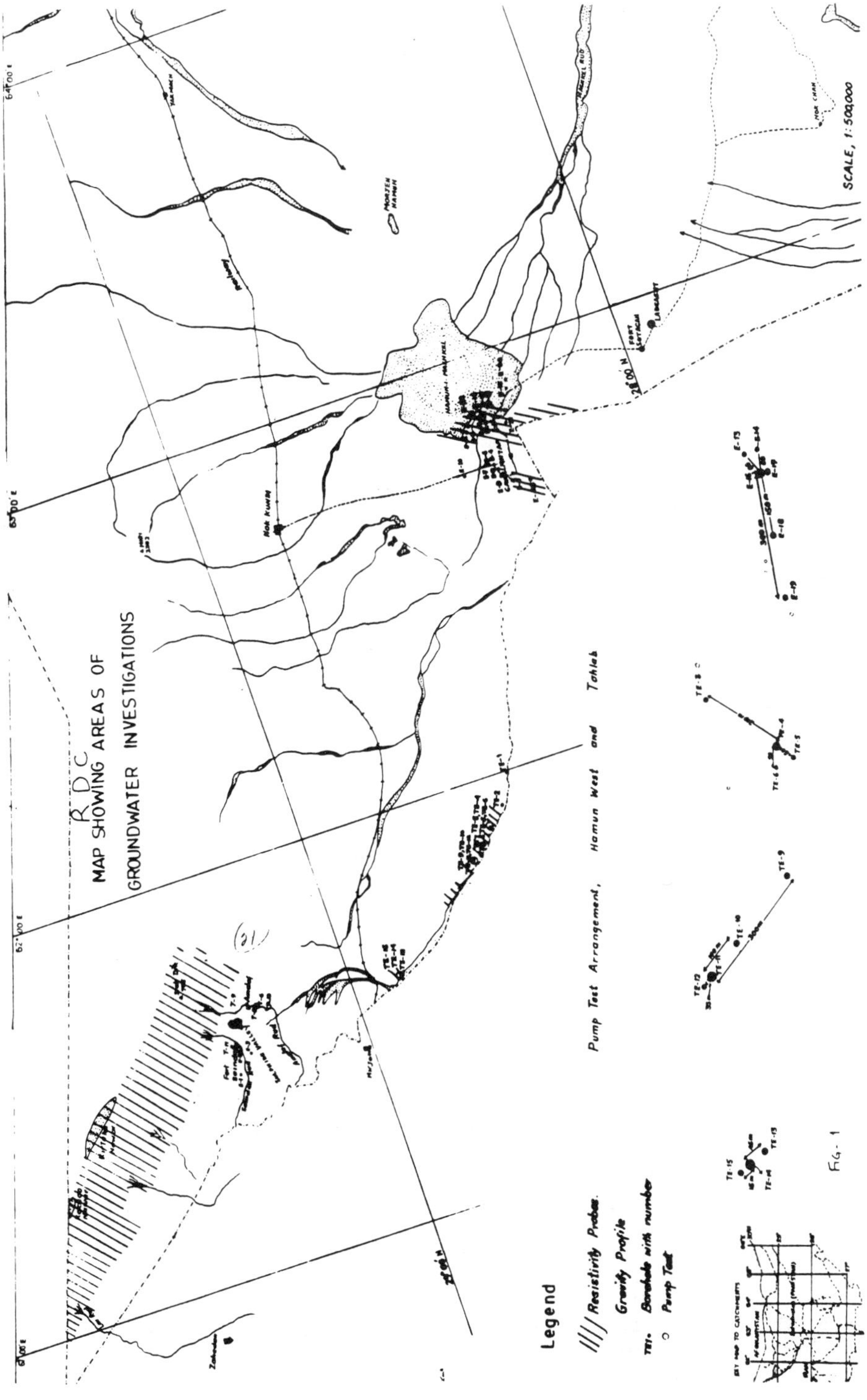
R D C
MAP SHOWING AREAS OF
GROUNDWATER INVESTIGATIONS
Pump Test Arrangement, Hamun West and Tahlab
Legend
Resistivity Probes
Gravity Profile
Borehole with number
Pump Test
SCALE, 1:500,000
FIG. 1

### Topography

The Sulphide Valley, (elevation about 1000 M above mean sea level) named because of the extensive sulphide mineralization, extends north-south and is surrounded by low rounded hills, up to 1300 m high. The Valley is about 2 Km wide in the south and gradually tapers to about ½ Km in the north. The total length of the Valley is about 7 Km and it lies about 45° through Saindak Syncline which strikes NW-SE.

## GEOLOGY

Copper in Saindak Sulphide Valley is associated with porphyry stocks, which have intruded volcano-sedimentary rocks of Oligocene age and are exposed in the main Amalaf Syncline running NW-SE. A total of five porphyry stocks of quartz, diorite composition have been explored in the Sulphide Valley in a zone running north-south. The Oligocene Volcano sedimentary rocks comprise mainly silstones, tuffs, and dacite. Some agglomerate developed on the outer side of the Amalaf Syncline also falls in the copper zone.

Three main copper mineralized porphyry stocks carry mineralization of economic potential. These have rings of alteration zones developed around the centres of the porphyry stocks. The alteration zones varying from an intensely mineralized core of potassium silicate through moderately mineralized quartz sericite, to feebly mineralized propylite zone are well developed over the east and south ore bodies. A well marked pyrite front defining the limits of copper mineralization encircles the deposits and is seen at the base of the hills forming a rim around the deposits. A zone of retrograde sericite alteration has been superimposed covering most of the east ore body and part of the south and north ore bodies.

Copper mineralization is mainly in the form of disseminations and along fracture planes. A good stockwork mineralization is developed in the south ore body. Mineralization in the north ore body is confined to replacement type bodies developed within the Skarn zones in contact with porphyries.

The porphyry stocks in Saindak have been affected in a series of late sulphide carrying tourmaline veins which at places have taken the shape of pipes. However, no significant copper mineralization has been found within the pipes.

## EVALUATION PROGRAMME

Not having had any experience of proving a porphyry copper deposit, we requested UNDP to make available to us on short term assignments, the services of geologists with porphyry copper exploration experience. Services of two geologists, Dr. John A. Wolfe and Dr. R.H. Sillitoe and one chemist Dr. F. Munizaga were made available to RDC by UNDP. Dr. Wolf visited Saindak first, confirmed that it was a porphyry deposit and for evaluating the deposits, specified a grid of 200 m x 120 m over the whole of the Sulphide Valley, a total drilling of about 50,000 m. Later on, Dr. Sillitoe visited the area and with RDC geologists prepared an alteration zone map of the Sulphide Valley. As a result, the total drilling required to prove the deposits was reduced to 30,000 meters. This illustrates the direct savings that can be achieved by alteration zone mapping of a porphyry copper prospect.

Both Dr. Wolfe and Dr. Sillitoe paid second visits to the project and carried out logging of the core and their own evaluation of the grade and tonnage of ore. Their evaluation agreed well with that done by RDC geoogists. The visits by UNDP experts provided extremely valuable training to RDC geologists who can now undertake the whole porphyry copper exploration and evaluation work independently.

### Reserves

As a result of more than 30,000 meters of core drilling and over 100,000 chemical determinations, the three ore bodies (north, south and east) were proved with the grades and tonnages shown in Table I.

**Summary of Ore Reserves**
**(Sulphide Valley elevation - 946 meters A.M.S.L.)**
**Reserves at various cut-off grades**

| Cut-off grades | Proved reserves (upto 476 meters AMSL) (Million tons) | Probable reserves (between 608 - 476 meters AMSL) (Million tons) | Total reserves (upto 476 meters AMSL) (Million tons) | Average expected grades |
|---|---|---|---|---|
| **North Ore Body** | | | | |
| 0.30 percent | 15 | 4 | 19 | 0.498 percent copper |
| 0.40 percent | 10 | 2 | 12 | 0.587 percent copper |
| 0.2 grams gold/ton | 9 | ... | 9 | 0.514 grams gold per ton ore |
| **South Ore Body** | | | | |
| 0.30 percent | 54 | — | 54 | 0.488 percent copper |
| 0.35 percent | 41 | — | 41 | 0.538 percent copper |
| 0.40 percent | 27 | — | 27 | 0.639 percent copper |
| 0.50 percent | 18 | — | 18 | 0.725 percent copper |
| 0.40 grams gold/ton | 28 | 11 | 39 | 0.70 grams gold per ton ore |
| 0.30 grams gold/ton | 54 | — | 54 | 0.49 grams gold per ton ore |
| 0.009 (Moly.) | 29 | — | 29 | 0.019 percent |
| **East Ore Body** | | | | |
| 0.30 percent | 207 | 57 | 264 | 0.388 percent copper |
| 0.35 percent | 118 | 34 | 152 | 0.430 percent copper |
| 0.40 percent | 74 | 15 | 89 | 0.50 percent copper |
| 0.45 percent | 32 | 1.0 | 34 | 0.572 percent copper |
| 0.50 percent | 20 | 0.59 | 21 | 0.621 percent copper |
| 0.009 (Moly.) | 55 | — | 55 | 0.017 percent molybdenum |

## PROBLEMS IN PROVING SAINDAK DEPOSIT

### Site Infrastructure

In the early stages of planning, it was realized that the isolated location of the area, the paucity of water and habitation and extreme climate were the main physical handicaps that had to be overcome. Since it was proposed to drill about 30,000 m in 18 months' time, the project had to be carefully planned.

It was estimated that there would be about 300 persons in the field consisting of geologists, mining engineers, drilling engineers and supporting crew, workshop, canteen, and office staff. It was realized that unless comfortable living conditions were provided, round the clock drilling all through the year could not be undertaken.

In view of the remoteness of the area, brick construction for the camp could not be undertaken. Since clean washed sand is plentiful in the dry stream beds, a cement concrete camp with sharing rooms for 2, 4, 6 and 8 persons was constructed with a large canteen which could also be used as a club and movie house. Cold storage facilities for perishable food items were provided and the camp was linked with Head Office at Karachi and the supply base at Quetta with a wireless.

Special cooking facilities had to designed by building a pressurized fuel tank (parafin) and connecting it to 8 burners for cooking. Large hot plates and special thick bottom cooking utensils were designed for preparing the meals.

A workshop for the maintenance of machinery, a power house to supply power 24 hours a day, a sample preparation laboratory and a core library were also constructed. Because of the remote location of the area, a well stocked store of spare parts was also established. A minimum of 3 months stocks of supplies for locally available spares and of 6 months for imported spares was developed.

In designing the camp accommodations, due consideration was given to the fact that the investigations may indicate a sub-marginal ore body and the camp may have to be abandoned. Since Saindak is located close to the Pakistan-Iran and Afghanistan borders the camp was so designed as to be of use to the various government agencies connected with manning of the border posts like the Customs, Immigration, Militia etc. As it turned out, the deposits have proved to be of economic interest and a full feasibility study is now underway.

## DRILLING

The greatest amount of physical effort was involved in the drilling operations. After careful scrutiny, it was decided to import five Longyear-38 wireline core drilling machines and one Ingersol Rand T-4 drill master. The main drilling work was done by Longyear machines and the T-4 was used for water wells as well as some test holes. Each Longyear machine was manned by one man incharge and three drilling assistants for each shift. Since three shifts were run all through, staff for four shifts was provided for each drill machine to allow for leave. To supervise the work of the drillers, each machine was under the overall charge of a drilling engineer. Three mechanics were posted for maintenance of the machines on a round the clock basis.

Since RDC did not have any staff of its own, trained drillers were taken on deputation from Geological Survey of Pakistan. We recruited diploma holders in mechanical engineering to be trained as shift foremen and graduates in science as drilling assistants. On completion of one year's training, written, oral and practical examinations were conducted for each category before confirming the recruits in their posts. It was made possible for lower staff to be promoted to higher grades by passing the examinations, conducted by our own engineers. In addition to permanent jobs, substantial monetary benefits were given to those qualifying in the examinations.

As a whole it was found that one year's training was enough to make a good driller, though performance varied widely from person to person.

With this staff we averaged a drilling speed of 11 meters per machine shift. Since the work was done continuously, average daily progress was 11 meters per shift for the actual drilling time. However, average for the whole time spent in the operation also counting the time spent in shifting rigs from hole to hole, repairs, and maintenance was 5 m. per shift per rig.

### Core Recovery and Storage

Core recovery on an average was better than 95%. The core was arranged in wooden boxes with a capacity of 6 meters core in each box. Three inch wide lengthwise partitions were provided in the boxes so that each length of core could be kept separately, duly marked on the side, as well as wooden blocks kept in the box. All the core boxes were clearly marked and the core of each hole was separately stocked on steel racks specially made for the purpose. This method of core storage makes it possible to examine any core length of any hole whenever required.

## SAMPLE PREPARATION LABORATORY

It was estimated that with five wireline drill machines working round the clock at least 70 meters of core would be available each day. Since our chemical laboratory was located at Karachi - a distance of over 1500 Km, it was necessary to set up a sample preparation laboratory on site.

### Sample Preparation Procedure (Drill Site)

Cores from bore holes are brought to the laboratory in wooden core boxes which accomodate 6 meters of core. Samples are taken of 3 meter core lengths which are marked on the core boxes. All the core is logged by a qualified geologist. After the core is logged, it is split into two halves and relogged after splitting. One half of the core is preserved in the core library; the other half is crushed to 1 cm in a jaw crusher and the crushed material is riffled in a Jone's splitter. 500 gms of the crushed core is dried in a oven at 50°C for over 3 hours. It is then passed through a grinding mill which grinds it to 8 mesh.

This is further split through a Jone's Separator and about 125 gms of this material is pulverised to 80 mesh in a Tema disc mill. This pulverised material is again split and about 35 gms is sent to our chemical laboratory for analysis and the balance of about 90 gms is retained at Saindak for reference. On receiving analyses of samples, the core is once again examined to see if there are any major differences between the estimates made at the time of logging and the assay results.

In the beginning core splitting, crushing and sampling was done by graduate mining engineers and geologists. It was noticed that the routine nature of the operation was very boring to them and lapses might occur. As an experiment, three local boys with secondary level education were taken as trainees and taught core splitting and sample preparation. They picked up the work very quickly and showed remarkable devotion. Also, they meticulously followed the procedures. They were at all times under the supervision of a geologist or a mining engineer and it was like achieving a sort of an automation in sample preparation work.

## CHEMICAL LABORATORY

Since no chemical laboratory capable of handing such a large number of samples as would be generated by our core drilling programme existed in the country, a modern chemical laboratory with 2 atomic absorption spectrophotometer units, one emission spectograph spectrophotometer, Leco furnace for sulphur, ancillaries and chemicals, 10 chemists and 5 supporting staff to handle 50 samples with 5 determinations of each (Cu, CuO, Mo, Zn, Ag and Au) per day was set up in Karachi. Analytical procedures generally used in mining industry were followed. In all, over 100,000 determinations were made in 15 months and a check of over 370 analyses by the Institute of Geological Sciences, London showed excellent agreement.

The quality of analyses was maintained through constant precision checks by analysing 8% duplicate samples; 5% standard reference samples with each batch and by analysing blind samples. In addition, about 5% of the analyses were checked by laboratories of international repute. A statistical study of duplicate analyses obtained during the period of one year showed a precision of ± 5% at the 95% confidence level.

## GROUNDWATER INVESTIGATION

As soon as positive indications for the existence of a sizable copper ore body were found the search for an adequate source of water supply assumed new urgency. Accordingly a groundwater investigation cell was commissioned and charged with the responsibility to:

a) Find a source of sustained water supply of the order of 1 cu m./sec without regard to quality;
b) Establish expeditiously the source and to keep cost of water delivered at a minimal level.

These requirements add up to a tall order when it is realised that the copper deposit lies in hostile Lut Desert environment characterised by erratic and scanty precipitation (40 mm - 120 mm), exceptionally high rates of evaporation (exceeding 4000 mm a year), extremes of temperature (-5°C to + 50°C), and persistent dust storms of considerable magnitude. Despite these discouraging realities, the groundwater investigations were launched immediately and on the basis of available information and reconnaissance surveys of prospective regions were identified.

### Problems Faced by Groundwater Investigation

(a) In all 44 exploratory holes were drilled totalling 6,000 m. Ingersoll Rand, T-4 Drillmaster was used throughout these operations. This rig has good penetration in hard rocks as well as drift. However, since unconsolidated sediments tend to cave in, mud has to be used in various consistencies to hold them. This necessarily interferes with sampling and distorts the textural complexion of the cuttings reaching the surface. Although uncertainties were reduced, they could by no means be eliminated.

(b) Down-the-hole hammer drill cuttings are brought up by air pressure with or without mud. This leads to a time lag between the point of drilling and the surfacing of the cuttings. This lag increases with the depth of drilling. It is often impractical to correct it.

(c) Loose sediments cave in whenever rods are pulled out and even while drilling is in progress. This is time consuming and retards the progress of work. When finer fractions, particularly silty clays are encountered they choke the inlet valve requiring time consuming cleaning operations.

(d) To procure sufficient water for mud preparation and drilling operations in an arid zone presents serious difficulties and there are frequent holdups due to lack of timely water supplies.

(e) The standard T-4 equipment was capable of drilling to a depth of 200 meters only and often much less. This did not permit deep enough drilling with the result that all our test holes have only partially penetrated the aquifer.

(f) Non-availability of turbine or submersible pumps has forced us to employ compressed air pumping. This has limited discharge to under 600 gpm even under the most favourable conditions. Relatively low discharge, in turn, has led to poor drawdown in the pumped wells and insignificant drawndown in observation wells.

(g) Because of dry conditions at the surface and of very low resistivity of the subsurface formations, the results of resistivity probes are unreliable. Comparison of interpreted geoelectric layering from depth probes carried out adjacent to boreholes yield no clear relationships between resistivity and soil types. Down hole resistivity logging has indicated that this is almost certainly due to variation in salinity of the groundwater in otherwise comparable horizons. Thus probes have not proved helpful in establishing a clear relationship between soil types and resistivity.

(h) Gravity profiles running north from Hamum to Nokkundi and beyond have suggested the existence of a graben-like structure but the picture is by no means clear. Certain points raised by the study could only be confirmed or discarded by deep holes (at least 300 m) for which existing facilities are inadequate. Efforts will be made to drill deeper holes as soon as facilities are available.

### Summarised Conclusions

(a) Adequate quantities of water are likely to be available from surface storage in Mashkel Rud Valley (over 250 km); from Mashkel Delta (over 200 Km); from Hamun West (over 150 km); from central Tahlab (75 km) and possibly western Tahlab (35 km). Thus availability of water is not in doubt though more drilling is indicated. At present Tahlab appears to hold out the most promise with a lead of 35 to 75 kms and with a head of 350 - 375 m.

(b) RDC intends to explore an alternate source of possibly adequate supply in Kirtaka Nawar area 40 - 42 kms from Sulphide Valley.

(c) While these figures may be used to make tentative estimates of costs involved in delivering Tahlab water to Sulphide Valley the final selection of site for well-field development may be deferred till results of Kirtaka investigations is known.

## SPECIAL INCENTIVES

In view of the isolated location of the Project area certain special incentives were introduced for the staff. These were:

### Rota Leave

It was felt that normal 33 days earned leave per year allowed to Public Sector Corporation employees was not encugh because a lot of time would be spent in travelling to and from the place of duty. Also, since the work on the Project was carried out around the clock all through the year, no Sundays or other holidays were allowed to the staff. A system of compensatory leave called Rota Leave was therefore introduced. Under the system, for every three months continuous duty at site 25 days leave (21 + 4) was allowed. The actual leave was 21 days and 4 days were allowed for travelling to and from the site.

This system, when introduced, was found to have a drawback in as much as staff members proceeding on Rota leave usually over-stayed and applied for earned or medical leave. This resulted in upsetting of the Rota schedule made out in advance and some of the staff members could not utilise Rota on a regular basis. It was therefore necessary to lay down that no other leave could be taken with Rota leave unless prior sanction of the Project Manager was obtained before proceeding on leave. Also, to provide additional incentives to the staff members monetary compensation could be obtained with the consent of the Project Manager. This enabled the work to proceed on a regular basis despite absenteeism because some of the staff members preferred to receive compensation for their Rota leave rather than use it. However, earned leave was not compensated for and staff was forced to use this leave so as to provide them with much needed rest and recreation.

## Transport Facilities

The Corporation also introduced a weekly bus service of its own to Quetta and back because the normal buses running on the route take almost two days for the one way journey. The Corporation assumed the responsibility of transporting, free of cost, all staff from the site to Quetta and back.

## Provisions

Special arrangements had to be made for the supply of provisions and for this purpose a special 5-ton truck was purchased and equipped to transport rations and perishable food items.

## Messing

Subsidised messing to all the members of the staff was provided. A central canteen was constructed and the same food was served to all categories of staff but on differential rate of payment. For staff getting a basic salary of Rs.500/- or less the monthly messing rate was Rs.100/- per month; for those with basic salary between Rs.501/- to Rs.999/- a messing rate of Rs.130/- per month was fixed and for those above the basic salary of Rs.1000/- or above per month the rate was Rs.200/- per month. The Mess was managed by an elected committee and all canteen staff was paid for by the Corporation. Utensils, crockery, cutlery, furniture, fuel and transport expenses of ration was on Corporation's account. A ice cream cone machine was also installed in the mess.

## Recreation Facilities

Recreation facilities for out-door games like football, volley ball etc. were provided. In addition, 35 mm sound projectors were provided so that two movie shows a week could be arranged. Films from distributors were hired in Karachi and flown to Quetta for despatch to the site. Normally titles of the available films were sent to the site beforehand and films selected by them were supplied.

### Medical Facilities

A qualified doctor was appointed and a two bed hospital cum dispensary was also set up.

### Emergency Measures

An old air strip about 25 km from the site was renovated for emergencies so that in case of accidents staff could be evauated to the hospital. As it turned out, the air strip was not used for emergencies but visiting geologists did use it for reconnaissance surveys.

As a whole, it is a matter of satisfaction that we have not only proved a porphyry copper deposit but have, in the process, developed expertise to undertake similar assignments without any external assistance.

# THE CREATION OF GHOST TOWNS IN ARID OR DESERT ZONES

By John Carman

All of us have a clear idea of what a ghost town looks like, chiefly because they have been prominently featured in Western movies. Gaunt, weatherbeaten, crumbling buildings, tumbleweed rolling down a dirt road, the spirits of people long departed still faintly haunting the place. There is a sadness to the scene stemming from realization that it represents loss, waste, disruption of lives and destruction of a community. Once thriving businesses almost overnight become valueless -- the swinging doors of Black Mac's Drinking and Dancing Emporium swing with the wind. The dentist, the doctor, the lawyer -- all have to relocate and start once again the arduous struggle to build up a practice. As business closes down and services are cut, those who might care to stay on have no choice but to join the trek. All lose to a certain degree, some lose almost totally.

Usually, the ghost town has been a by-product of the exhaustion of mines. In deserts normally these will be found in remote areas which can support no other type of economic activity. In any case for the purpose of this paper, the ghost town will be defined as one which was originally established for the servicing of a mining enterprise.

Now let us consider the pro's and con's of ameliorating or eliminating the wastage and sociological stresses involved in the death of towns.

## TRAILER CAMPS

At first glance, establishment of trailer parks appears to offer interesting possibilities. Certainly it would save money and there would be salvage value when the mine closed down. However, for a number of reasons it is proposed that meeting housing needs with trailers would be the worst possible way to try to solve the problem.

1. Savings would not be as substantial as one might think. Even a trailer park has to be supplied with water, electricity, sewage disposal roads. Additionally, in a day when in many places it is difficult to attract and hold good miners, a large part of the cost of establishing a new community is in the provision of common services and facilities. Shopping mart, bank, specialty stores, hospital, schools, playing fields, parks, arenas and community centre -- these are costly particularly as they have to be of high quality.

2. Far too many people who would arrive in a mobile home would be of the "drifter" type; a few months her to get a stake and then on again.

3. Trailer living by-and-large must involve more than a little psychological stress not to mention plain discomfort from over-crowding. In general, an unhappy family life does not produce a good worker.

4. In many instances limitations of space would lead to the decision to not have children. Thus the powerful anchor of children in school, playing Little League baseball or in other parts of the world soccer, would be lost.

## PLANNED DEVELOPMENT OF A MINING AREA

Where a number of orebodies are located within a relatively small area, in theory a centralized community could be created to service each in turn. As Mine A becomes exhausted Mine B swings into action. Thus the life of the area would be greatly prolonged. It could be done and probably is being done in USSR among other places under the control exerted in the Centrally Planned Economies. But where there are still vestiges of laissez faire, one questions that the owners of Mines B and C will defer commencement of return of investment by twenty or forty years as the case may be. Also, in the developing world even with mines which are wholly owned by governments, all the pressures will be to maximize near future gains. The problems confronting most of such government are so immense and complex, tomorrow will almost always be sacrificed before the stark exigencies of today.

## REDUCING THE SCALE OF OPERATIONS

The question may be posed whether it would be advantageous from a national as well as from a company point of view to restrict rate of mining to a moderate scale so as to extend the life of the mine to, say, 60 to 100 years instead of perhaps, 15 to 25.

With most mining companies, policy is to maximize production if only because of uncertainty as to what the future holds in a very uncertain world. In other words, get it while the getting is good. Increasingly often the company has no choice.

1. In a growing number of places in the world governments are now offering exploitation concessions of as little as 15 years and rarely more than 25. That the agreement offers possibilities of extensions has to be assigned a book value of about $1.00 by prudent management.

2. Also, governments have numerous other ways of applying pressure for speed-ups rather than stretch-outs.

3. More and more the characteristics of the orebody, particularly grade, will be the major determinant of scale of operations.

4. Financing houses will usually view with a decidedly jaundiced eye any scheme which increases costs per unit of output.

5. In the U.S.A. a number of states tax reserves in the ground, in part to obtain revenue but also with the intent of preventing the mining companies sitting on such.

From the viewpoint of governments extending the life of mines will usually be attractive only when the social costs involved in closure will outweigh the loss of revenues. Extreme examples of such instances are:

1. Continuing operations of the Portobello gold mine in Ecuador about twenty years after the foreign owners walked away because reserves were almost exhausted and it simply was not an economic operation anymore. Nevertheless, today work still goes on, heavily subsidized by government, of course, because a community numbering several thousands is involved. Losses in recent years have amounted to a million dollars or more which is no mean price to pay for side-stepping creation of a ghost town in a poor country.

2. Howard-Goldsmith[1] described the ultimate in closed circuits which is probably still operating at Colomb-Beshar which lies deep in the Algerian Sahara. There in the late 1950's the French developed a coal mine to fuel an on-site generating station which was important to a nation in desperate need of energy. Before the transmission line was built, great reserves of oil and gas were discovered. Colomb-Beshar no longer had any economic justification but as late as 1970 the mine was being worked simply to supply the town with power and provide employment.

As for the mining companies the main advantages would be:

1. Reduced scale would often cut financial requirements to the point where funding could be done from internal sources. This would be a favorable point solely in times of tight money when financing can be obtained only at exorbitant rates. Normally, the mining concerns seek the maximum percentage of loan capital so as to increase the leverage of the equity holders.

2. Mining personnel would be offered the opportunity of employment throughout their working life and thus labor turnover and training of new personnel would be considerably reduced. However, as will be described there are other means of ensuring these desirable results.

## USE OF ALREADY EXISTING TOWNS

Where the mine is located in a physically hostile setting and yet lies within reasonable distance, let us say, 50 miles or less of an established town serious consideration should be given to constructing only those facilities required for mining operations on the site. The work force would be domiciled and serviced in the aforementioned community. Generally, people would arrive at their place of work by vehicular transportation. In other instances, when distances are well in excess of 50 miles, recourse can be had to practice which is becoming common in Northern Canada, namely, airlifting. Under this system personnel work like dogs, seven-day week and twelve hours per day with a commensurate rest period when returned to their permanent domicile.

---

1. H.G. Howard-Goldsmith *"The Mineral Picture of Africa"* paper given at the U.N. Seminar on Mineral Economics, Ankara 1971.

## DOING IT ON THE CHEAP

"Historically, the company town has exerted maximum political, economic and social control over its inhabitants, while commonly offering minimal housing, facilities and amenities".[2] In an earlier and usually brutally exploitive day such towns were almost always an abomination in the industrial world. Mean shacks, outdoor plumbing, sordid bunkhouses referred to generally as "ram pastures", the company store, frequently employed as an effective and profitable means to impose de facto enslavement. Where operations were so isolated that management had to live in proximity to the serfs, contrasts were appalling. The psychological wounds visited upon both parties to such an unequal contract had to be extremely damaging. Nevertheless, these were the norms in many mining camps in more than a few parts of the industrial world up to World War II.

Conditions far worse prevail in the Third World, particularly with mines owned and operated by nationals. In places where families actually live in the mouths of abandoned adits or in adobe shacks constructed by the bare hands of the inhabitants, ghost towns have to represent the most minimal of financial losses.

Today, in those important mining countries where unions are strong and unemployment is low the converse holds. A high standard of living has to be guaranteed or continuously there will be a shortage of skilled labor, not to mention administrative and supervisory personnel. The old bunkhouse has been replaced by comfortable apartments and emphasis is placed upon well-planned and built separated houses, each on more than ample land. The company store, now a supermart, frequently is a loss leader on the company's books. A former manager of a fairly large mine in Northern Canada was queried as to his major problem. Mining operations, milling operations, labour relations? None of these but instead it was the need to explain and defend the price increase of two cents on a quart of milk or a loaf of bread. No, doing it on the cheap is not a workable solution in the industrial world. And one could hope with considerable feeling that with the new mines to come into production in the developing nations, there will be sufficient enlightenment on the part of governments and owners to ensure that the horrors of the past are not visited upon the future mining community. Whether the solution proposed later for eliminating ghost towns in the developed lands will work in the Third World is unfortunately questionable.

## SUGGESTED BLUE PRINT FOR THE FUTURE

A number of proposals of doubtful merit have been briefly considered for reducing the social and financial losses represented by ghost towns. Instead, with imagination, vision and early planning, it is strongly proposed that such towns can be relatively easily eliminated in almost all desertic environments of the industrialized countries. No new mine should be permitted to come into production without measures being taken to largely ensure a life for the community long beyond the day when the mine closes. After all, it is rather paradoxical to think of abandoned mining communities in the same deserts which are being inundated by immigrants seeking a place in the sun. As is evidenced by Arizona and other states with decidedly arid climates, the desert, excluding tundra, of course, is a very pleasant place to live. Sun-

2. *"Colstrif, Montana: The Modern Company Town"* Coal Age Vol. 80, No. 5 May 1975, New York, pp. 54-7

shine, cool nights, relatively clean air unless there is a smelter nearby, space and vistas; these things are golden to many who lived for long in "concrete jungles". Particularly to the old, a retirement home near Tucson is infinitely preferable to braving the streets in which furtive figures move as darkness descends, the grey dankness and cold of winter and the frequently stupifying heat of summer of a New York or Washington.

It is obvious that what is being proposed is that new mining towns should be set up with the idea that in the end they will become at least partially retirement centres. To do so will require in the first instance good land. Fertility will rarely be a problem because soils will not have been leached but space there will have to be. Space for the "good life" and for growth for reasons which will become clear later. Accordingly, in those instances where the mine is located in surroundings which are very rugged, consideration should be given to establishment of the town in the nearest available place with gentle topography and limiting of construction at the mine to those edifices directly involved in the extraction and treatment processes. Of course, there has to be a limit in regard to distance but it is proposed that there are few orebodies in the deserts of this world which lie more than 40 or 50 miles from tracts suitable for community development.

The communities envisioned can be depicted by two recent examples of mining town planning. Neither are located in deserts but as is the general rule with mines nor are they in the midst of Gardens of Eden.

At Colstrip, a major strip mining operation for coal has been set up in the range country of Southern Montana. The housing breakdown is 226 single family detached houses, 60 multi-family units and 180 spaces in a mobile home park. With regard to the latter for reasons expressed earlier the writer hopes that few people venture in. Accommodations are of a high standard and rates are relatively low. Rent of a $30,000 house amounts to $210 per month including maintenance, water and sewer, garbage collection, landscaping, range and refrigerator and insurance upon the structure.

The philosophy behind not sparing cost in setting up the company town has been that it is recognized that poor living conditions lead to labour unrest whereas good housing and other facilities reduces turnover and improves morale which reflects favorably on job performance.

A considerably more interesting and advanced approach to creating something approaching the ideal company town, the town of the future, the town which can easily convert to a retirement place in the desert has been described at length by Clarke and Grimes when treating with development of the Ruttan Mine in Northern Manitoba.[3]

The mine involved has an indicated life of twenty years. In that geological environment it is probable that it will survive in good health for considerably more time. The town site will be located 15 miles away. This is significant. Typically, in the past everything was clustered around the headframe. Men walked to work. In this instance, it is evident that aesthetics have pointed out that it is much more pleasant to live near the banks of a river and it is far easier to construct on bottom land than in the rocky bush.

---

3. W.F. Clarke and D.R. Grimes, *"Planning and Construction of Leaf Rapids - A New Town with a New Concept"* CIM Bulletin, Vol. 68, No. 756, April 1975, Montreal, pp. 84-93

From the beginning the company realized that it did not have "in house" capacity to plan and construct a modern community sufficiently attractive and workable to lure people into the northern wilderness and keep them. Accordingly the job was contracted to a major subdivision developer. Initially, it was planned to have 600 single family units on lots measuring at least 60 by 100 feet and multiple units to accommodate 300 families. All electrical wiring was buried.

In 1971 a startling event took place. The Manitoba Government decided to build the town itself using the town developer referred to above. This was not an indication that the Government eventually wanted to be the proud owner of a ghost town but instead recognition of a trend which is bound to grow stronger; namely, homes owned largely by employees and facilities by firms and individuals servicing the community. In other words, Leaf Rapids has no intention of dying even though located in an environment which is far from hospitable. Significantly, if one thinks of people carrying on after the mine closes down, basically as a retirement community living of savings, public and company pensions, most deserts have charms which are rarely if ever encountered in the rocky forests of Northern Canada.

The Government decided to make a major change by constructing 600 apartments and 300 houses instead of the reverse. There was bitter disagreement as the majority of employees wanted their own plot of land. Apparently, they prevailed because as of 1975 there were 119 townhouses and 74 apartments being rented. Also, there were 196 family units and 67 duplexes which were owned by the employees. Additionally, there were 35 mobile homes, a fact which will elicit no further comments.

Among the terms are a ceiling with municipal taxes of $450 for a single family unit costing $24,000 and a minimum lot size of 6000 feet. The cost of land cannot exceed $60 per frontage foot.

It is planned to have a work force which is 80-90 per cent married. In effect, a community which is stable and which will grow old in rhythm with the depletion of the mine. It is evident that with such policy and planning, turnover at the Ruttan Mine will be insignificant, a highly vital factor in this age of increasing technological complexity constantly demanding higher skills.

To complete the picture, the town boasts a school running to Grade 12 for 625 students and which almost certainly is very well staffed indeed. There is a 40-room hotel complete with all the amenities which go to make "a home away from home". With an arena, curling rink, gymnasium, sauna and a cultural area composed of theatre, library and exhibition centre, there is obviously no lack of recreational and intellectual opportunities. The commercial area contains the supermarket, department store, allied shops, coffee shop, bank and business offices. The hospital and municipal offices and council chambers complete the roster of facilities and amenities enjoyed by the community. All of the foregoing are contained within one building of 210,000 square feet which has been the product of a great deal of creative and imaginative architecture. Homes and the town centre are placed in settings which respect the need for greenbelts, playing fields and variety, anything to get away from the deadly checkerboard monotony and total lack of character of the company town of the past.

Let us now consider transferring this example to some spot in New Mexico or Arizona. In order to ensure a community which will be viable into the indefinite future, what steps will be required?

1. Mandatory home ownership by the 80-90 per cent of the work force which is married. Of course, the company would have to provide easy terms -- low down payments, modest interest rates and mortgages with a life approximating that envisioned for the mine but not to exceed thirty years.

2. Professional services such as those of the lawyer, doctor and dentist to be obtained by offering facilities for setting up private practices. As the number of such people would be small and collusion regarding price fixing not totally unknown in such professions, the isolation of most mining towns would dictate that both mine management and town authorities would have to exercise considerable vigilance to prevent rapid enrichment of the few at the expense of many.

3. All retail establishments should be operated and owned by concerns or people with an established track record. Needless to say, contractual arrangements would have to be made which would protect the community and ensure more than acceptable servicing.

4. As the end of mining operations looms, steps must be taken to incorporate smoothly as much as possible of its physical plant into privately owned businesses for supporting the community; i.e., those lathes, grinders and metal working tools which any town needs reserved with payment to be made when the last skip dumps its load at surface and the vehicle maintenance facilities which could be utilized by the local garages treated in the same manner. More important, very considerable extensions of land should be prepared for farming. The cessation of mining will release comparatively vast amounts of water unless the operation has been living at the expense of the water table, something which is happening at an alarming rate in too many places.

5. The community will grow older. Men will reach retirement age. They will own their homes free and clear, and have sufficient income to live out their years in comfort and decency. Most will elect to stay right where they are when where they are happens to be the "Promised Land" for urbanites in the crowded cities of the rest of the country. Accordingly, there must be establishment of sinking funds to constantly finance creation of new accommodations to house the newcomers who replace those who have drilled off their last round, walked their last beat and settled down to reducing their handicap on the golf course or producing a prize flower.

6. It is suggested that plans should be based on having the mining company construct everything but purely operational facilities on the basis that the end product will be a self-sufficient town into the foreseeable future and one that will be owned and governed at the earliest possible date by the employees and those who serve them.

7. Finally, when mining operations cease, outsiders should be allowed to purchase homes as they become empty because of death or departure.

# INDUSTRIALIZATION IN ARID AND SEMIARID LANDS

W. C. Ellet

## SUMMARY

For the better part of three decades, industrialization has been taking place in many of the arid and semiarid areas of the United States. While this can be an inspiration for many nations with similar regions, the United States' experience is by no means a model for the rest of the world in this regard.

The United States does share a common experience with many arid parts of the world, in that settlement in the arid and semiarid sections was due to the mineral and petroleum wealth. On the other hand, many of the arid settlements were originally nothing more than way stations servicing the infrastructure system.

Industrialization of the arid areas was facilitated by a well established infrastructure that was in place well before the industrialization started to develop.

After the revolution, the Continental Congress provided for public education in the territories, and as the pioneers moved west, the schools went with them. In 1862, the Congress established a system of public higher education for the entire nation. From these institutions came the research and trained manpower which would one day develop the industries of the arid and semiarid regions of this country.

The U.S. experience is not without relevance for the rest of the arid areas of the world. The Government has an historical statistical series on the use of water by industry type. It is to that particular matter that this paper is addressed.

Since the end of World War II, there has been a significant industrialization of many of the arid and semiarid areas of the United States. Much of this industrialization has come from industries not necessarily associated with the extraction and processing of minerals and petroleum.

In all probability, most of these industries might not have developed had it not been for the infrastructure put in place to tap the mineral and agricultural wealth of the west. To reach the west from the east, it was necessary to route much of it through the arid lands of the United States.

Many of the initial settlements in the harsher lands were the consequence of the construction of this infrastructure, and subsequently, its operation and maintenance. That some of these wayside places would become centers of industry in the twentieth century was the result of certain other forces, rather than any concerted planning.

That, over time, most of these areas would be populated with relatively literate people was due to the foresight and wisdom of our forefathers. The Continental Congress, the Congress of the Articles of Confederation, ordained the establishment and means for financing public schools in the territories. Thus, as the pioneers moved west to farm, construct the infrastructure, or seek minerals, the public schools followed them.

Seventy-seven years later, the Congress of the United States enacted the Morrill Act, which established the great public universities of this nation. Prior to this act, public universities in this country numbered less than half a dozen. By the turn of this century, with the exception of the old universities in the east, established well before the founding of the nation, the land grant universities were beginning to dominate higher education in America. Under the terms of the Morrill Act, thirteen million acres of the public domain were given to the states for the establishment of mechanical and agricultural colleges. Passed in 1862, in the midst of the U.S. Civil War, this act provided the South, the Middle West, and the Western states not only with institutions of higher education, but also major centers of research, from agriculture to higher energy physics, by the early twentieth century. Even though the United States has been blessed by nature with a tremendous abundance of natural resources, the nation would not be where it is without these two fundamental acts of legislature.

Some of the foremost universities of this country are located in the arid and semiarid parts of the United States. Without the Morrill Act of 1862, these educational institutions would, in all probability, not exist. To state it another way, the winning of the West was not due so much to the cowboy, or the federal troops and the Indian Wars, as to the land grant colleges. Today, several of these western universities are not only prestigious in the United States, but in the world, as well. The number of Nobel laureates in these institutions is proof, to some extent, in itself. It should be recognized that this development has ocurred, by and large, only in the past half century. Nevertheless, it would probably not have developed without the support of the water side of the United States. My fellow citizens of the arid and semiarid areas of the West may disagree, but I think history proves the point.

In the early part of the twentieth century, commencing with Theodore Roosevelt's administration, entire major programs were begun in the west with regards to water conservation. With the exception of a hiatus under the Harding and Coolidge administrations, the programs have been continued into the present day. These projects have had a multifold set of objectives - flood control, agricultural irrigation, industrial water requirements, and plain human need for water. It is too early to evaluate the present presidential polarization to water projects. To make any judgements on this program is wholly premature.

Simply stated, the industrial development of these areas of the United States owes much to the innovations of the industrial citizens, local industries, and indigenous political institutions. Nevertheless, without the policies and impetus of the more water-rich parts of the United States, this industrial development would not be where it is today. It would be less than candid not to admit that much of the initial incentive was the western wealth in minerals and petroleum that had a major impact on the industrial development of the arid and semiarid parts of the United States. The development of industries unassociated with mineral resources, including food industries, would not have evolved without the federal policies noted above.

Although we can reject the developments in the arid and semiarid parts of the United States as an analogy for other parts of the world with similar climactic problems, the empirical experience is very much relevant to these other areas.

Due to the fact that, over the years, the United States Government has collected statistics on water use by various types of industries, there is a data base available to other nations on the types of industries which can operate effectively in arid and semiarid regions. Nonetheless, it is probably impossible for other regions of the world to fully emulate the United States experience of industrialization in the Southwestern part of the nation. The lack of a continental support system is in itself the margin of difference. For obvious reasons, such industrialization must be taken within the context of resources available to the individual nation, both human and physical.

The requirements for both infrastructure and an educational system as prerequisites for industrialization are so obvious and well-known, the subject need not be treated here. There is, however, a fact of life which should be stated, and has been too often neglected in the past, particularly by the industrialized nations with regard to less industrialized countries.

Industrial development requires technology. In the case of many arid and semiarid lands, this means some kind of technology transfer from the industrialized countries. Unfortunately, technology transfer is not as simple as some would have it. Since technology is knowledge, it behooves a nation considering industrialization to recognize that it requires knowledge to buy knowledge. The asymmetry in knowledge about knowledge between the industrialized nations and the less industrialized ones can only be corrected by the latter building the national expertise and institutions which will permit the acquisition of appropriate technologies. Fortunately, today many of the arid and semiarid nations have citizens with excellent technical backgrounds. Unfortunately, many countries do not have the appropriate institutions to effectively utilize such specialists.

Be that as it may, the real key lies in the selection of industries which most efficiently utilize water. It is in this regard that the U.S. experience is of substantial benefit to countries who must carefully husband their water resources.

Thanks to the United States census of manufacturers, this nation does have extensive statistical data of water use by major industrial groups. Moreover, the statistics have been kept in terms of volume of water used as well as by the net water lost in the industrial process. Last, but not least, the statistics have an additional economic measure in that data has been maintained with regard to the value added.

The two industries which require the greatest volume of water, and have the greatest water loss, are the primary metals and chemical and allied products. With respect to the former, we are speaking of such elements as blast furnaces and basic steel production, iron and steel foundries, primary nonferrous metal production, nonferrous rolling and drawing, and nonferrous foundries. Chemical and allied products are usually subdivided into industrial chemicals, plastic materials and synthetics, drugs, soaps, cleaners, toilet goods, paints, and organic and agricultural chemical material.

Of the two major industrial categories in the United States, primary metals is third in rank, and chemical and allied products is fourth. It is no wonder these industries appear attractive to developing countries, arid or not. It should be recognized that these industries expend a great deal of water, and they also require a substantial energy input.

Not far behind the above two industries in their water requirements and expenditures are paper and allied products, and petroleum and coal products.

Paper and allied products consist of pulpmills, paper mills, paper board mills, paper board containers and boxes, and building paper mills. As to the value added, this classification ranks eighth. On the water required, the value added, I believe, would have to be higher to be an attractive industry for either arid or semiarid lands. The case could be made for a semiarid region, depending on the other types of industries in the country. In that region, due consideration might be given to such elements as boxes, containers and building paper.

While the United States classification system separates them, petroleum and coal products are actually, in many respects, closely associated with chemicals and allied products. The category contains only petroleum refining, and paving and roofing materials. The value added here exceeds that of chemical and allied products, but together they rank fourth and fifth, respectively. Consequently, it is no wonder that nations blessed with petroleum, or petroleum and coal, would seek to develop such industries. Granted, an argument can be made that this is not a wise economic decision in terms of water use. Nevertheless, the decisions have been made, and there is little, if any, possibility they will be reversed. In this regard, I am not just talking about some Middle Eastern nations, but my own as well. Even with our water problems in California, I cannot at this time see our closing down our petrochemical industry in favor of the middle and eastern parts of the United States.

At the other end of the spectrum, the lowest users of water are: furniture and fixtures, tobacco manufacturers, leather and leather products, and instruments and related products.

Furniture and fixtures has one of the lowest values added. This category involves such elements as household furniture, office furniture, and partitions and fixtures. The value added cannot be taken at its face value, as it is a United States statistic. This business is highly competitive in the United States, and is very sensitive to changes in the domestic economy.

Tobacco is concerned with cigarettes, cigars, chewing and smoking tobacco, and tobacco stemming and drying. Tobacco has two redeeming virtues in that it is the lowest water user, while having the highest value added, percentage-wise, in the United States. The problem for the arid land is the cost and availability of the raw material.

Leather and leather products embrace tanning and finishing, boot and shoe cut stock, bindings, footwear (except rubber), leather gloves, luggage and handbags. As with the furniture industry, the value added measure is meaningful. It is extremely low. It indicates a condition peculiar to the United States.

The instrument classification covers such items as engineering and scientific instruments, measuring and control devices, lenses, medical instruments, photographic equipment and supplies, watches and clocks. Value added in the United States is numbered after tobacco.

Slightly higher users of water are such industries as stone and clay products, textiles, food and kindred products, and lumber products.

Stone and clay comprise a host of industry types, such as flat glass, glass, glassware, pressed and blown, cement, structural clay products, pottery, concrete, gypsum and plaster products, and cut stone and stone products. Value added, on the whole, is relatively high, considering it is quite competitive in the United States.

The textile industry, for reporting purposes, is comprised of weaving mills, cotton, weaving mills synthetics, weaving, finishing mills, wool, knitting mills, floor covering mills, and yarn and thread mills. Value added in the United States ranks it along with stone and clay products. However, there is a distortion in that floor covering brings the ranking value added up much higher than one would anticipate. Since most American automobiles have carpets, and most homes built in the last thirty years have wall-to-wall carpeting, the value added here must be used with some circumspection.

Food and kindred products covers meat and dairy products, preserved fruits and vegetables, grain mill products, bakery products, sugar and confectionary, fats and oils, and beverages. As with the last two industries, the value added here compares very favorably. Although there are many large food processors, there are thousands upon thousands of smaller firms, as well. Consequently, it is a highly competitive business.

In the middle, between the intensive and the less intensive water industries, are electronics, transportation equipment, non-electrical machinery, rubber and plastic products, and fabricated metal products. With the exception of instruments noted above, these are all industries, with exceptions, that require fairly substantial technological input in one manner or another.

Electronics includes electrical distributing equipment, electrical industrial apparatus, household appliances, electrical lighting and wiring equipment, radio and television receiving equipment, communication equipment, computers, and electronic components and supplies. As might be expected, the value added is of a relatively high percentage. Generally speaking, the business is very competitive. It is also an industry which has demonstrated continual growth for many decades.

Transportation equipment includes motor vehicles and equipment, aircraft and parts, ship and boat building and repair, railroad equipment, and guided missiles. The added value is extremely high, and many nations consider becoming involved in transportation manufacture in one form or another.

Non-electrical machinery encompasses engines and turbines, farm and garden machinery, construction machinery, metal working machinery, general industrial machinery, and refrigeration and service machinery. Added value here is not as high as the several recently enumerated, but, on the whole, it is satisfactory. Generally, with a few exceptions, it is a very competitive business in this country.

Included in rubber and plastic products are such items as tires and inner tubes, rubber and plastic footwear, hoses and belting, reclaimed rubber, and fabricated rubber products. The value added as a percentage is very high. Although the tire portion of the industry is not highly competitive, the remainder is. Since the value added on a tire is quite high, caution must be used with regard to value added on many of the other segments of the industry.

Fabricated metal products includes metal cans and containers, cutlery, hand tools, hardware, plumbing and heating equipment, fabricated structural products, screw machine products, metal forging and stamping, metal services, and ordanance and accessories. The value added here is relatively low, compared to others in the middle of the water-use spectrum.

In selecting any of these industries, a number of considerations must be taken into account before a decision is made. First, does a reasonable domestic market exist? Does the establishment of the industry enhance and supplement existing industries, thereby making them more competitive in the world market? Does an overseas market exist for the industry, particularly in adjacent areas? Is there the technical expertise within the region to evaluate whether or not a technology transfer is feasible? Will the new industry reduce unemployment, or place too much strain on an already existing manpower pool? What will be the skills required to staff the industry, and what will be the training time required to bring the plant up to full operating capacity?

It is assumed that the means of financing have already been ascertained, and the logistics and infrastructure are in place, or will be, well before the plant is ready to operate. Unless all of the above are determined and planned for, the availability of water takes on little meaning. Obviously, the matter is one of the key foci, but a failure to come to grips with these basic issues will only result in water waste.

Percentage of Total Water Intake, by U.S. Manufacturing Industries by Major Industry Groups

| SIC Code | Industry Group | Percentage |
|---|---|---|
| 20 | Food and Kindred Products | 5.4% |
| 21 | Tobacco Manufacturing | .03 |
| 22 | Textile Mill Products | 1.2 |
| 24 | Lumber and Wood Products | 1.1 |
| 25 | Furniture and Fixtures | 6.04 |
| 26 | Paper and Allied Products | 16.1 |
| 28 | Chemicals and Allied Products | 27.8 |
| 29 | Petroleum and Coal Products | 8.5 |
| 30 | Rubber and Miscellaneous Plastic Products | 1.03 |
| 31 | Leather and Leather Products | .05 |
| 32 | Stone, Clay and Glass Products | 1.46 |
| 33 | Primary Metal Industries | 32.9 |
| 34 | Fabricated Metal Products | .7 |
| 35 | Machinery, Except Electrical | 1.1 |
| 36 | Electric and Electronic Equipment | .66 |
| 37 | Transportation Equipment | 1.6 |
| 38 | Instruments and Related Products | .25 |
| 39 | Miscellaneous Manufacturing Industries | .08 |

Percentage of Total Water Intake Discharged by U.S. Industries by Major Industry Groups

| SIC Code | Industry Group | Percentage Discharged |
|---|---|---|
| 20 | Food and Kindred Products | 93% |
| 21 | Tobacco Manufactures | 80 |
| 22 | Textile Mill Products | 90 |
| 24 | Lumber and Wood Products | 77 |
| 25 | Furniture and Fixtures | 100 |
| 26 | Paper and Allied Products | 95 |
| 28 | Chemicals and Allied Products | 94 |
| 29 | Petroleum and Coal Products | 90 |
| 30 | Rubber and Miscellaneous Plastic Products | 93 |
| 31 | Leather and Leather Products | 100 |
| 32 | Stone, Clay and Glass Products | 88 |
| 33 | Primary Metal Industries | 96 |
| 34 | Fabricated Metal Products | 94 |
| 35 | Machinery, Except Electrical | 96 |
| 36 | Electric and Electronic Equipment | 93 |
| 37 | Transportation Equipment | 94 |
| 38 | Instruments and Related Products | 95 |
| 39 | Miscellaneous Manufacturing Industries | 100 |

Percentage of Total Water Lost by U.S. Manufacturing Industries by Major Industry Groups

| SIC Code | Industry Groups | Percentage of Total Lost |
|---|---|---|
| 20 | Food and Kindred Products | 10% |
| 21 | Tobacco Manufactures | 0 |
| 22 | Textile Mill Products | .2 |
| 24 | Lumber and Wood Products | 4 |
| 25 | Furniture and Fixtures | 0 |
| 26 | Paper and Allied Products | 12.6 |
| 28 | Chemicals and Allied Products | 29.3 |
| 29 | Petroleum and Coal Products | 14 |
| 30 | Rubber and Miscellaneous Plastic Products | 1.2 |
| 31 | Leather and Leather Products | 0 |
| 32 | Stone, Clay and Glass Products | 3 |
| 33 | Primary Metal Industries | 20.4 |
| 34 | Fabricated Metal Products | .7 |
| 35 | Machinery, Except Electrical | .7 |
| 36 | Electric and Electronic Equipment | .4 |
| 37 | Transportation Equipment | 1.6 |
| 38 | Instruments and Allied Products | .1 |
| 39 | Miscellaneous Manufacturing Industries | 0 |

# INDUSTRIAL DEVELOPMENT IN ARID REGIONS

Zvi Pelah[1]

The lecture will describe the special problems of development of the Negev, taking into consideration environmental aspects such as climatic conditions, including low humidity combined with a high rate of radiation, and wind direction channeling effects. The industrial zones will be discussed from these points of view.

The quality of land is also one of the important factors weighing in the economic decisions leading towards the erection of industrial plants. The major problem of water sources are discussed. The development of water sources includes the transportation of potable water from the northern parts of Israel, desalination processes (including desalination by atomic power) and the construction of a saline water conduit from the Mediterranean to the Dead Sea. The impact of such projects on industrialization and on the quality of life in the desert will be discussed.

The planned infrastructure of roads and the building of a new international airport and aviation industry, will be surveyed. A brief summary of the agricultural achievements will be given including the utilization of brackish water for the growth of industrial crops. Energy sources will be outlined, including coal deposits, natural gas, solar energy, wind energy, and hydroelectric power to be obtained from the saline conduit, (which will allow the establishment of inland atomic power plants for water desalination and electricity). The saline conduit will supply the cooling water needed for these power plants as well as for industrial ventures around them.

The lecture will emphasize the importance of Ben-Gurion University of the Negev as a focal point of scientific endeavour, a center for higher education and generation of new ideas for the industrialization of the region.

The medical problems encountered by people living in a hostile environment, and their solution will be discussed.

---

[1]Director, Research and Development Authority, Ben-Gurion University Beer Sheva, Israel

The above-mentioned points indicate the complexity of the problems of life in the desert and industrialization of a desert region. After elaboration of these problems, a description will be given of the industries in the Negev, their history and present status. Industries will be classified into: 1) Industries near sources of raw material; 2) Manpower oriented industries; 3) Polluting industries.

The economic feasibility of industry in the Negev will be discussed as well as its impact on population growth in the Negev.

The author will try to draw conclusions from the Israeli experiment applicable to other parts of the world.

Arid and desert land comprise a large portion of the land surface of our planet. Reclaiming the desert, making it less hostile to human life, has become a major goal of humanity. I am going to discuss the Israeli experiment in utilizing the Negev desert for industrial development, hoping that our experiences, achievements, and plans for the future, might contribute to the quality of life in other hostile desert environments.

The late David Ben-Gurion - first prime minister and founder of the State of Israel, was the main driving force in the Israeli government's resolution to resettle, develop and industrialize the Negev. It was his vision that started the real drive. He said: "Without the settlement of the South and the Negev, there could be no security to the State, and we will not reach economic independence. This resettlement can be achieved only by changing the order of nature; I am sure that this is within the capacity of science and the pioneering initiative of our people. This miracle will become a reality through scientific pioneering".

The Negev desert comprises 62% of the land area of Israel; it is sparsely populated even now, supporting only 6.5% of Israel's population. It is rather obvious that in order to meet the challenge of a densely populated Negev, the necessary infrastructure must be supplied e.g. roads for efficient transportation, educational institutions, medical centers of high quality, recreation facilities, industrial zones, commercial centers, and an advanced agriculture.

The productive activity of people may be divided into three main categories:

Primary activity - Utilizing raw materials directly from nature - e.g. mining, fisheries, agriculture and forestry. The location of primary industry is determined by the location of the raw materials.

Secondary activity - deals with processing of primary natural materials into other useful products - a change which adds to their economic value. Location of such industries is ideally between the source of the raw materials and the consuming market. The governing factors will be minimal production costs (including transportation costs) and maximum profit.

Tertiary activity - commerce and services - should obviously be in the vicinity of urban centers.

The Negev desert - sparsely populated, and having some valuable natural resources, was until several years ago, dependent on primary industry for the majority of its economic activity. This activity consisted of the mining of phosphate rock, copper mining (and production of copper cement), sand (and its spin-off - production of glass bottles), Kaolin (and the ceramic industries). The Dead Sea is Israel's largest and most important natural resource and its exploitation forms the base for a variety of allied industries, e.g. production of $KCl$, $MgCl_2$, $Br$, etc. In the last few years, new industries have been erected:

1) a plant for the production of bromine and bromine compounds.
2) a phosphoric acid production plant
3) a periclas plant - producing MgO of high purity suitable for refractory bricks and pharmaceuticals.

Phosphate rock utilization has also become a basis for important industrial activity, annually producing about three million tons of phosphate for export. It may be understood from the foregoing that the availability of and the high cost of transportation were the determining factors in the choice of industrial locations.

An industrial plant is the intermediary between the market and the raw materials producers. This function is performed by:

1) Import of raw materials, 2) processing and modifying them according to market needs, 3) marketing the finished product.

These factors show the importance of an optimal location, which takes into consideration the cost of all parameters involved, and which, in all probability determines the economics and viability of the particular plant and the industry as a whole. Industries are rather diverse in character, and therefore may be divided according to their location coefficient - defined as their dependence upon a specific location.

A high location coefficient would be assigned to an industry highly dependent on location; location factors include: raw materials, market, transportation requirements, size of plant, availability of skilled workers, complementary industries at the site, energy sources, ecological considerations and last but not least - political reasons. Some industries clearly have a low location coefficient, such as the mobile industries. One major example of such an industry in Israel is the textile industry, which does not require proximity to markets or raw materials because the transportation component in the overall production cost is minimal. Due to the Government's policies, the textile industry was encouraged to locate in the Negev. This type of industry needs large numbers of unskilled laborers, and was therefore a good solution at a time of acute unemployment among the new unskilled immigrants who came to Israel from underdeveloped countries. Up to seven years ago, about 40% of the manpower in the Negev was still employed in these industries. The trend however, now is towards new, more sophisticated industries, requiring skilled labor.

Taking into account what was mentioned earlier, it is obvious that the Negev Desert is the only land reserve of the State of Israel, available for further development and absorption of immigrants. The development of any desert is dependent on energy sources, (either conventional or novel) development of water sources, industrialization and modern agriculture. It is of interest to note that the Negev was not always undeveloped. Two thousand years ago this was a province of the Nabateans. Rainfall in the southern Negev is from 50 to 100 mm a year, a meager quantity by any standard, but still the accumulated about was utilized by building elaborate water harvesting and spreading system, using low dams, which enabled them to cultivate tiny fields and terraces. With water, the desert bloomed. Only later, with the destruction of the system, these fields were again abandoned. The crucial problem of water for development and taming of the desert is well known.

The Israeli Government and its science advisors realized from the beginning that utilization and settlement of the Negev would have to be accompanied by a vigorous program of road building, construction of power lines and development of local water sources or the transport of water into the desert. The realization of this infrastructure brought about the conversion of large unproductive tracts of arid land to productive and prospering farmlands. Public funds financed industrial development through governmental industrial cooperatives whose subsidized budgets enabled them to penetrate into the very heart of the desert, and form the nucleus of new urban settlements.

As mentioned before, the first industrial ventures were raw material oriented, afterwards came the textile industry supported by Government development loans and encouraged by special tax benefits, in order to create the much needed jobs for the unskilled new immigrants. In the early fifties, a small chemical plant was established in Beer Sheva, for the same purpose - to create jobs for the growing population. It is interesting to note that this small plant for the production of D.D.T., and formulations of D.D.T. containing kaolin (a reason for its establishment in Beer-Sheva?) has developed into a chemical giant, the largest chemical company in Israel, producing a full line of organo-phosphorous and carbamate insecticides for the world market, as well as several types of polyester resins, photographic chemicals and others. All this from a meager beginning with a few unskilled workers, "Machteshim" now employs over 1200 people, with a scientific staff of over 130 researchers and development engineers.

The location of "Machteshim" in Beer Sheva, considered earlier to be a deterrent to its development, became an asset. Even Beer Sheva, with a present population of over 110,000 (as compared to 200, thirty years ago, or 50,000, ten years ago) is now too crowded for "Machteshim". Growth of the city is forcing further development on the city's outskirts. The growing awareness of ecological problems and quality of life in the world as a whole and in Israel in particular, has started to have an impact on Israeli industry. Polluting industries - be it pollution of water, or air or noise pollution - which are located in the center of populated areas, must now plan for their future taking ecological factors into consideration.

In Israel, a small country, the problem of distance from ports and population centers is not so acute, so that planning the transportation infrastructure for easy access to sea ports and air ports, will enable economical transfer of ecologically problematic industries to the open spaces of the Negev. Thus it may be seen that the open, undeveloped desert lands can be a real asset of the region.

The Negev Desert, with all its difficulties, has several real advantages:

1) The land is cheap and uninhabited
2) The soil is impermeable to water
3) The region has a very high level of insolation which represents an inexhaustible energy source.
4) Low humidity.

One example of converting these shortcomings into assets is the Dead Sea Works which uses evaporation ponds to produce its line of products.

The same is true for other industries producing industrial wastes containing a high concentration of inorganic salts. The problem of handling and disposing of these wastes is eliminated by their evaporation. Hundreds of thousands of cubic meters of these wastes could be treated annually by evaporation in shallow ponds of a few tens of hectares. Once every few years, the accumulated inorganic material would be removed from the ponds, thus making them fit for reuse. The desert begins one or two kilometers south of Beer Sheva. It is an ideal site for industries with ecological problems which prevent them from being established in urban areas.

I have mentioned the case of "Machteshim Chemical Works" in Beer Sheva. It was outside the city when built some 25 years ago, in an industrial area, together with other industries. The accelerated urbanization process brought these industries into the city limits. Thus these industries and especially "Machteshim" and "Bromine Compounds Ltd." became a real ecological nuisance. Further development was hindered, and therefore, planning started for a new industrial complex, to be erected far enough from Beer Sheva or any other settled area so as not to be an ecological nuisance, but without losing the advantage of being near a large urban center, with its manpower reserve, cultural activity, scientific center and parent plant.

A study of international experience with industrial parks, showed that the success of such a park is dependent upon a critical mass of allied industries utilizing common facilities. The study pointed out the feasibility of the idea. A special committee was appointed with the goal of finding the right site, taking into consideration topographical, climatic and hydrological factors, as well as the economical problem in creating the infrastructure of roads for transportation and the lay-out of the necessary services for "Machteshim" and other industries for the first decade, as well as for the more distant future. The industrial park was planned to be erected in three steps, the third step bringing development to completion. In the first stage four industrial plants are now being built; within half a year or sooner, production will start. The plants are: "Machteshim South" as the main plant, "Bromine Compounds", "Plantex" (to be moved partly from the center of Israel) and "Zion Chemical Company". The Government of Israel supports this project, which is looked upon as a model park, to be copied, if successful, on other regions of the Negev.

The Negev desert lies between two seas, the Mediterranean and the Red Sea, and therefore could be used as a land bridge between the seas. A thorough study indicated the great development potential of the Negev desert, as the real industrial backbone of Israel. Therefore the region is now being surveyed from all aspects, to provide information for formulating the guidelines for technical, organizational and legislative planning. This planning will ensure the social and technical foundation required.

The goal is to provide employment for a population of 1 million inhabitants in the nineties, a goal which is certainly attainable taking into consideration the local advantages, and the formulated development program which comprise:

1. a) natural resources - minerals from the Dead Sea, the phosphates, pure sand, kaolin, copper ores, etc; b) The crude oil pipeline from Eilat to Ashdod which passes through the Negev and should become a factor for the erection of petrochemical industries in the future; c) Large unpopulated areas, suitable for polluting industries; d) Low humidity and high solar radiation; e) Significant difference in elevation between the Mediterranean and the Dead Sea.

2. In order to ensure economic stability, the program recommends diversification of industrial ventures, reducing the textile component in the overall employment picture and striving towards science based industries. It is of importance to mention that agriculture will play a far greater role in the Negev, due to the implementation of results of advanced research leading towards speciality crops e.g. Jojoba, which will provide the basis for a new cosmetic and pharmaceutical industry, and the production and processing of medicinal plants.

It is true that water is the major problem, and in order to increase Israel's potential water supply, especially in the Negev, unconventional methods are and should be utilized. These include reducing the rate of evaporation from water and land surfaces, enhancing rainfall, trickle irrigation, water desalination using atomic and solar energy, utilization of fossil water, recycling of sewage water, etc.

3) In order to enhance the value of local raw materials, the trend is to produce finished products with higher market value.

The optimistic evaluation presented here, is the result of the master plan for the Negev - a plan which includes a number of central projects:

a) The conduit between the Mediterranean and the Dead Sea will allow the utilization of the difference in elevation between the two bodies of water for the production of hydroelectric power, and will make it possible to build atomic power plants in the Negev. The conduit will supply all the necessary cooling water for the piles, and for other power stations that will use the deposits of shale oil found in the Negev. It will supply the cooling water for the large industrial plants to be established, without draining the meager supply of potable water.

Building of atomic power plants will trigger new industries - such as the processing of phosphates and uranium extraction, and a full line of sophisticated industries - dependent on the atomic plants.

There is no question that the conduit might be utilized in the formation of artificial lakes similar to Lake Mead in Nevada. Such a venture will change the monotonous scenery of the desert, and open the way towards establishing tourist and recreational centers. Centers of this type are a must for an urbanized civilization.

b) Coal and petrochemicals - there are now two petrochemical centers in Israel - one in Haifa and one in Ashdod. In the 1980's these two centers will reach their saturation point. It is only logical to assume, that the third center should be established in the Negev, where the crude-oil pipe line from Eilat passes. Another reason would be the coal found in the Negev. A petrochemical industry in combination with the inorganic industry based on local raw materials - not to mention the other chemical plants - would allow the establishment of an industrial center of international proportions.

The Israeli government has decided to build the second international airport in the vicinity of Beer Sheva. According to the resolution, an aeronautical industry will be developed, with a projected employment of over 15 thousand workers. Such an industry needs a large variety of supporting industries, and highly skilled personnel, in various disciplines.

The Ben-Gurion University of the Negev, and its Applied Research Institute, is even now the scientific, academic and educational center of the Negev. It not only furnishes the scientific personnel for the sophisticated industries of the present and future, but also acts as a cultural meeting-place. Its impact is felt all over the region, by raising the educational level of primary and secondary schools. In the few years since its establishment, it has helped the existing industries and led them towards excellence. The Univeristy is, as a matter of fact, one of the main factors playing a major role in the future development plans, by the inspiration, enduring and pioneering spirit of its academic staff.

In conclusion, I would say, that generally speaking - the human element is the decisive factor in making the desert bloom, and the scientific center is the focal point for those gifted individuals having the necessary combination of vision and implementation.

# Use and Reuse of Water by the Mining Industry of Arizona

Walter W. Fisher
Arizona Bureau of Mines
University of Arizona

## INTRODUCTION

The objectives of this paper are to describe water use practices in the Arizona mining industry, to demonstrate how water is recycled and delineate the major areas of consumption. This will be accomplished by analysis of process flowsheets and water balances for the major water consuming operations.

The importance of mining to Arizona's economy is illustrated by the fact that mining accounted for 27 percent of the state's income in 1974 [1]. Copper which is the major resource mined in Arizona contributed 91.5 percent of the income from mining in 1974 [1]. Since the metals molybdenum, gold, and silver are recovered as by-products of copper mining, their value is included with copper. The mining of all other metals accounted for only 0.5 percent of mining income while sand and gravel production contributed 2.7 percent to the total income.

It is clear that the copper industry is the dominant mining consumer of water in the state. In fact, about 80 percent of the water consumed for mining is used by the copper industry [2]. Most of the remaining 20 percent is consumed by sand and gravel operations. Since copper production plays such an overwhelming role in Arizona, this discussion will be limited to copper mining and processing.

## WATER BALANCE CALCULATIONS

A simple diagram illustrating a general water balance for a mining operation is given in figure 1. The basis for this water balance is that the water lost through evaporation, retention in the product and tailings, seepage, and transfer must be replaced with new water. The sum of the losses (or the new water) is the water consumed by the operation. This approach was used in the preparation of the flowsheets and water balances presented in this paper.

In general, mining operations do not have firm data on water losses in the mining and processing systems. The new water required by an operation, the

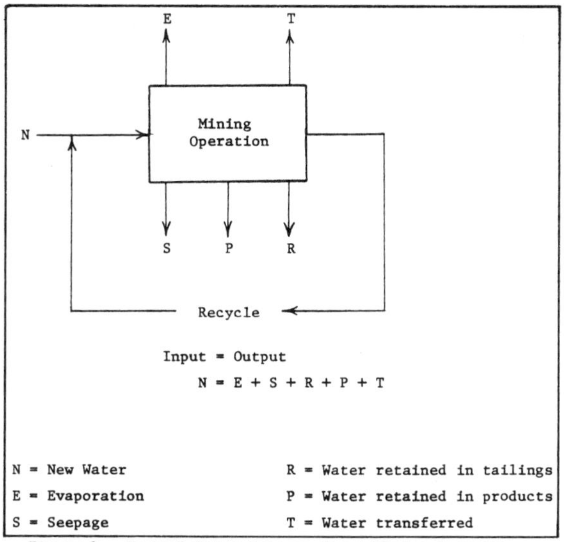


Figure 1. General Water Balance for a Mining Operation.

water recycled within the operation, and the water transferred out of the operation are closely monitored. The water lost through evaporation, retention in products and tailings, and seepage must be estimated. In this paper the evaporation rate from bodies of water was estimated to be 5 gallons per minute per acre [3]. Product and tailings were estimated to contain 15 percent water. Seepage was estimated to be the difference between new water supplied to the operation and the water lost through evaporation, retention, and transfer. In some types of leaching operations seepage was assumed to be negligible. Groundwater (G) that percolates into a part of an operation was included as new water.

The water balances that will be presented represent either a specific mining operation or a composite based on several operations. Each mining operation has unique problems that may cause its water balance to deviate from the examples discussed in this presentation. The examples are meant to show trends rather than absolute values and to illustrate the areas for future water conservation.

## WATER USE IN THE PRODUCTION OF COPPER

### INTRODUCTION

Most mining operations use water as conservatively as possible and recycle water extensively. A summary of water use for the production of copper is given in table 1. Since concentration, leaching, and smelting require the largest water inventories and account for most of the water consumption, this discussion will be limited to those operations.

Table 1. WATER USE IN THE PRODUCTION OF COPPER.

| Operation | Unit Processes | Water Use | Magnitude |
|---|---|---|---|
| Mining | Drilling | Material transport | Small |
| | Blasting, Loading, Hauling | Dust control | |
| Concentration | Crushing and sizing | Dust control | Large |
| | Grinding, Flotation, Tailings disposal | Material transport | |
| Leaching | Extraction, Washing, Concentration and Purification, Recovery | Solvent | Large |
| Smelting | Roasting, Concentrate melting, Converting, Casting | Cooling | Large |
| | Emission control | Contaminant removal | |
| Refining | Electrorefining | Solvent | Very Small |
| Wire Bar or Rod Production | Melting, Casting | Cooling | Very Small |

MINING AND CONCENTRATION

The production of copper begins with the mining of copper ore from an open pit or underground mine. In Arizona about 78 percent of the copper ore is mined from open pits with the remainder coming from underground mines [1]. The major use of water for mining is dust control and most of this water is lost to evaporation.

Ore from the mine is crushed and sized to produce a feed to the grinding circuit that is less than ½ inch (1.3 cm ) in size. Water used in this operation for dust control and cooling continues to the grinding circuit and only a small amount is lost due to evaporation. The crushed and sized ore undergoes grinding to liberate the sulfide mineral particles from the host rock and produce a product that contains 80 percent smaller than 65 mesh (0.02 cm ). Water is used in the grinding process to eliminate dust and facilitate the transport of materials as a slurry. Additional water is added after the ore is ground to produce a slurry feed to the flotation process containing about 30 percent solids.

In the flotation process small additions of frothing and collecting reagents are added to the ore slurry to promote froth formation and condition the sulfide mineral particles so that they will attach to air bubbles. Air is injected into the vigorously agitated pulp in the flotation cell where sulfide mineral particles become attached to air bubbles, float to the surface, and are recovered as a concentrate. The flotation process separates the sulfide minerals from the gangue and forms a concentrate of the sulfide minerals. A typical ore may contain from 0.3 to 1.0 percent copper. The flotation concentrate will contain from 18 to 30 percent copper.

The sulfide concentrate is thickened, filtered, and sent to a smelter for copper recovery. The tailings are thickened and the clarified water is recycled to the flotation process. The thickened tailings are sent to a tailings pond for disposal. After the solids have settled out of the slurry in the tailings pond, the clear water is recycled to the process.

The flowsheet and water balance for a typical mine and sulfide concentrator are given in figure 2. The water balance for this operation was calculated using the data and assumptions listed in table 2. Water losses amounting to

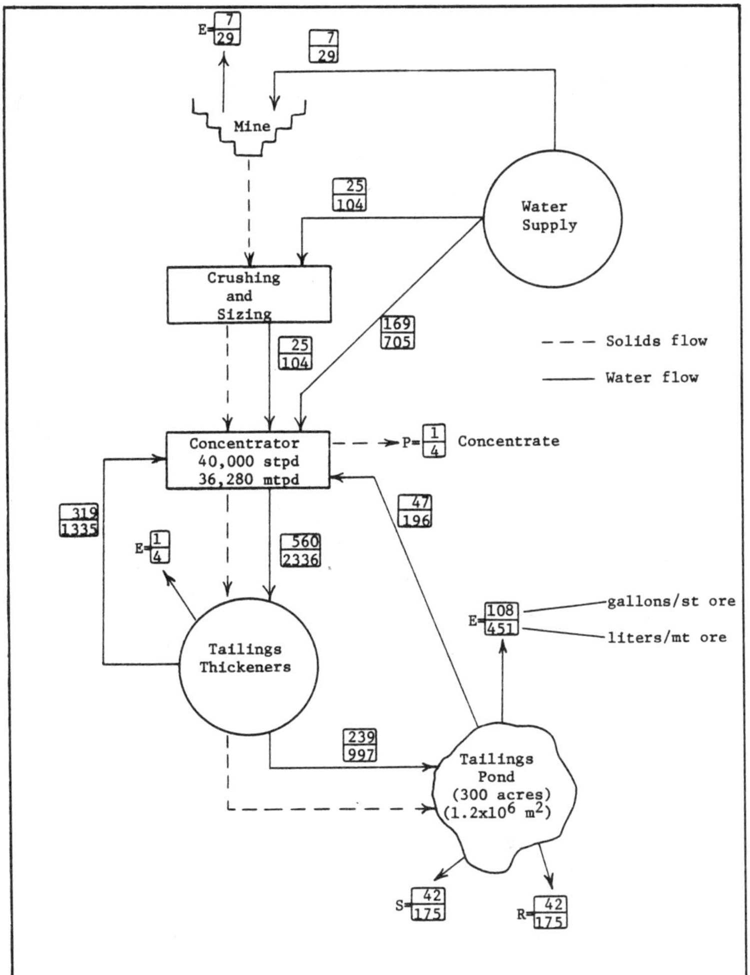


Figure 2. Flowsheet and Water Balance for a Typical Copper Sulfide Mining and Concentrating Operation.

less than one gallon per short ton of ore were neglected. Water supplied to the mine was assumed to be lost through evaporation.

The total water required for mining and concentrating one short ton of ore is 568 gallons. Water recycled in the system amounts to 65.4 percent of the total. Water lost through evaporation, retention, and seepage amount to 20.4, 7.7, and 7.5 percent, respectively. The major losses of water in the concentrator system are evaporation from the tailings pond, retention of water in the tailings, and seepage of water from the tailings pond. In other words, the major water consuming operation is tailings disposal.

The loss of water through evaporation in the mine and retention in the tailings cannot be improved significantly due to physical restrictions. Additional water could be conserved for recycle by reducing evaporation losses from the thickeners and tailings pond and by preventing seepage from the tailings pond.

Table 2. DATA AND ASSUMPTIONS USED FOR CALCULATING THE WATER BALANCE IN FIGURE 2.

| Item | Value |
|---|---|
| Concentrator capacity | 40,000 stpd |
| Ore grade | 0.5 percent Cu |
| Copper recovery | 90 percent |
| Concentrate grade | 25 percent Cu |
| Concentrator pulp density | 30 percent solids |
| Thickener discharge pulp density | 50 percent solids |
| Tailings thickeners -2 | 400 ft. diameter |
| Tailings pond area | 300 acres |
| Evaporation rate | 5 gpm/acre |
| Water retention in product and tailings | 15 percent |

## DUMP LEACHING

The mining of copper ores requires the removal of vast tonnages of low-grade ore and barren overburden. The low-grade ore is placed in special dumps that are leached to recover copper over a long period of time. The site for a leach dump is selected so that the base is impervious and the natural topography provides easy solution recovery. A dam is constructed below the base of the dump to collect solution. After a dump has been laid down, the surface is leveled and prepared for solution application. Solutions are applied to the dump surface by spraying, ponding, dripping, or injection. As the leach solution percolates down through the dump, it dissolves copper and other metals. At the bottom of the dump the pregnant leach solution follows the original drainage system, flows out of the dump, and is recovered behind the collection dam.

Copper can be recovered from the pregnant leach solution by cementation or by solvent extraction and electrowinning. Cementation is the electrochemical replacement of a more noble metal ion in solution by a more active metal. Copper is recovered by cementation on scrap iron to produce an impure copper product that is usually treated along with sulfide concentrates in a smelter.

In the solvent extraction process, copper is selectively extracted into an organic liquid phase leaving unwanted aqueous species in the leach solution. The copper is then extracted (stripped) back into a new aqueous solution to form a purified, concentrated copper solution for metal recovery. Copper is recovered from the pure, concentrated solution by electrowinning to produce cathode copper that does not require additional refining.

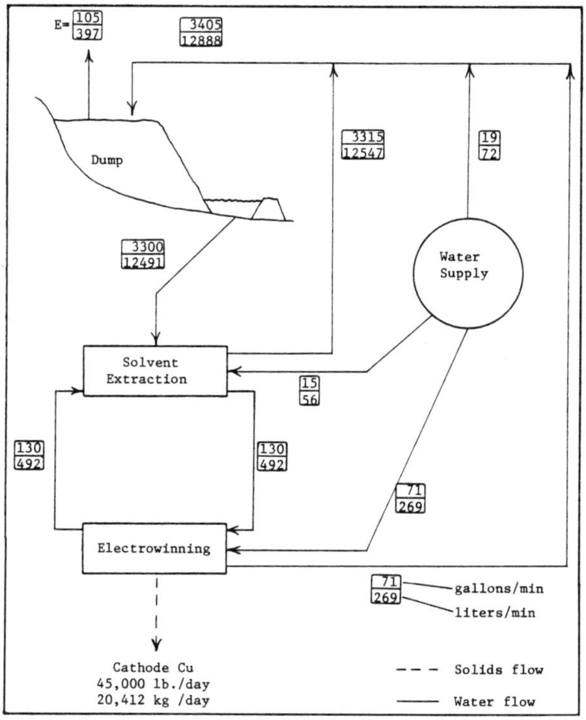


Figure 3. Flowsheet and Water Balance for a Typical Copper Dump Leach with Solvent Extraction and Electrowinning.

Figure 3 is a flowsheet and water balance for a typical dump leach-solvent extraction-electrowinning operation [4]. Since a single dump is operated for many years, water retained in the dump is lost only during the first leach cycle and is neglected because the volume is very small compared to the total water use. Dumps are built to minimize solution seepage and, therefore, the seepage loss is assumed to be negligible. The major water loss in this operation is through evaporation.

The total water inventory required to produce one pound of copper is 109 gallons. The amount of water consumed to produce a pound of copper is 3.36 gallons. About 97 percent of the water in the dump leach system is recycled. The major loss of water in this system is by evaporation from the surface of the dump where the leach solution is introduced.

IN SITU LEACHING

When an ore deposit is mined by underground mining methods, a large amount of copper-bearing rock and ore must be left in place. Usually this material is rubblized due to collapse of the rock overlying the mine and is suitably fractured for recovery of the copper by leaching. In other cases a copper deposit may be too small to be mined profitably by conventional techniques. The deposit may be small enough to create rubblization and fracturing of most of the rock and ore by conventional explosives. In either case copper-bearing rock and ore rubblized and fractured in place can be leached by applying leach solution to the surface of the rubblized material. The leach solution percolates down through the column of broken ore and rock, dissolves copper and is pumped back to the surface or exits the deposit at a lower elevation and is recovered behind a collection dam. The pregnant solution produced by in situ leaching can be treated by cementation or solvent extraction-electrowinning to recover the copper.

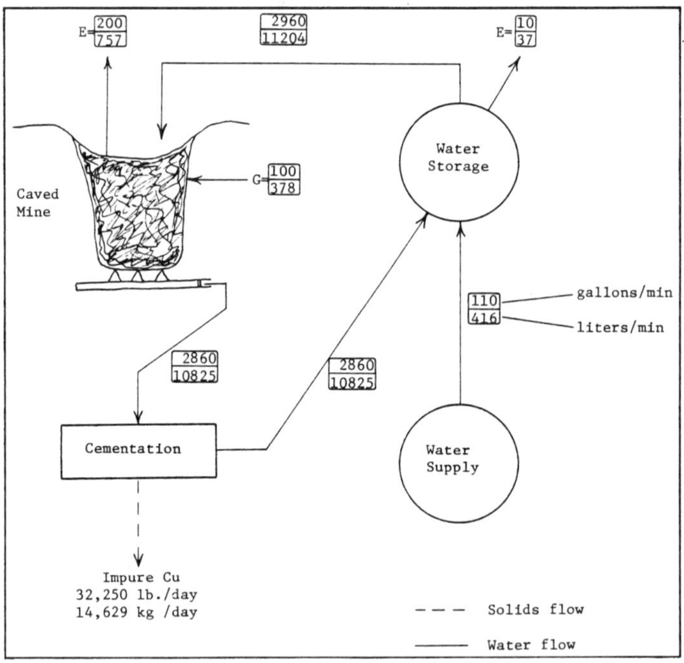


Figure 4. Flowsheet and Water Balance for Copper In Situ Leaching with Cementation.

Figure 4 is a flowsheet and water balance for a typical in situ leach-cementation operation [4]. The water retained in the ore material is neglected since it is only lost during the initial leach cycle. Most in situ leach sites are within rock that has a very low permeability and consequently water lost by seepage can be neglected.

In addition to the fresh water added to this in situ leaching system, a small amount of groundwater (G) percolates into the fractured ore body. The total water inventory required to produce one pound of copper amounts to 136 gallons. The water consumed in the production of a pound of copper is 9.38 gallons. About 93 percent of the water in this operation is recycled. Again, the major loss of water is through evaporation at the surface where solution is applied.

HEAP LEACHING

Some copper deposits consist of mostly oxidized copper minerals that are easily leached to extract the copper. One of the methods for leaching this type of ore is heap leaching. Heap leaching is similar to dump leaching but on a much smaller scale. Oxide copper ore is placed on a prepared, impermeable surface in heaps about 20 feet in depth containing about 100,000 tons of material. The heap is positioned to use the natural drainage system for solu- collection and storage. The surface of the heap is prepared for solution application and solutions are applied to the heap surface by spraying, ponding, dripping, or injection. The leach solution percolates down through the ore, leaches the copper from the oxide minerals and is recovered behind the collection dam below the heap. Copper is recovered from the pregnant leach solution by cementation or solvent extraction and electrowinning. When most of the copper has been extracted from a heap, a new heap is constructed on top of the first heap and leaching is continued. This procedure is continued until the total height of the heaps reaches 150 to 200 feet.

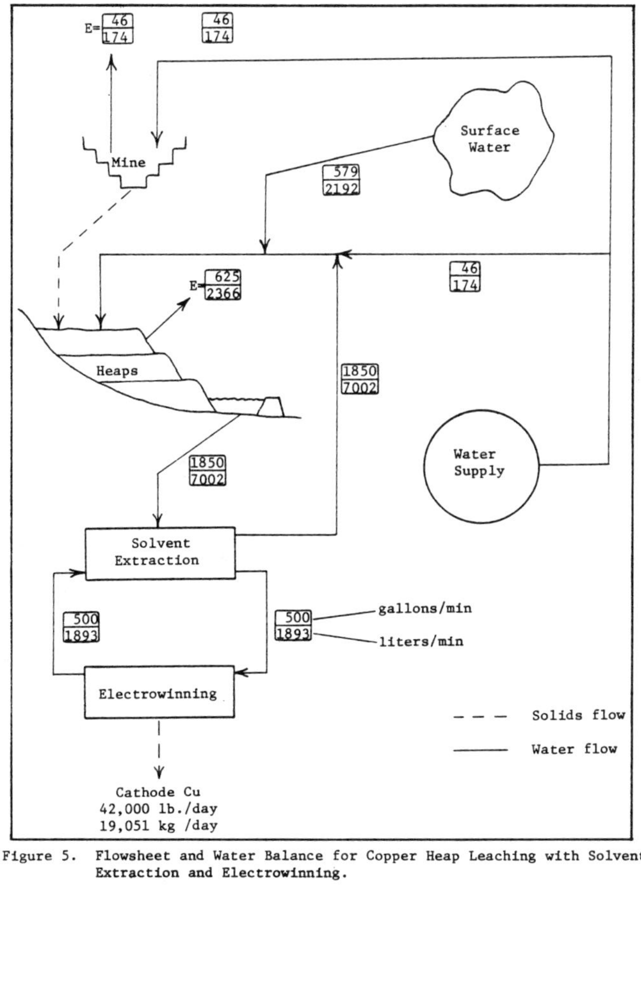


Figure 5. Flowsheet and Water Balance for Copper Heap Leaching with Solvent Extraction and Electrowinning.

Figure 5 is a flowsheet and water balance for a typical heap leach-solvent extraction-electrowinning operation [4]. Water retained in the heap and seepage are neglected for reasons discussed under dump leaching. Since new material must be mined for additional heap construction, some water is consumed in the mine for dust control. The major water loss is through evaporation on the heap surfaces.

The total water inventory for the production of one pound of copper is 86 gallons. The amount of water consumed to produce one pound of copper is 23.0 gallons. The water recycled in this heap leaching system is about 73 percent. Evaporation from the heap surfaces is the major loss of water in the operation.

## AGITATION LEACHING

Another method for copper recovery from oxide copper ores is agitation leaching. In this approach the ore is crushed, ground, and subjected to leaching in stirred reactors. The leached ore slurry is washed by counter current decantation (CCD) and the thickened tailings are sent to a tailings pond for disposal. The pregnant leach solution is treated by solvent extraction and electrowinning to produce cathode copper.

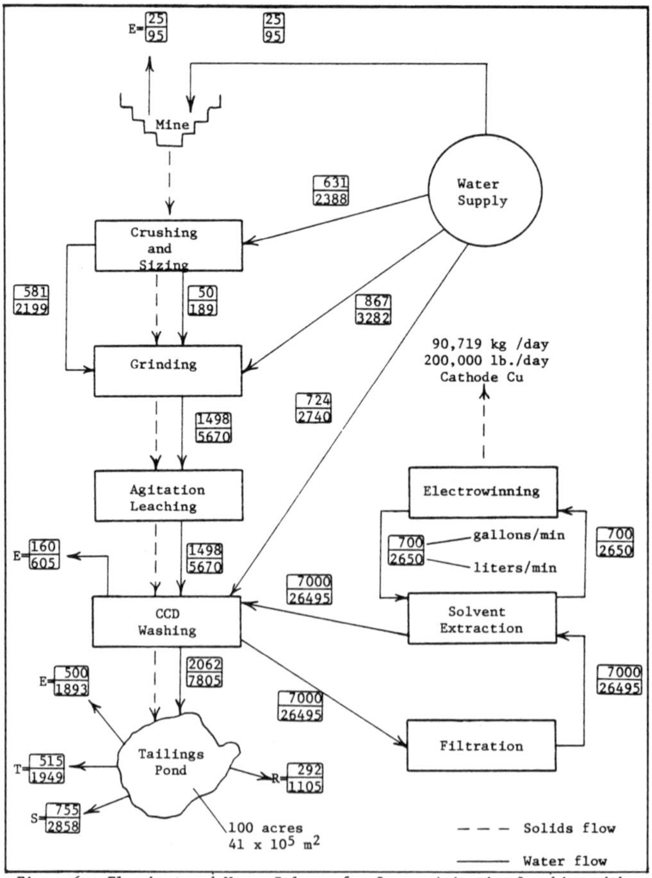


Figure 6. Flowsheet and Water Balance for Copper Agitation Leaching with Solvent Extraction and Electrowinning.

Figure 6 is a flowsheet and water balance for an agitation leach-solvent extraction-electrowinning operation [5]. The water inventory required to extract one pound of copper in this system is 66 gallons. The production of one pound of copper consumes 12.5 gallons of water. About 76 percent of the water is recycled in this agitation leaching system. Another 5.6 percent of the water is transferred to another operation. The major water losses from this operation are evaporation from the mine, CCD thickeners and tailings pond, retention of water in the tailings and seepage from the tailings pond. Water lost in the mine and retained in the tailings cannot be reduced because of physical factors. Water could be conserved by eliminating seepage and reducing evaporation in the tailings pond.

## SMELTING AND REFINING

The final step in the production of copper is the smelting and refining of sulfide concentrates and cementation precipitates (cement copper). Concentrates may be roasted as a first step to remove part of the sulfur. Then, concentrates and cement copper are melted and treated in an electric or reverbatory furnace to form matte (molten copper-iron sulfide). The molten matte is transferred to a converting furnace where the sulfur is oxidized and

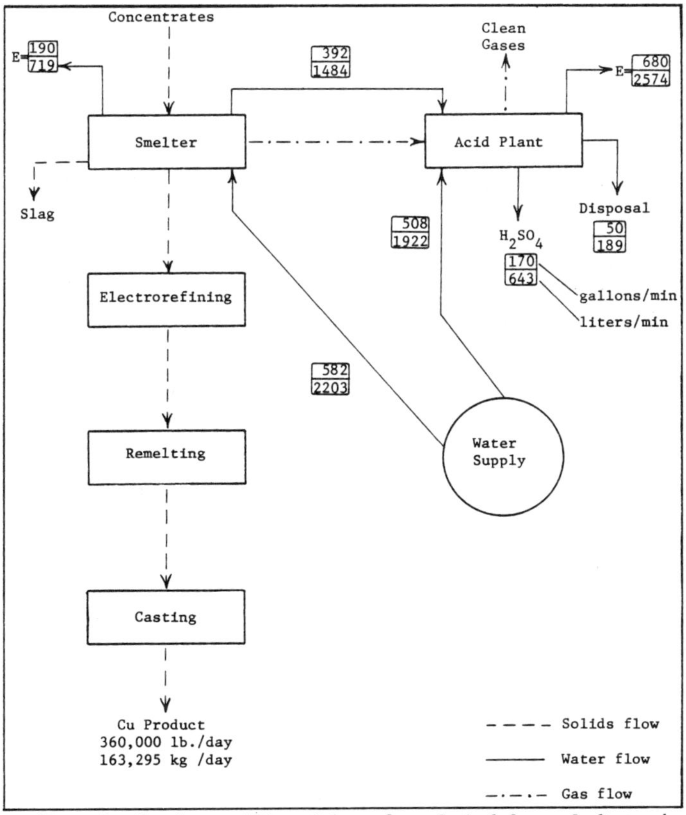


Figure 7. Flowsheet and Water Balance for a Typical Copper Smelter and Refinery with Pollution Controls.

removed as sulfur dioxide gas, the iron is oxidized and removed as an oxide in the slag and the copper is reduced to form molten copper. The molten copper is cast into anodes which are refined to a high purity by electrorefining. The refined copper is melted and cast to form wire bar, wire rod, or other marketable products. Most modern copper smelters utilize the heat content of furnace gases to produce steam for generating electricity and driving blowers. In order to simplify the presentation, this part of the smelter operation is not included in the discussion.

Water is used in the smelter for furnace and mold cooling, cooling of furnace gases, and recovery of contaminants in the furnace gases. Water is used as the solvent in electrorefining but the losses are very small. Figure 7 is a flowsheet and water balance for a typical copper smelting and refining operation [4,5]. The total water inventory for the production of one pound of copper is 4.36 gallons. Most of this water is consumed through evaporation or formation of sulfuric acid. The addition of pollution control equipment to smelters has greatly increased water consumption during smelting.

## SUMMARY

Water use statistics for the operations discussed in this paper are listed in table 3. The total water consumed for the mining, concentration, smelting, and refining of one pound of copper is 26.7 gallons. The water consumed by leaching, cementation, smelting, and refining to produce one pound of copper is 13.8 gallons. A pound of copper produced directly by leaching, solvent extraction, and electrowinning consumes from 3.4 to 23.0 gallons of water. This large variation is due primarily to differences in evaporation rates caused by location, altitude, leaching surface area or tailings pond area.

Table 3. SUMMARY OF WATER USE STATISTICS FOR SELECTED COPPER PRODUCTION PROCESSES.

| Operation<br><br>Operation | Total Water, gal./lb. Cu (1/kg Cu) | Water Consumed, gal./lb. Cu (1/kg Cu) | Percent Water Recycled | Percent Water Consumed by Evaporation |
|---|---|---|---|---|
| Mining and Concentration | 63<br>(526) | 22.3<br>(186.0) | 65.4 | 57.7 |
| Dump Leach-SX-Electrowinning | 109<br>(909) | 3.4<br>(28.4) | 96.9 | 95+ |
| In Situ Leach-Cementation | 136<br>(1135) | 9.4<br>(78.4) | 93.5 | 95+ |
| Mining-Heap Leach-SX-Electrowinning | 86<br>(718) | 23.0<br>(191.9) | 73.4 | 95+ |
| Mining-Agitation Leach-SX-Electrowinning | 66<br>(551) | 12.5<br>(104.3) | 75.6 | 39.5 |
| Smelting and Refining | 4.4<br>(36.7) | 4.4<br>(36.7) | | 79.8 |

The largest consumption of water is by evaporation from thickeners, tailings ponds, and leaching surfaces. A lesser amount of water is lost due to seepage from tailings ponds and some leaching operations. Each of these water consuming areas are targets for future water conservation.

## REFERENCES


1 U. S. Bureau of Mines, Minerals Yearbook, 1974.

2 Arizona Water Commission, Alternative Futures - Phase II - Arizona State Water Plan, Phoenix, Arizona, 1977.

3 Gilkey, M. M. and R. T. Beckman, Water Requirements and Uses in Arizona Mineral Industries, U. S. Bureau of Mines, IC 8162, 1963.

4 University of Arizona, Department of Mining and Geological Engineering, Unpublished file data.

5 Arizona Bureau of Mines, Unpublished file data.

# WATER REQUIREMENTS FOR PETROLEUM PRODUCTION

William E. Bye[1]

## ABSTRACT

The water requirements for Petroleum Production will be addressed under four distinct categories as follows:

- Water Use for Personnel
- Water Use for Drilling Muds
- Water Use for Enhanced Oil Recovery
- Water Use for Associated Processes

In water resources development project studies, it is essential that an industrial water base be established. That is, there must be a reasonable estimate formulated of the quantity of water used at the present time by each of the various categories of petroleum production. The best source of information in making such a base study is the industry itself. The information relating to individual water uses was obtained from various reference sources using the most recent information available and a more or less complete census of industry sources.

## INTRODUCTION

Water uses related to petroleum production would follow in approximately the same sequence as the development of an oil field. It requires people to locate, drill and produce an oil field. A well is drilled and if the well is productive it will be completed and stimulated. At a certain stage in the life of an oil well, it may be necessary to develop additional processes to recover more oil, either by secondary or tertiary recovery methods. Throughout the development of an oil field, water is required to complete each stage. This paper will provide some estimates that may be used as a guide in estimating water needs during the development of an oil well.

---

[1]Environmental Engineer, U.S. Environmental Protection Agency

To develop water needs for personnel, you must consider the essential elements of water demand including the average daily water consumption and the peak rate of demand. The average daily water consumption must be estimated to determine the ability of the water source to meet continuing demands over the critical periods when surface flows are low and the ground water tables are at minimum elevation. To estimate the quantity of stored water needed to sustain the demand during these critical periods, the peak demand rates must be estimated. These rates are used to determine plumbing and piping sizing, pressure losses and storage requirements necessary to supply sufficient water during periods of peak water demand.

Many factors influence the water use of a given system. For example, the fact that water under pressure is readily available stimulates its use for watering lawns and gardens, washing automobiles, operating air conditioning equipment and for performing many other useful but nonessential activities in the home and industry. Modern kitchen and laundry appliances, such as food waste disposals and automatic dishwashers, contribute to a higher total water use and tend to increase peak demands. Table I(1) presents a summary of average water use which may be used as a guide in estimating water needs. Some adjustments may be required to meet specific requirements in localized areas.

The rate of water use of an individual water facility will vary with the activity in the home or with the industrial operational program. Rates are generally the highest near mealtimes, during laundry periods and shortly before bedtime. During the intervening daytime and night hours, water use may be virtually nil. Thus, the total amount of water used by a facility may be distributed over only a few hours of the day during which the actual use is much greater than the average rate indicated in Table I.

Table I. Planning Guide for Water Use

| Types of Establishments | Gals./day |
|---|---|
| Airports (per passenger) | 3-5 |
| Apartments, multiple family (per resident) | 60 |
| Bath houses (per bather) | 10 |
| Camps: | |
| Construction, semi-permanent (per worker) | 50 |
| Day with no meals served (per camper) | 15 |
| Luxury (per camper) | 100-150 |
| Resorts, day and night, with limited plumbing (per camper) | 50 |
| Tourist with central bath and toilet facilities (per person) | 35 |
| Cottages with seasonal occupancy (per resident) | 50 |
| Courts, tourist with individual bath units (per person) | 50 |
| Clubs: | |
| Country (per resident member) | 100 |
| Country (per nonresident member) | 25 |
| Dwellings: | |
| Boardinghouses (per boarder) | 50 |
| Additional kitchen requirements for nonresident boarders | 10 |
| Luxury (per person) | 100-150 |
| Multiple-family apartments (per resident) | 40 |
| Rooming houses (per resident) | 60 |
| Single family (per resident) | 50-75 |
| Estates (per resident) | 100-150 |
| Factories (gallons per person per shift) | 15-35 |
| Highway rest area (per person) | 5 |

(continued)

Table I. (continued)

| | |
|---|---|
| Hotels with private baths (2 persons per room) | 60 |
| Hotels without private baths (per person) | 50 |
| Institutions other than hospitals (per person) | 75-125 |
| Hospitals (per bed) | 250-400 |
| Laundries, self-service (gallons per washing, per customer) | 50 |
| Livestock (per animal) | |
| Cattle (drinking) | 12 |
| Dairy (drinking and servicing) | 35 |
| Goat (drinking) | 2 |
| Hog (drinking) | 4 |
| Horse (drinking) | 12 |
| Mule (drinking) | 12 |
| Sheep (drinking) | 2 |
| Steer (drinking) | 12 |
| Motels with bath, toilet, and kitchen facilities (per bed space) | 50 |
| With bed and toilet (per bed space) | 40 |
| Parks: | |
| Overnight with flush toilets (per camper) | 25 |
| Trailers with individual bath, no sewer connection (per trailer) | 25 |
| Trailers with individual baths, connected to sewer (per person) | 50 |
| Picnic: | |
| With bathhouses, showers and flush toilets (per picnicker) | 20 |
| With toilet facilities only (gallons per picnicker) | 10 |
| Poultry: | |
| Chickens (per 100) | 5-10 |
| Turkeys (per 100) | 10-18 |
| Restaurants with toilet facilities (per patron) | 7-10 |
| Without toilet facilities (per patron) | 2 1/2-3 |
| With bars and cocktail lounge (additional quantity per patron) | 2 |
| Schools: | |
| Boarding (per pupil) | 75-100 |
| Day with cafeteria, gymnasiums, and showers (per pupil) | 25 |
| Day with cafeteria but no gymnasiums or showers (per pupil) | 20 |
| Day without cafeteria, gymnasiums or showers (per pupil) | 15 |
| Service stations (per vehicle) | 10 |
| Stores (per toilet room) | 400 |
| Swimming pools (per swimmer) | 10 |
| Theaters: | |
| Drive-in (per car space) | 5 |
| Movie (per auditorium seat) | 5 |
| Workers: | |
| Construction (per person per shift) | 50 |
| Day (school or offices per person per shift) | 15 |

Much of the information in Table 1. is not appropriate to water uses related to petroleum production, but was included for general interest and use.

Simultaneous operation of several plumbing fixtures will create maximum peak rate of water delivery for a water system. For example, a shower, an automatic dish-water and a flush valve toilet all operating at the same time could produce a near critical peak in water usage. Naturally, all of these facilities are not usually operating together. However, if they exist and are supplied by the same facility, there is always a possibility that a critical combination could result. Table II summarizes the rate of flow expected for certain plumbing and other fixtures.

Table II. Rates of Flow for Certain Plumbing, Household, and Farm Fixtures

| Location | Flow Pressure[1] lbs/sq. in. (psi) | Flow Rate gal./min. (gpm) |
|---|---|---|
| Ordinary basin faucet | 8 | 2.0 |
| Self-closing basin faucet | 8 | 2.5 |
| Sink faucet, 3/8 inch | 8 | 4.5 |
| Sink faucet, 1/2 inch | 8 | 4.5 |
| Bathtub faucet | 8 | 6.0 |
| Laundry tub faucet, 1/2 inch | 8 | 5.0 |
| Shower | 8 | 5.0 |
| Ball-cock for closet | 8 | 3.0[2] |
| Flush valve for closet | 15 | 15-40 |
| Flushometer valve for urinal | 15 | 15.0 |
| Garden hose (50 ft., 3/4-inch sill cock) | 30 | 5.0 |
| Garden hose (50 ft., 5/8-inch outlet) | 15 | 3.33 |
| Drinking fountains | 15 | .75 |
| Fire hose 1-1/2 inches, 1/2-inch nozzle | 30 | 40.0 |

[1]Flow pressure is the pressure in the supply near the faucet or water outlet while the faucet or water outlet is wide open and flowing.

[2]Wide range due to variation in design and type of close flush valves.

The second water requirement would be to provide fluids to drill the well.

The drilling fluid, or drilling mud, performs a number of essential functions on a drilling well, most of which are well known. Among these are cooling the bit, lifting drill cuttings to the surface, maintaining sufficient pressure to hold back the flow from the formations penetrated, and sealing the hole to prevent the mud, or mud filtrate, from flowing in significant amounts into the formation. To satisfy all these requirements, the mud must be compounded to control the chemical and physical properties to meet the varied and changing sub-surface conditions. The science of drilling fluid control is highly developed and its on-site application is essential to the success of a well.

The most commonly used drilling fluids are water-based, and there are undoubtedly hundreds of examples of water-based muds. Table III and IV list the constituents in two of the more common types of water based muds.

Table III. Clay-Base Mud

| Component | Concentration (lb./bbl) |
|---|---|
| Water | 200 to 340 |
| Bentonite | 15 to 30 |
| Lignite | 1 to 6 |
| Lignosulfonate | 2 to 10 |
| Sodium Hydroxide | 0.5 to 1.5 |
| Barite | 0 to 500 |

Table IV. Polymer Mud

| Component | Concentration (lb/bbl) |
|---|---|
| Water | 300 to 345 |
| Bentonite | 0 to 10 |
| Salt (NaCl, KCl) | 10 to 100 |
| Sodium or Potassium Hydroxide | 0.1 to 0.3 |
| Polymer (Starch, Polyacrylamide) | 0.5 to 5 |
| Bactericide (Paraformaldehyde) | 0.1 to 0.5 |
| Barite | 0 to 300 |

Rule of thumb estimates are generally not considered to be scientifically acceptable, however, in the water requirements for drilling muds, the industry contacts generally agreed that the average, normal well, with no lost circulation problems will require approximately two to three barrels of water per foot of hold drilled.

Table V indicates the concentration of various components used in oil muds. Oil muds are customarily prepared from diesel oil using some type of surfactant to emulsify water. The filtrate reducer and gellant can be oil-dispersible versions of the lignite and clay used in the water-base system, while the barite is the same as that used in the aqueous system.

Table V. Oil Mud

| Component | Concentration (lb/bbl) |
|---|---|
| Diesel Oil | 150 to 230 |
| Water | 35 to 50 |
| Calcium Chloride | 15 to 25 |
| Emulsifier (Soap, Poluamide) | 5 to 20 |
| Filtrate Reducer (Amine Lignite) | 0 to 10 |
| Gellant (Amine Clay) | 2 to 4 |
| Barite | 0 to 500 |

When a well is drilled every cubic foot of cuttings removed must be replaced with a barrel of drilling fluid. A typical water based mud is made of approximately 90-95% water. The amount of water will decrease as the amounts of weighting materials increase. The water content of the drilling fluid can be reduced to as low as 50-60%. Oil base muds, specifically invert muds, may have 20-35% water in the fluid phase. It was interesting to note that in checking with industry representatives, the exact amount of water required was not generally considered. The most common reference is "water as required" and specific areas with unique individual drilling conditions will dictate the amount of water needed.

The amount of water required for associated processes in oil production, which include cementing and stimulation of the well are fairly well established, as a set amount of water is required to mix the cement and known amounts of water will be specified in the stimulation process. The length of casing and the type and amount of cement required will vary according to the conditions and what has been encountered during the drilling of the well.

The API classification of cements and suggested depths, water requirements and temperatures where these different types should be used are shown in Table VI.(2)

Table VI. API Cement Classification and Applications

| API Classification | Mixing Water gal./sk | Slurry Weight lb/gal | Well Depth ft. | Static* Temp. F |
|---|---|---|---|---|
| A (Portland) | 5.2 | 15.6 | 0-6,000 | 80-170 |
| B (Portland) | 5.2 | 15.6 | 0-6,000 | 80-170 |
| C (High Early) | 6.3 | 14.8 | 0-6,000 | 80-170 |
| N (Retarded) | 4.5 | 16.2 | 6,000- 9,000 | 170-205 |
| D (Retarded) | 4.5 | 16.2 | 6,000-12,000 | 170-260 |
| E (Retarded) | 4.5 | 16.2 | 6,000-14,000 | 170-290 |
| F (Retarded) | 4.5 | 16.2 | 10,000-16,000 | 230-320 |

*These depths and temperature conditions are based on casing-cementing conditions as taken from API RP 10B and should be considered as approximate values.

The most common method of mixing cement is to add water to dry Portland cement called "neat cement" but in most cases a neat cement will not do the job and a more sophisticated cementing material will be required. This can be obtained by adding various additives to the neat cement to obtain the desired type of cement.

These additives will vary the amount of water required per cu.ft. of material. Table VI lists the most common additives used in cements today.(3)

Generally, on land based operations fresh water is used to mix the cement slurry. On offshore installations, sea water may be used if fresh water is not available. Ocean water or water from the Gulf of Mexico contains small percentages of sodium, magnesium and/or calcium chloride (20,000-40,000 ppm) and will reduce the setting time of all types of cementing compositions.

Well stimulation practices are conducted to increase the production and/or injection rate of a producing or injection zone. Two methods have been used extensively for stimulation - fracturing and acidizing.

The size of fracturing treatments varies from a few thousand gallons to over a million gallons of fluid pumped while fracturing the formation. The massive fracturing treatments being performed today are designed for 2,000 ft. of penetration in each direction from the well. The created fracture is propped open by a fracturing sand or propping agent and acts as a wedge to hold the fracture open and provide a highly conductive channel from the reservoir to the wellbore of the transmission of fluids.

The second method of stimulation is acidizing which can be done two ways. The acid is either injected at high rates and pressure fracturing the formation, or it is injected at a rate and pressure low enough to force the acid into the pore space of the producing formation, creating permeable drainage channels into the well bore.

The water requirements for stimulation treatments, either fracturing, acid-fracturing or acidizing will vary according to the size of the treatment. Acids are made from concentrated acids diluted with water and fracturing fluid which can be either oil or water. In a large massive fracturing job, economics dictate the use of water as the costs and hazards of pumping over a million gallons of oil would be prohibitive.

Within the producing element of the oil industry, the greatest need of water usage is for enhanced recovery. Enhanced recovery includes secondary recovery, pressure

Table VII. Cement Additives

Cement Accelerators
Purpose: Reduce waiting on cement
Increase early strength

1. Calcium
2. Sodium chloride
3. Sodium silicate
4. HA-5
5. Gypsum cement

Cement Retarders
Purpose: Increase thickening time
Retard setting

1. Calcium lignosulfonates
2. Modified lignins
3. CMHEC (carboxymethylhydroxethyl cellulose)
4. Saturated salt water

Lightweight Additives
Purpose: Reduce slurry weight
Increase yield
Lower costs

1. Bentonite
2. Pozzolans
3. Gilsonite
4. Expanded perlite
5. Diatoamceous earth
6. Hydorcarbons (emulsions)

Special Cements

1. Gypsum cement
2. Resin cement
3. Diesel oil cement
4. Pozzolan-lime cements
   (for high temperatures)

Heavyweight Additives
Purpose: Increase density
Restrain pressure

1. Ilmenite
2. Barite
3. Sand (40-60 mesh)
4. Saturated salt water

Low Water Additives
Purpose: Protect sensitive formations
Prevent premature dehydration
Better squeeze jobs

1. CMHEC (carboxymethyl-hydroxethyl cellulose)
2. Latex
3. Acrylic
4. Organic polymer blends

(continued)

Table VII. (continued)

Lost Circulation Additives
Purpose: Restore circulation
Increased fill-up in column
Reduce costs

1. Gilsonite
2. Granulated nut hulls
3. Cellophane flakes
4. Expanded perlites
5. Shredded fibres
6. Bentonite
7. Bentonite diesel oil

Special Additives

1. Radioactive tracers
2. Activated charcoal
3. Paraformaldehyde - sodium chromate
4. Silica flour

maintenance and tertiary recovery activities. In each of these activities, water is injected into the producing horizon to assist in the recovery of oil which otherwise would be left in the reservoir. In these activities, the water is injected into wells completed within the producing horizon, and the oil is pushed toward producing wells through which it is brought to the surface.

The water used in these operations may be depending upon availability and compatibility with water in the reservoir, either fresh or saline. Prior to the design and installation of surface facilities handling the water to be injected, careful studies should be made to determine whether the proposed injected fluid is compatible with the water existing in the reservoir. When dissimilar waters are mixed, there is often a chemical reaction which results in the formation of precipitants or the formation of scales. A laboratory analysis can usually pinpoint such tendencies and the time when these conditions can be anticipated. At times, chemical treatment will not adequately prevent the formation of precipitants or scales, and it is necessary to find some alternate source of water for injection. The analytical program should include the result of the compatibility study of the produced water with the "make-up" water at the surface. It is sometimes necessary to dispose of produced water in a separate operation and to use only "make-up" water for injecting into reservoirs for recovery enhancement.

On the average, the amount of water required for a complete recovery enhancement program from the beginning of injection until the depletion of the reservoir is two displaceable pore volumes of water. If the laboratory analysis indicates that the produced water may be recycled in the water injection process, the amount of "make-up" water will usually be about 50% of the total water required. The pore volume of the reservoir may be determined by the following formula:

$$PV = 7758 \ B/AF \times O(1 - Swi - Sor),$$

where

V = pore volume
B/AF = bbls. per acre ft.
O = porosity

Swi = interstitial (or connate) water
Sor = residual oil saturation

Not all oil producing reservoirs are amenable to enhanced recovery. Each individual reservoir must be examined separately. If the recovery mechanism is an effective bottom water drive, the injection of water would not be economically justifiable. Other factors may also preclude the injection of water as a means of enhanced recovery.

A study of a typical sandstone oil-bearing reservoir will demonstrate the relative amounts of water required for enhanced recovery by water flooding. A review of 215 sandstone reservoirs under flood now in the United States indicates that the average porosity of a sandstone reservoir is 22.1% and interstitial or connate water is 28.6%. If a residual oil saturation (after flooding) of 25% is assumed, the amount of water required for injecting 2 pore volumes on an acre-foot basis is 1592 bbls. If the oil bearing reservoir covers 5,000 acres and is on the average 40 feet thick, the total amount of water required will be 318,400,000 bbls.

A similar examination of a typical carbonate oil-bearing reservoir of 5,000 acres and 40 feet thick, would require 238,800,000 bbls. of water to inject 2 pore volumes in an enhanced recovery operation. A review of 81 carbonate formation enhanced recovery operations in the United States reveals that the average porosity is 14.4% and the interstitial or connate water is 21.6%. Again, for estimation purposes a residual oil saturation after flooding of 25% is assumed.

## CONCLUSIONS

It is evident that the one key requirement for petroleum production is water and in some cases it must be fresh water. However, it becomes more evident that in drilling, completion and enhanced recovery operations more non potable water is being used in these processes. It is critical in most cases to conduct compatability studies to determine if water other than fresh water will work as efficiently and will not damage the formations. As technologies improve, more produced and saline water can be used which reduces the demands of fresh water supplies for this purpose and conserves them for drinking and irrigation purposes.

## ACKNOWLEDGMENTS

I would like to thank Mr. Victor J. Kimm, Deputy Assistant Administrator; Mr. Alan Levin, Director of State Programs; and Mr. Thomas E. Belk, Chief of the Office of Water Supply, Environmental Protection Agency. Without their assistance and approval this paper could not have been presented.

I would also like to thank Mr. Seth Abbott, Jr., Atlantic Richfield Company, Dallas, Texas, for his efforts in compiling the data for the enhanced recovery portion of this paper.

## REFERENCES

(1) U.S. Environmental Protection Agency, Water Supply Division, "Manual of Individual Water Supply Systems, U. S. Government Printing Office, Washington, D. C. 20402, Stock No. EP 2.8:N2 Reprinted 1975.

(2) "Recommended Practices for Testing Oil Well Cements," API RP 10B, Ninth Edition, January, 1960.

(3) "Oil Well Cementing Practices and Materials," Phil C. Montgomery and Dwight K. Smith, The Petroleum Engineer, May and June 1961.

# THE USE OF SULPHUR AS A BUILDING MATERIAL IN DESERTS.

A. Ortega - Architect - UN Adviser to the UAE and
B. Lefebvre - Architect - UAE Consultant.

## AVAILABILITY OF SULPHUR

In the Arabian Peninsula and particularly in the United Arab Emirates, sulphur will be produced in large quantities following implementation of the plans for the construction of liquid natural gas plants and recovery from petroleum refineries. The recovery of sulphur in the U.A.E. will be about 230,000 tons a year in the case of the LNG plant at Das Island. The proposal for a LNG plant on the Abu Dhabi mainland, with twice the capacity of the Das Island plant, will imply the recovery of another 450,000 tons of sulphur. The prospective proposals for similar plants - Dubai's firm initiatives are now taking concrete shape; and Sharjah in a more tentative way - and the spectrum of oil refinery proposals, taken together, could add up, on a conservative basis, to a total recovered sulphur potential of the order of 1 million tons by 1980, in addition to any by-product production of sulphur from mining or other industrial operations.

There are three ways of obtaining sulphur in the arid region of the Arabian Peninsula.

1) Native sulphur.
2) Sulphur recovered from LNG plants; and
3) Recovery of sulphur from petroleum or mineral refining operations.

The native sulphur, occurs in deposit some 50 miles from Jebal Dhana in Abu Dhabi and there are several salt domes in four of the northern Emirates that may be associated with the existance of sulphur deposits.

## WHAT TO DO WITH SULPHUR

Sulphur is a basic material for the chemical industry, particularly for producing sulphuric acid, a primary material for large-scale industrialisation.

Even if the industrialisation in the U.A.E. calls for sulphuric acid in vast quantities, the sulphur production outmatch all foreseeable needs. Other uses of sulphur are in the production of fertilizers and insecticides.

Judging from experience in other sulphur producing countries, the amount of sulphur used in chemical, agricultural and industrial sectors is less than fifty per cent of their production. Therefore there is a need to find new uses for sulphur.

One of these new uses is to utilize sulphur as a building material.

Sulphur can be used to fabricate a strong and durable building material; using 1 part molten sulphur to 4 parts aggregates (sand and gravel) eliminating completely the use of water.

This characteristic is very important in the context of desert regions.

Sulphur concrete has many interesting advantages:

1. It is a fast setting high-strength concrete, it cures in a matter of minutes.
2. Good water impermeability, insoluble in water.
3. Adheres strongly to a wide range of materials.
4. Poor conductor of heat and electricity.
5. The techniques for mixing and casting are simple.
6. Sulphur can be stored out doors for indefinite time without any damage.
7. There is no need to use graded or washed sand and aggregates in the making of sulphur concrete since sulphur is not attacked by salts. Hence sea sand and aggregates can be readily used.
8. Melting sulphur uses very little energy, less than one per cent of that of cement, since the sulphur melting point is a few degrees above boiling water and can be endlessly recycled.

In recent years, interest has been also shown in the use of sulphur as a binding agent. Investigations have shown that, in place of cement, sulphur could be used as a binding material for joints in blockwork.

The Southwest Research Institute in San Antonio, Texas, USA, and San Carlos University in Guatemala City, Guatemala, have used sulphur for coating and joining building blocks and concrete pipes. It was also demonstrated that by impregnation with molten sulphur, a significant improvement in the strength and impermeability of concrete elements can be achieved.

The uses of sulphur in construction are just now being looked at in several parts of the world, but the arid regions rich in oil have the advantage of being able to produce sulphur in large quantities.

PROPERTIES OF SULPHUR

→ a) Sulphur is a non-matalic, chemical element. It was first a classified as an element in 1777 by Lavoisier.

→ b) Sulphur weighs 2 tons per cubic metre; its melting point is between 114 degree C and 119 degree C and its boiling point 444 degree C. Sulphur ignites at 245 degree C.

→ c) A compressive strength of sulphur concrete of 189 Kg/cm2 can be obtained by using 16 per cent by weight of sulphur. But compressive strengths up to 600 Kg/cm2 have been reached by selecting the type and size of the aggregate.

→ d) Sulphur itself is relatively brittle but the use of a plasticizer will overcome this problem.

→ e) Sulphur ignites at 245 degree C and produces sulphur dioxyde (SO 2). However, when it is burning, it does not soar like a wood or plastic fire and is very easily detected because of the particular odor of sulphur dioxyde. In order to make sulphur concrete self extinguishing, it can be mixed with fire resistant chemicals like "Dicyclopentadeine". Sulphur fires produce a low flame and are easily put out by dry chemical sprays or simply "choked" by throwing small quantities of sand on it.

→ f) Sulphur concrete can be casted to produce building components of precise dimensions allowing the manufacture of inter-locking blocks that do not required any morter or special skill to be laid.

WHAT DO WE NEED

Workshop

A workshop area of approximately 30 square meters is sufficient as working space. The area should be covered but can be open sided as to provide as much ventilation as possible. An area adjacent to the workshop should be accessible for delivery of raw materials such as sulphur, gravel, sand, pyrites, etc. as well as an area for storing finished products.

## MELTER MIXERS

### CONCRETE MIXER

The melter used for the first stage of the project was a regular one bagger concrete mixer. With a cast iron bottom under the bucket a propane gas burner was attached for heating it up while it rotated.

★★

★★

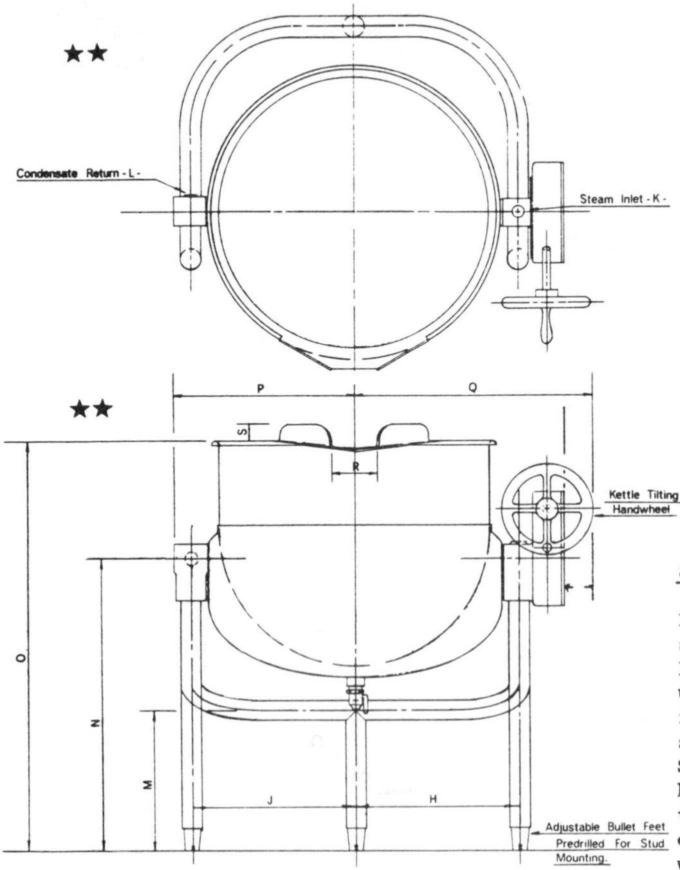


### STEAM KETTLE

In a second stage a steam kettle of 170 litre capacity was used as an alternative for melting and mixing sulphur.
Steam providing a better control by thermostat and a more efficient use of energy were the basis for this selection.

* MOULDS

* Moulds should preferably be made of metal; steel or aluminium.

They can also be made out of hard woods. Metal permits the heat to dissipate much faster, hence accelerate unmoulding time, approximately ten minutes after pouring; this time can be reduced further if the mould is dipped into cold water. Wood moulds take a longer time and should be limited to simple forms where size tolerances are not critical such as bricks.

Based on the design of an interlocking concrete block in the Philippines by Bruce Etherington. Modified steel moulds were produced in order to satisfy the moulding techniques of sulphur - concrete. They are made of steel plates approx. 6mm thick and weight about 35 kilograms each. The blocks are design following modular coordination sizes 2M x 2M x 4M. (20 cm x 20 cm x 40 cm) and 2M x 2M x 2M (20 cm x 20 cm x 20 cm). In both case there are allowed tolerances of 2 mm. Respectively the blocks weight 13 and 8 kilogrammes.

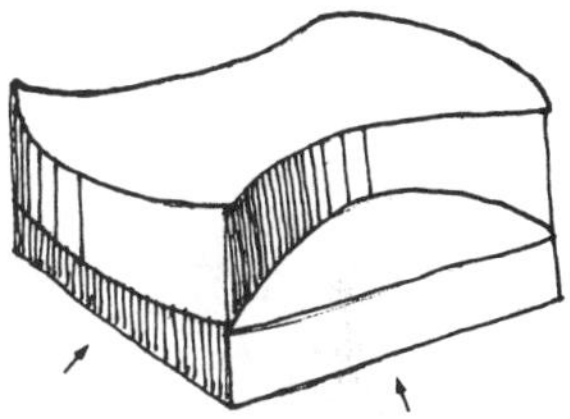

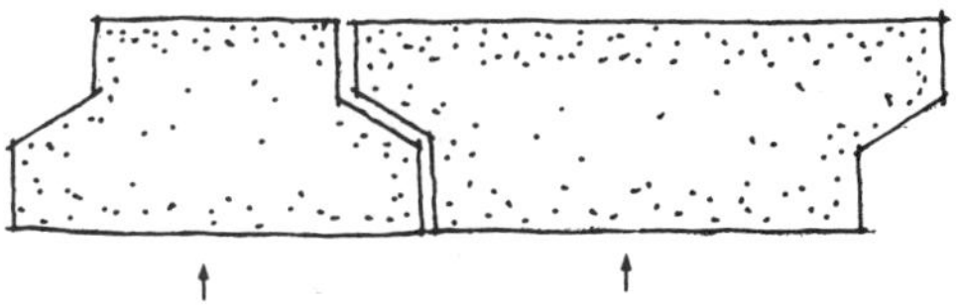

Two other steel moulds to produce a modular paving block 2M x 2M (20 cm x 20 cm) that can be used to make sidewalks, driveways and roads were manufactured. This design was developed by the United Nations "Projecto Experimental de Vivienda" (PREVI) in Lima, Peru in 1970, using a sand/cement mixture.

Manpower.

A group of four persons was found to be the minimum number in a team for the production of wall and paving blocks.

Although labor requirements will vary according to the size of any workshop, it can be broken down as follows: one person mixing and pouring, two persons assembling, greasing and disassembling moulds and one person providing raw materials.

HOW TO DO IT

Melting, Mixing and Moulding

Proportions of sulphur and aggregates must be by weight. To obtain a workable mix approx. 30% - 35% sulphur must be mixed with 70% - 65% sand.

With experience the person in charge of mixing learns to evaluate the proportion of the ingredients by "eye". It is best to start by melting the sulphur and then when liquid to add the aggregates (sand, gravel, pyrites, etc).

If the cold aggregates are added too fast in large quantities they can lower the temperature of the mix below 120 degree C and sulphur will then solidify.

Sometimes overheating is also encountered. When this happens the mix loses its fluidity and becomes viscous. The temperature is then closer to 180 degree C and should be lowered by regulating the heat source, or by adding cold aggregates.

Two disadvantages that previously retarded the use of sulphur as a building material were: sulphur is relatively brittle and can burn.

In recent years the development of aditives (unsaturated hydrocarbons) eliminated those drawbacks.

One of these is Dicyclopentadiene used at a level of approximately 2% by weight in combination with sulphur in coatings, for impregnating concrete and in sulphur concrete.

Sulphur, by itself is brittle and crystalline. To overcome this, Dicyclopentadiene at levels of 2-5% can be used.

When Dicyclopentadiene and sulphur are heated together complex reactions occur which are temperature dependent. For example, at a temperature of 125 degree C, the reaction is similar to the vulcanization of rubber. At 140 degree C, polysulfides form similar to Thiokol rubber. At temperature over 170 degree C, hydrogen sulfide forms, which is toxic. Hence temperatures this high are to be avoided.

Sulphur concrete can be produced in many colours resulting from the combination of the yellow of sulphur and the natural colour of the aggregates used. Other combination of colours can also be achieved by adding to the mix metal oxide colours.

## MOULDING

Moulds have to be greased before assembly to prevent sticking, and to make unmoulding easier. Different kinds of oil can be used. We utilised motor oil. This is done very easily with a paintbrush.

Once the moulds are assembled and the sulphur mix is ready they can be filled. Plastic beakers were used for this operation. These pitchers do not melt under the heat and sulphur residues left are easily cleaned out. Gloves have to be used and protective glasses to avoid burns from splashing of the mix.

Moulds during pouring can be vibrated by hand or by knocking on the table to ensure flow into all parts. When sulphur cools a contraction of the mix happens. It might be necessary after a mould has been filled to come back and add some more mix to make sure the moulds are completely filled. When pouring an open sided mould, like the paving block, large size aggregates can be added (25mm to 50mm). These add to the strength of the block and diminish the quantity of mix required to fill the mould. One half of the block weight can be of these stone aggregates. This method of casting reduces greatly the cooling time, the stone absorbing much of the heat of the mix. For example without stones the cooling time of the paving block is 10 minutes and with the stones is only 2 minutes. Another method of shortening cooling time for moulds without stones is to immerse them into water.

Generally if metal moulds are used, unmoulding can be done approximately 10 minutes after pouring. Care should be taken not knock the blocks since they are still fragile. A block takes about 1 hour to cure and reach full strength. They can be used right away, which is a great advantage.

Diassembled parts should be cleaned and oiled again before reassembly.

Since the curing time is short, a relatively small number of moulds are sufficient. A continous production can be achieved with 4 hollow block moulds, 2 paving block moulds. Pouring takes about 2 minutes, by the time the last mould is poured, the first is ready to be poured again.

One of the advantages of sulphur concrete is that it can be recycled. Broken blocks can be remelted again to form new ones.

## HOW TO BUILD

To demonstrate the uses of sulphur-concrete building components an addition was built to the existing office building of the Department of Studies and Research on the experimental grounds of the Ministry of Public Works and Housing.

First a strip footing foundation was laid in reinforced concrete. This was the only stage of the construction that needed a measuring tape. The first layer of the sulphur blocks was laid on cement mortar to make sure that they were levelled and leaving door openings.

Once the mortar was set, the laying of 450 interlocking and mortarless blocks took 15 man-hours of unskilled workers.

The resulting walls, with good thermal insulation properties, need no mortar, plastering or painting, they are waterproof and do not required any future maintenance. The finish is marble-like and can be easily washed.

## FINAL CONCLUSION

THE USE OF SULPHUR AS A BUILDING MATERIAL IS A VERY PROMISING TECHNIQUE TO BE USED IN DESERT AND ARID REGIONS.

A finished house in the Philippines.

Dubai - April 1977

# QUALITY PROBLEMS OF BUILDING MATERIALS IN THE ARAB PENINSULA.

A. Ortega
Architect
U.N. Adviser to the U.A.E. Government

In Desert climates such as that of the Arab Peninsula, building problems are more difficult to solve than in temperate zones. In such regions, the choice of aggregates may be resticted or be in short supply or of low quality and their contamination with sulphates, chlorides or saline water is common.

The climatic conditions in the Arab Peninsula need to be taken carefully into account for the production of concrete. The summer climate consists of great heat combined with high humidity. From May to October, temperature ranges between 35 degree C and 50 degree C during the day. The humidity has an average maximum of over 88 per cent. The daily average hours of sunshine vary between 9 hours and 12 hours according to the season.

Aggregates need to be chemically stable so that they will neither react chemically with cement, nor be affected by outside influences, such as the penetration of humidity in the concrete.

Also high porosity weakens the durability of concrete exposed to moist air, especially in proximity of the sea, as it reduces the effectiveness of reinforcement covering and facilitiates steel corrosion. When chloride contents are high, it is not possible to leave them in contact with steel.

Sulphates in concrete are undesirable, as they can develop dangerous salts that may contribute to fast deterioration of concrete products.

In the UAE and some other areas of the Arabian Gulf, both sands and aggregates often contain a high percentage of gypsum, and this can have a serious effect on the quality of concrete in which they have been used.

Aggregates constitute about 75% of the total volume of concrete, and thus their influence is extremely important. Aggregates should not contain injurious amounts of clay, silt, mica, organic matter, and chemical salts. The chemical salts produce bad effects, caused by their reaction with the cement, or by preventing hydration.

The factors influencing the selection of both fine and coarse aggregates are generally their physical properties, grading, the distance that have to be hauled, and the degree of quality variability within the quarry. Samples for chloride and sulphate analysis should be taken from the various types of stone distinguishable on a visual basis, as well as on a representative sample of the crushed aggregate.

The source of aggregate employed for concreting, should be known to the user, and the supplier should be able to produce up-to-date data on the sulphate and chloride contents. A change in grading of sand supplies may indicate a change in source and this should prompt a check on the salts content. Some supervision of the winning of sand used for aggregate should be made to avoid the use of deposits in which salts have accumulated through evaporation of salt water.

Where a new source of rock is used for concrete production, it should be ascertained that it is compatible with Portland Cement, and will not give rise to an alkali-aggregate reaction. Concrete suffering from this interaction between aggregate and cement is very liable to crack through the expansive formation of an alkali silicate gel, which may be seen as a white exudation at the surface.

Marine aggregates in the Arab Peninsula contain large amounts of hollow and flat shell fragments. Crushed shells, like any other flake material, if present in sufficient quantity, may adversely affect the work ability of the concrete. Both hollow and flat shells in the coarse aggregate can materially reduce the strength of the concrete in which it is used. Natural sands available in the desert areas may have salt contents higher than in sands dredged from the sea. Where significant evaporation of sea water can take place, for example on beaches above high water level, the saline content of the sand can be higher than in sand dredged from the sea.

Harmful effects of impurities in the raw materials will be then kept to a minimum. In considering the effects of soluble salts on concrete, it should first be established whether salts will be introduced into the concrete, with the raw materials, or whether the available aggregates and mixing water are salt-free, and secondly whether or not the concrete so made will be placed in a salt-free or salt-laden environment. Of these four combinations, most reported investigations have been concerned with, either salt-free concrete placed in a sulphate-laden environment, or concrete containing chloride placed in a salt-free environment. In the Arab Peninsula, the possibility exists of having both problems: chlorides and sulphates as contaminants of the aggregates and mixing water and also for the environment to contain these salts both below and above ground level. The increased risk of sulphate attack of concrete is due to the presence of sulphate in the raw materials, and the increased risk of corrosion of reinforcement is due to the presence of chloride, either in the raw materials or in the atmosphere.

Loss of strength and chemical resistance of the set concrete can also occur at high temperature and in the presence of moisture. This is due to conversion of the two principal calcium aluminate hydrates to alternative forms resulting in an increase in porosity.

Sulphate content is usually expressed as percentage of $SO_3$ (sulphur trioxide). The most common sulphates to be found as impurities in aggregates and waters used for mixing concrete are those of calcium, sodium and magnesium, and to a much lesser extent potassium.

Difference between these sulphates lies in their relative solubilities. The maximum solubility of calcium sulphate or gypsum yields a solution with only 0.12 per cent $SO_3$, whereas the other sulphates will go completely into solution. Sulphate in the presence of water can have an adverse effect on set Portland cement, and the risk will be greater with the more soluble sulphates.

If a ground water contains more than 0.12 per cent $SO_3$ there is clear evidence that one of the more soluble sulphates is present. When the percentage of $SO_3$ in a water extract is known, we get some idea of whether the sulphate is present as gypsum or one of the more soluble salts.

Under tests on concrete immersed in concentrated solutions of magnesium and sodium sulphate, the disintegration of concrete can be brought about relatively quickly. The expansion of concrete slabs which rest on damp sulphate-containing fill probably provide the most rapid practical demonstration of the effects of sulphate attack.

The attack on concrete by acid sulphate is sometimes encountered in certain places due to bacteriological activity. Results in the gradual erosion of the concrete are found in concrete sewers, particularly in severe climates like the one of the Arab Peninsula and can be distinguished from the fact that it takes place well above the normal flow level in pipes and on the walls of manholes.

The principal troubles arising from the presence of chloride in set concrete are the increased of corrosion of reinforcement and the migration of chloride to the surface of the concrete, spoiling the wall of ceiling surfaces. An increase in the chloride contents occur in confined seas in tropical areas. Coastal ground water supplies are often contaminated with chloride by seepage of sea-water, but brackish waters also occur well inland due to the presence of soluble salts in the surrounding rocks and sands. Use of the latter material as aggregates for concrete provides another source of chloride to be avoided.

Corrosion is usually due to a failure of the concrete to maintain its normally high alkali environment around the steel, and rusting and serious spalling and disruption of the concrete result. Corrosion of reinforcement in concrete containing no soluble salts can also occur if the concrete cracks, or the depth of cover of the reinforcement is inadequate.

There is evidence that the rate of corrosion of reinforcement increases with temperatures over 30° C and also that it reaches a maximum at between 70 and 80 percent relative humidity. At this humidity level, both moisture and oxygen can gain access to the steel through any porosity in the concrete. At lower humidities there will be less moisture available and at higher humidities, admission of oxygen will be diminished. In the Arab Peninsula there are about 240 days per year when temperature exceeds 30° C.

The safe level of chloride inclusion in concrete varies with the type of concrete and the environment in which it is placed. An added hazard is that all along the Arabic coast, salt-laden dew can be deposited during the night, the salinity being derived from sea spray and from minute dust particles carried by the wind. This is particularly important in relation to the storage or reinforcement prior to use, and every measure should be taken to ensure that loose salt and rust particles are removed from the steel before placing the concrete. Storage of steel directly on desert sand can also be a source of rapid rusting.

Sulphates can completely destroy the binding power of cement and could be a danger to the safety of any concrete structure. The damage caused by a sulphate attack may only be apparent after several years.

The introduction of the more soluble salts such as the chlorides or sulphates of sodium, potassium and magnesium into concrete may result in the production of an unsightly white efflorescence on the surface of the concrete as the mixing water dries out, and in some conditions it may lead to repeated crystallisation so that surface erosion can occur. When the surface of the concrete is painted or plas-

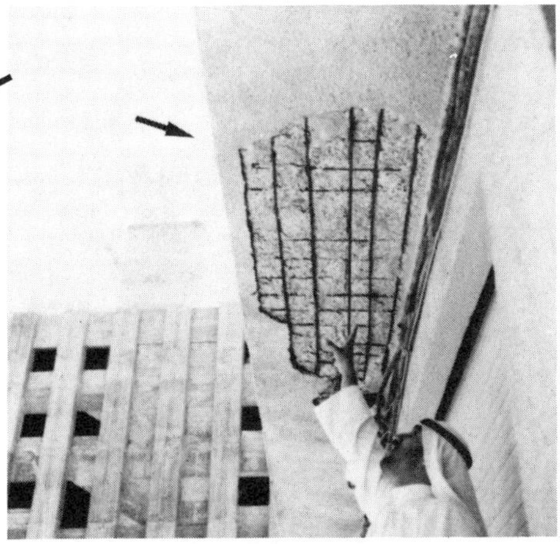

The photographs show corrosion problem of the steel reinforcement due to chloride attack. This always happens when the sand used is contaminated with marine salts, principally sulphates and chlorides. The higher the chloride content the more quickly rusting occurs as well as with the increase in temperature and humidity.

Chlorides are soluble and can be removed from the aggregates by washing, however large quantities of drinking water is required.

The amount of chloride which can be tolerated in a concrete mix is not easily determined. It needs to consider the conditions of each site.

THERE IS NO WAY TO PREVENT CONCRETE FROM DETERIORATING WHEN THE AGGREGATES CONTAIN SULPHATES AND CHLORIDES.

The photographs show the disintegration of a structure in Dubai, built six years ago. In many cases when the quantity of chloride in a concrete mix exceeds 10,000 ppm reinforcing bars are found to have been reduced to powder within a period of 6 months.

Another problem with the use of sea-sand in cement is the unsightly efflorescence that appears on the surface destroying the adherence of the paint film. This happens particularly in periods of high humidity. Concrete blocks manufactured with sea-sand will have this problem.

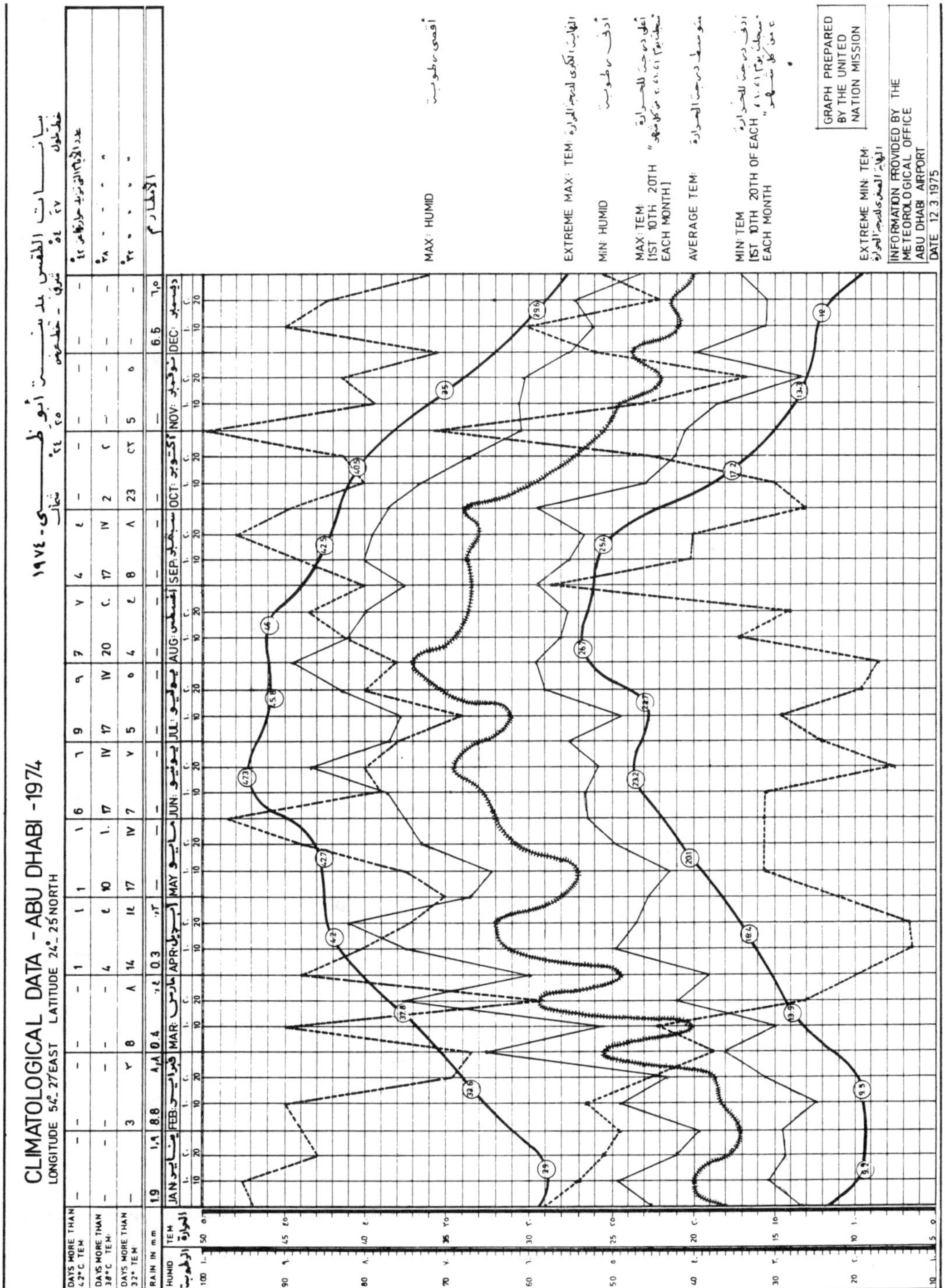
CLIMATOLOGICAL DATA - ABU DHABI -1974
LONGITUDE 54° 27' EAST LATITUDE 24° 25' NORTH
DAYS MORE THAN 42°C TEM
DAYS MORE THAN 38°C TEM
DAYS MORE THAN 32° TEM
RAIN IN mm
HUMID
TEM
JAN
FEB
MAR
APR
MAY
JUN
JUL
AUG
SEP
OCT
NOV
DEC
MAX: HUMID
EXTREME MAX: TEM
MIN: HUMID
MAX: TEM [1ST 10TH 20TH EACH MONTH]
AVERAGE TEM
MIN: TEM [1ST 10TH 20TH OF EACH MONTH
EXTREME MIN: TEM
GRAPH PREPARED BY THE UNITED NATION MISSION
INFORMATION PROVIDED BY THE METEOROLOGICAL OFFICE ABU DHABI AIRPORT
DATE 12.3.1975

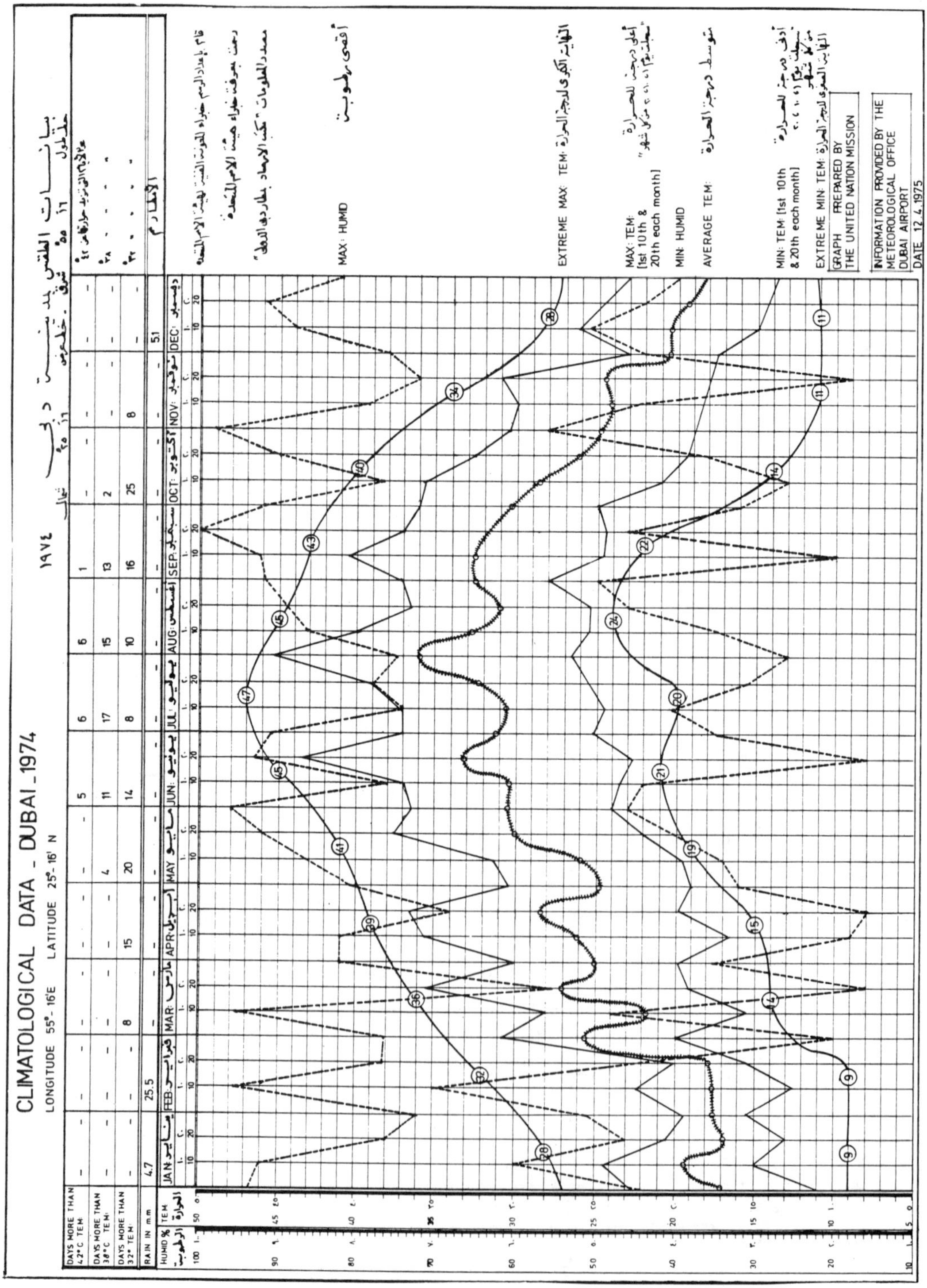
CLIMATOLOGICAL DATA _ DUBAI _ 1974
LONGITUDE 55°- 16'E LATITUDE 25°- 16' N
MAX: HUMID
EXTREME MAX: TEM:
MAX: TEM: [1st 10th & 20th each month]
MIN: HUMID
AVERAGE TEM:
MIN: TEM: [1st 10th & 20th each month]
EXTREME MIN: TEM:
GRAPH PREPARED BY THE UNITED NATION MISSION
INFORMATION PROVIDED BY THE METEOROLOGICAL OFFICE DUBAI AIRPORT
DATE 12.4.1975

tered before the migration of salt to the surface has finished, these surface coatings may easily become dislodged. There is little that can be done about this without redecorating. It is recommended that the salt or salt-laden surface finish should be removed without the application of water otherwise this would tend to wash salt into a porous backing. If the walling material is dry and will remain always dry, further migration of salts from the wall may not occur after redecoration.

Sand is found naturally in immense quantities in the Arab Peninsula. However, to find suitable sand for concrete is not an easy task. There is not any available survey yet about suitable sand quarries in the region.

Sand is fine granular aggregate resulting from the natural disintegration of rock or from crushing of rocks. The sand usually consists of hard quartz grains.

Aggregates fall into two main categories^ those derived from natural rock, weighing more than 1,700 kg/m3 and porous ones with a weight less than 900 kg/m3 for use in light-weight concretes.

In the Arab Peninsula, there is a large amount of natural rock in the mountain ranges and adjacent gravel plains that could be used for the production of concrete, but in many locations the rocks have a mineral composition that contains serpentinite, vermiculite, chlorite^ materials that when used as aggregate in concrete, expand excessively. This swelling facilitates moisture penetration reducing the effectiveness of the reinforcement covering and produces steel corrosion.

## CONCLUSIONS

There is an urgent need in desert climates to improve the quality of concrete work which requires¨

1. To stop the use of high contaminated sand for manufacturing concrete. ½Contaminated sand corrodes reinforcing steel and disintegrates the concrete.¼
2. To limit the admissable concentration of sulphates and chlorides in the concrete aggregates to maximum content of 3000 parts per million.
3. To create laboratory facilities for quality control of building materials particularly to carry out acceptability test on local materials for sulphates and chlorides.
4. To carry out surveys of suitable sand and aggregate quarries and to provide information about the properties for acceptable sand. ½There is no way to promote the use of proper sand without this information.¼
5. To improve cement handling, protecting it from atmospheric moisture and to reduce the time of storage to a maximum of 6 months. ½There is a great loss of strength when cement is stored for long periods.¼
6. To design a concrete mixture determining the correct amounts and sizes of each of the ingredients including the proper water-cement ratio. ½Incorrect proportion of ingredients produces a weak and porous concrete and a high water-cement ratio produces cracks in the concrete and excessive shrinkage.¼

7. To pay attention to the climatic conditions in the region, particularly to the high temperatures and intense sun radiations, when the placing and curing operations take place. (Fresh concrete exposed to direct sunlight during summer months may lose up to 60% of its strength when it is not properly cured.)

# ASPECTS OF THE DESIGN AND CONSTRUCTION PROCEDURES FOR LOW-COST ROADS IN ARID AREAS

M. Larcher

The increasing need for a better road system into outlying arid areas of the world, and especially of developing countries, poses many problems for the responsible planner and the highway engineer.

These problems are basically associated with economic and social considerations, since it is one of the objectives of any highway authority to provide a satisfactory network at a cost consistent with the traffic needs. This objective is more difficult to achieve in desertic areas, due partly to environment and partly to remoteness.

High costs and difficulties due to remoteness are inevitable, but economy can generally be achieved by the selection of a standard design and construction procedure differing in some aspects from that which might be acceptable and used elsewhere.

For example, the specifications of geometric standards, materials, methods of working, and testing procedures which would be perfectly appropriate in Europe and North America, have to be revised and adapted when applied to desertic areas. The systems and procedures used in road construction industries of developed countries are efficient within their own environment but have also to be revised prior to being considered for application to road construction in arid areas.

The use of locally occurring materials, the quality of which is outside normally accepted limits, and the practice of providing compaction at natural moisture contents much below optimum where water is scarce and costly to obtain, are just two examples where significant economy may be possibly achieved, both in terms of capital and energy expenses.

In recent years, a great deal of attention has been given to the idea of developing so-called "intermediate" techniques to reduce the construction and maintenance costs of roads in desertic areas by intensive use of local material and resources, dry compaction of the soils to reduce the water and energy consumption, and by an adequate choice of design and stage construction and maintenance policies.

The roads under consideration are mainly but not necessarily limited to those with low level traffic projections, say less than approximately 20-30 veh./day, half with axle loads approaching the legal limits. These roads cross wide arid areas where ground water resources are poor and scarce, and which are characterized by long travel distances which must be covered at high speeds and therefore require reasonably good surfacing conditions.

For the type of roads under consideration, the empirical "trial and observation approach" can reasonably be used to deal with the problem of optimum design selection of stage construction policy and adequate construction and maintenance technology.

It is the purpose of the annexed paper to present very briefly some results of a recent experience gained by the U.N. services dealing with technical studies for roads in arid areas, which could be brought to the attention of planners and engineers who are confronted with similar problems.

The recent improvements observed in the equipment performances, and especially in the field of vibratory compaction, demonstrate the possibility of achieving required soils' densities at a low operating cost, even in very thin surface layers and at a low natural moisture content, say approximately between 1 and 2% of the dry specific weight of the soil. The compaction of road-making materials can be effectively carried out using moisture content well below optimum, if the choice of the specifications of the vibratory compaction equipment is correct. Among the features of vibratory compactors, which make them able to meet the increasing demand for economic operation, and to meet technical requirements of high densities with dry material, we shall mention:

- Hydrostatic drive to ensure smooth, jerk-free reversing and an infinitely variable speed control;
- Hydrostatic vibrator drive with infinitely variable frequency and centrifugal force to match the various soil materials and site conditions;
- Large roll diameter to reduce the horizontal material displacement and shearing strength;
- Vibrator system generating vertically directed compaction forces only;
- Vibrator system providing infinite control of the impact force without reducing or adjusting the operating frequency.

However, with some unfavourable material characteristics, resulting from compaction at low moisture content well below optimum, dry densities less than the maximum, modified Proctor density will be achieved. Even in this case, the effect of these lower densities on soil strength is often offset by the lower moisture content, and acceptable Californian Bearing ratio are achieved.

Where possible, the most effective and cheapest method of providing compaction is by stage construction. The benefits achieved by this procedure are in the compaction provided by the traffic, even with low-level traffic, the ability to correct shape and local failures prior to the final stage of sealing, and if the compaction has been provided at any moisture content other than equilibrium, the time span between stages provides for the moisture content to approach an equilibrium value.

While the compaction aspect of the soils in arid areas could be considered in itself as the major issue of the annexed paper, this one presents a number of other concepts showing that adequate design in association with stage construction and relevant technology make possible, when applied to roads in desertic areas, significant savings without sacrificing the surface conditions improvement compared with the present conditions of most of the existing track and earth roads in the same areas.

One of the main constraints stems from the imperative necessity, at the initial stage of construction, to have an equipment maintenance unit standing by to keep the surface reasonable, repair local failures, and take advantage of favourable conditions when they occur (rainy season).

The solution of a single-lane paved surface has been considered as a stage construction method. However, the increasing concentration of the loading repetition requires an increase in the pavement thickness, which means, when compared with the full width paving plus maintenance costs, that there is not necessarily an overall saving with this type of solution. In addition, it is more dangerous in terms of traffic safety.

It is felt that the annexed paper could be of interest to engineers concerned with the construction and maintenance of less heavily used roads in desertic areas.

Although described as being of an interim nature, several of the observations and suggestions appear to be relevant to present practice.

## Améliorations de l'état de surface des liaisons par route et pistes en milieu désertique<br>Diminution des coûts d'études de construction et d'entretien<br>Réduction des consommations en eaux souterraines et en énergie<br>Rôle décisif des Techniciens et des populations locales dans la conception et la réalisation des aménagements routiers

### PRÉAMBULE

Les études techniques de projets routiers à faibles projections de trafic en régions arides comportent, outre les rubriques spécifiques habituelles à ce type d'opération, la recherche des conditions optimales d'emploi des matériaux locaux et des faibles quantités d'eaux souterraines disponibles en vue de la construction d'une route économique.

A la lumière de plusieurs expériences récentes, nous tenterons ci-après de dégager quelques idées susceptibles d'influencer l'orientation future des études et des modalités d'exécution de travaux d'aménagement routier en pays désertique. Ceux-ci intéressent en premier lieu le cas général d'itinéraires à faibles perspectives de trafic pour lesquels on peut accepter des caractéristiques routières réduites afin de faciliter le désenclavement de populations disséminées sur des territoires immenses, mais aussi l'aménagement de routes plus circulées.

Au cours de notre exposé, nous identifierons en particulier les trois groupes de facteurs suivants:

- Economies et efficacité des études
- Economies apportées par l'adaptation adéquate des normes géométriques et des structures de chaussées à l'environnement des conditions désertiques.
- Economies en consommation d'eau et d'énergie dans le compactage à densité élevée sans eau des couches de sols de remblais et de chaussées.

Ces économies peuvent être amplifiées par une participation locale accrue à la conception et à la réalisation des aménagements.

## ASPECT ÉCONOMIQUE ET SOCIAL DES TRAVAUX D'AMÉNAGEMENT ROUTIER EN RÉGION DÉSERTIQUE

L'importance de l'amélioration de la qualité des communications routières dans la réduction des disparités régionales est bien connue et justifie que les décisions relatives aux aménagements de réseaux de desserte des régions désertiques soient prises même si le taux de rentabilité de l'investissement est marginal, compte tenu de l'ensemble des effets attendus positifs ou négatifs, économiques, sociaux et autres, quantifiables et non quantifiables.

## ASPECTS PARTICULIERS DE LA TECHNOLOGIE ROUTIERE EN PAYS DÉSERTIQUES

Nous aborderons successivement les rubriques suivantes:

1. Trafics
2. Tracé et photointerprétation
3. Le problème de l'eau et les solutions possibles
4. Tracé et terrassements en zones ensablées
5. Etudes géotechniques et matériaux locaux
6. Essais de déflexion
7. Compactage vibrant au laboratoire
8. Le choix du matériel de compactage
9. Abaissement des teneurs en eau avant compactage des sols
10. Chaussée en terre battue sous climat désertique
11. Choix et modalité d'emploi des sols en chaussée non revêtue
12. Teneur en eau et sols plastiques sous climat désertique
13. Lutte contre l'apparition de la tôle ondulée
14. Structures de chaussées revêtues (Trafics supérieurs à 30 V/j)
15. Profils en travers des chaussées revêtues
16. Le choix du niveau minimal d'aménagement routier
17. Travail de nuit et réduction de consommation d'eau
18. Passage des vallées d'oued
19. Conclusion

## 1. TRAFICS

Le présent mémoire se réfère essentiellement au cas économiquement défavorable des routes à très faible projections de trafic; moins de 20 véhicules par jour, même parfoins moins de 10 véhicules par jour, en supposant que la moitié de ce trafic est composée de poids lourds, avec 13 tonnes de charge par essieu, les pneumatiques étant gonflés à la pression de 7 kms/cm2.

Ce cas est difficile à solutionner car les calcules économiques classiques aboutissent à des taux de rentabilité très bas pour les investissements que nécessite la construction d'une chaussée revêtue, niveau d'aménagement considéré souvent minimal pour obtenir une amélioration durable et significative de l'état de surface de la chaussée, en même temps que la réduction des sujétions d'entretien courant. Le problème est d'autant plus aigu que les distances à aménager sont importantes et l'éloignement des sources d'approvisionnement considérable. Pourtant, l'amélioration de la qualité des transports de surface est bien l'une des conditions à satisfaire dans le but de désenclaver ces régions déshéritées, abaisser les coûts de transport, multiplier les échanges et amorcer le processus de participation des populations locales à l'amélioration des conditions de vie que connaissent leurs homologues des régions plus favorisées.

Nous évoquerons ci-après certains aspects de la technologie routière apparue au cours d'études techniques routières récentes en pays arides et dont l'exploitation peut contribuer, dans des conditions d'environnement analogues, à réduire de façon sensible le coût des études et des travaux conduisant à l'aménagement d'une route au tracé moderne et confortable, capable de supporter un trafic rapide d'une dizaine de véhicules lourds par jour, nécessitant un entretien faible pour conserver ces qualités pendant plusieurs années. On suppose en outre que si le trafic était affecté d'un coefficient multiplicateur égal ou supérieur à 3, il deviendrait économiquement possible et techniquement nécessaire de procéder à une étape de renforcement de la chaussée, celle ci ne nécessitant qu'un investissement additionnel plus léger puisque la géométrie et la structure de chaussée existante seraient alors en majeure partie incorporés et valorisés par le renforcement en cause.

## 2. LES TRACÉS - ETUDE PAR PHOTOINTERPRÉTATION

La plupart des régions désertiques ont été couvertes par photographie aérienne à diverses échelles: satellites à grande échelle (1/500 000) et aussi le plus souvent à des échelles variant du 1/40 000 au 1/80 000.

Une fois fixés les pôles d'aboutissement des itinéraires à aménager, l'une des premières mesures préconisées consiste à se procurer les photographies aériennes existantes le long d'un couloir couvrant les itinéraires en cause.

On réalisera une mosaïque à échelle si possible contrôlée sur points géographiques suivie d'une interprétation photogéologique soignée qui sera poussée simultanément dans les quatre directions suivantes: Hydrogéologie, Hydrologie de surface, identification des sols superficiels et ressources en matériaux de construction, reliefs significatifs pour un tracé routier.

Les travaux de photointerprétation au bureau seront complétés par une mission de reconnaissance au sol multidisciplinaire d'assez courte durée à laquelle participeront les Techniciens responsables de l'étude dans chacune des disciplines suivantes: Tracé, Hydrogéologie et Hydrologie de surface, Géotechnique routière. Des échantillons de sols seront prélévés au cours de la mission puis analysés au laboratoire afin d'identifier les séries reconnues par photointerprétation.

A l'issue de cette mission un rapport choix de tracé sera préparé avec dessin du ou des tracés étudiés sur photomosaïques.

L'importance de cette phase de l'étude est capitale car elle conditionne étroitement et de façon généralement définitive les autres phases de l'aménagement et influence l'économie du projet tout entier.

Une fois le tracé choisi par le maître d'oeuvre, il pourra être parfois avantageux de compléter les couvertures photographiques existantes par une nouvelle couverture au 1/20 000 sans préparation au sol le long de l'axe retenu dans le but de produire une photomosaïque sur laquelle sera reporté de façon plus précise le tracé proposé, après nouvelle étude de photointerprétation géotechnique détaillée.

Dans les zones topographiques difficiles, on pourra utiliser économiquement la couverture au 1/20000 afin de produire un photoplan équipé de lignes de niveau à l'échelle du 1/5000. Là encore, on pourra faire l'économie de la stéréopréparation au sol dans la mesure où l'Ingénieur routier s'intéresse davantage aux différences de niveau sur une section déterminée qu'aux cotes altimétriques exactes.

Muni des documents qui précèdent, l'Ingénieur Topographe local expérimenté en tracés routiers sera en mesure de procéder à une implantation directe de l'axe du projet avec lever altimétrique au sol du profil en long et si nécessaire, des profils en travers du terrain naturel.

Dans le cadre de la procédure que nous venons de décrire, il sera généralement nécessaire pour les pays en voie de développement de faire appel à des spécialistes étrangers pour effectuer les études préalables à l'implantation. Cependant, la participation étroite des Techniciens locaux à la totalité des étapes de l'étude apparait indispensable dans la perspective d'une relève progressive du concours étranger au fur et à mesure de la répétition de travaux semblables dans la même région, d'autant plus qu'une grande similitude géologique superficielle règne souvent sur de vastes étendues désertiques.

Dès l'étape implantation directe de l'axe sur le terrain la relève pourra être effectuée à la condition que les Techniciens locaux responsables aient été pleinement associés aux phases précédentes.

De façon générale, le modelé souple et la nudité des paysages désertiques facilitent la mise en place d'un tracé routier économique et confortable à vitesse de base élevée (de l'ordre de 100 KH) conforme aux normes internationales et ne nécessitant que des terrassements insignifiants. La ligne rouge sera en général fixée à environ 15 cms minimum au-dessus du sol naturel pour faciliter l'assainissement transversal et éviter les dépôts de sables éoliens.

## 3. LE PROBLÈME DE L'EAU ET LES SOLUTIONS POSSIBLES

a) Hydrologie de surface - Les régions désertiques sont marquées par une pluviométrie très faible et irrégulière tombant par averses de forte intensité, des lits d'oued généralement secs sauf au moment des pluies où ils deviennent le siège d'écoulements de courte durée à régime torrentiel. Au cours de l'étude de photointerprétation, le contour des bassins versants sera dessiné sur photomosaïque au 1/50000, et les traces d'érosion soigneusement identifiées.

b) Hydrogéologie - Les seules ressources en eau permanentes sont constituées par les nappes d'eau souterraine quand elles existent. Dans l'affirmative, le potentiel de débit de ces nappes peut varier de quelques litres minute à plusieurs centaines de m3 jour dans les cas exceptionnellement favorables.

Le problème de l'eau est fondamental en zone désertique et tout projet d'aménagement routier justifie la mise en oeuvre d'une étude systématique des ressources en eaux souterraines existantes à proximité de l'itinéraire si celle-ci n'a pas déjà été faite.

La phase photointerprétation géologique des vues aériennes sera utilement suivie d'une campagne de relevés géophysiques de resistivité et sondages électriques associés à la réfraction sismique, puis d'une campagne de forages et d'essais de pompage. Si les nappes d'eaux souterraines sont importantes et renouvelables, le problème de l'eau se ramènera à une étude de coût, et dans la fourchette des contraintes économiques, l'eau pourra être employée dans ce cas pour les travaux d'aménagement routier suivant les technologies traditionnelles susceptibles de consommer 500 m3 d'eau par km environ, soit 500 m3 par jour si l'avancement du chantier routier atteint 1 km/jour.

Bien souvent en zones désertiques, les nappes d'eau souterraines ne sont pas renouvelables ou offrent un potentiel de débit très inférieur aux besoins d'un chantier de travaux routiers. Ou bien encore les coûts de forages de puits profonds et de transport de l'eau sur de grandes distances (parfois supérieures à 100 km) grèvent lourdement le coût des travaux routiers.

Enfin et surtout, en pays aride plus qu'ailleurs, l'eau constitue un capital précieux et limités, indispensable à la vie, étroitement lié à toute perspective de développement: Ceci s'applique même au cas des eaux chargées en sols et impropres à la vie animale et parfois végétale. La salinité élevée de certaines nappes d'eaus souterraines ne constitue nullement un inconvénient pour compactage des sols à l'eau. Cependant, la réalisation de stations de dessalement de ces eaux n'est plus une utopie, et l'on peut affirmer que quelle que soit sa qualité, de façon générale, une politique d'utilisation prioritaire des eaux souterraines est nécessaire en région désertique plus qu'ailleurs. S'il existe une alternative économique, ce serait certainement une erreur de continuer dans ces régions à gaspiller l'eau pour des travaux de génie civil qui, comme nous le verrons peuvent si non s'en passer complètement, tout au moins en consommer beaucoup moins.

## 4. TRACÉ ET TERRASSEMENT GÉNÉRAUX EN ZONES RECOUVERTES DE SABLES ÉOLIENS

En règle générale, on cherchera à éviter les déblais car le sable est présent partout dans la plupart des régions désertiques et les passages routiers en déblais constituent autant de pièges à sable.

Le tracé suit donc autant que possible le terrain naturel en léger remblai en évitant les "sifs" de sables vifs par des tracés parrallèles à leur direction générale.

Si le franchissement d'un sif ou d'une dune est inévitable, le franchissement se fait à la perpendiculaire pour limiter la section vulnérable en se plaçant, le cas échéant, en léger déblai sur la crête avec talus transversal à 10/1.

Dans tous les passages considérés comme vulnérables au risque d'ensablement, le déblaiement du sable au bulldozer de part et d'autre du tracé peut constituer un investissement rentable. La réduction de vitesse au passage de ces zones constitue néanmoins une sage précaution pour améliorer la sécurité, car quelles que soient les dispositions prises, en période de vents de sable, des dépôts peuvent se produire sur la chaussée et causer des accidents. Chacune de ces zones fera donc l'objet d'une signalisation de limitation de vitesse en attirant l'attention des usagers sur la présence d'eventuels ensablements.

Une solution efficace de protection contre le sable est parfois constituée par la stabilisation latérale des sables superficiels à l'intérieur d'un couloir de largeur suffisante par aspersion de produits de fixation dérivés de la distillation des pétroles bruts. Cette fixation est assez onéreuse et les faibles trafics que connaissent de nombreuses routes et pistes en zones désertiques ne justifient pas toujours un tel investissement.

## 5. LES RECHERCHES DES MATÉRIAUX ET LES ÉTUDES GÉOTECHNIQUES AU LABORATOIRE

L'étude de photointerprétation mentionnée au paragraphe 1 ci-dessus aura fourni une carte à échelle du 1/50.000, soit mieux encore au 1/20.000 sur laquelle auront été identifiés, sur une largeur variable, les contours de formations géologiques superficielles.

A l'aide de cette carte, les recherches au sol de matériaux se trouvent grandement facilitées: On peut souvent se borner à identifier les propriétés géotechniques des sols type repérés à partir de la carte photogéologique et à conduire sur ces sols type une série d'essais très poussés au laboratoire. Ces essais comporteront l'exploration de leurs possibilités d'emploi en sols de fondations et en couches de chaussée sous climat désertique.

Dans les conditions désertiques, il n'est pas rare que l'identification d'une demi-douzaine de sols type par section de 200 ou 300 kms se révéle suffisante, et que l'extrapolation d'une bonne partie des résultats obtenus sur ces sols type sur les sols des sections de routes voisines soit possible, diminuant d'autant l'effort d'identification et de recherche à conduire sur ces sections voisines.

En dehors des essais classiques d'identification géotechnique (granulométrie, Limites d'Atterberg, essais Proctor-CBR, analyses chimiques), il sera utile d'étudier en laboratoire la réponse de ces sols aux essais de compactage vibrant aux mêmes points de teneur en eau que ceux utilisés dans le dessin des courbes densités Proctor-Résistance au poinçonnement CBR, en notant avec soin les teneurs en eau naturelles des échantillons au moment de leur prélévement puis après exposition à l'air libre et au soleil.

## 6. LES ESSAIS DE DÉFLEXION

Pour mesurer la portance des sols en place, par rapport à l'essai CBR qu'il complète, l'essai de déflexion présente plusieurs avantages: Il est plus directement lié aux efforts de flexion auxquels la chaussée sera soumise au passage des charges roulantes, il est facile à relever à la poutre BENKEIMANN, sans parler des appareils à grand rendement du type déflectographe LACROIX et courvimètre CEBTP. Ces instruments présentent l'avantage de fournir un relevé de mesures de déflexions et de rayons de courbure des déformations en "continu", mais il s'agit d'outils relativement délicats qui ne peuvent opérer que sur des surfaces propres et convenablement réglées, essentiellement sur chaussées revêtues.

## 7. LES ESSAIS DE COMPACTAGE VIBRANT SANS EAU AU LABORATOIRE

A la suite des expériences de compactage vibrant effectuées dans certains laboratoires géotechniques aux E.U. A. (table vibrante du Bureau des réclamations), en Suède (M. Lars Forssblad, Dynapac) et en USSR (Prof. Barkan, Institut des Transports, Moscou), en Australie et en Inde, il a été tenté d'amorcer au laboratoire l'étude de la réponse de plusieurs sols Sahariens au compactage vibrant.

A cet effet le moule à hausse renforcé Dynapac a été utilisé avec une dame vibrante constituée de divers instruments: La dama Dynapac TE 10, la dame Cleveland SCR 37, le marteau Skill-Roto Hammer, et enfin le marteau automoteur à percussion Cobra.

On a constaté avec les deux premiers dames, Dynapac TE 10 et Cleveland SCR 37, guidées l'une et l'autre par la hausse du moule, que les vibrations transversales parasites transmises au moule par la hausse nuisaient à l'obtention de densités élevées.

Au contraire, les densités obtenues avec la dame Skill Roto Hammer guidée manuellement ont été en moyenne meilleures. Les plus fortes densités ont été obtenues avec le marteau à explosion Cobra animé d'un mouvement linéaire de va et vient et guidé également manuellement: Quoique 30% plus légère, et avec une énergie de compactage inférieure de 50%, la dame Cobra conduit à des densités 10% supérieures en moyenne à celles obtenues avec les dames à vibrations circulaires SCR 37 et TE 10.

On a pu constater qu'il était possible avec toutes ces dames, mais bien plus facilement avec la dame à vibrations verticales COBRA, d'obtenir de bonnes densités en couches minces (3 cms après compactage) et même de dépasser jusqu'à plus de 5% la densité OPM, avec des teneurs en eau de l'ordre de 1%.

M. Lars Forssblad (Dynapac) a montré qu'il obtenait avec la dame TE 10 de bonnes corrélations de densités avec le compactage Proctor et meilleures encore avec le compactage vibrant sur chantier. Ces résultats favorables ont été possibles dans la zone des teneurs en eau voisines de l'OPM: La cohésion due à l'eau rend le sol moins sensible au décompactage causé par les vibrations transversales. Dans le cas du compactage à très basse teneur en eau (autour de 1%), la mobilité relative accrue des grains de sols facilite la désorganisation du compactage superficiel sous l'effet des percussions latérales (effet comparable à celui observé de part et d'autre de la génératrice de la bille vibrante d'un compacteur vibrant omnidirectionnel), spécialement si le réglage de la force centrifuge est à un niveau excessif qui provoque le décollement de la dame. On peut remédier à cet inconvénient en réduisant la force centrifuge ou en utilisant un instrument à vibrations exclusivement verticales.

Il apparait préférable au laboratoire comme sur chantier d'employer dans la zone des basses teneurs en eau des compacteurs à vibrations monodirectionnelles verticales pour réaliser des densités élevées en surface égales ou supérieures à l'OPM. A cette condition, l'examen des résultats comparés des essais peut montrer, à notre avis, les corrélations significatives qui existent entre la réponse d'un sol au compactage vibrant au laboratoire et sur chantier.

## 8. LE CHOIX DES COMPAÇTEURS VIBRANTS FACTEUR CLEF DU COMPACTAGE A BASSE TENEUR EN EAU

### Terrassements Généraux

Les travaux de terrassement généraux sont effectués depuis longtemps sans apport d'eau en différentes régions désertiques: Sahara (Communication Algérienne de 1969 à la conférence d'Addis Abéba), Australie (Chemin de fer TOM PRICE-PARABURDOO), Moyen Orient, etc.

En couches de chaussée, on en est resté au compactage à l'eau, au prix de dépenses souvent considérables de forages et de transport, et du prélèvement de quantités importantes d'eau dans des nappes au potentiel limité et pas toujours renouvelable.

Les travaux de compactage vibrant effectués tant au laboratoire que sur certains chantiers permettent de penser qu'il est aujourd'hui économiquement possible d'envisager le compactage de la grande majorité des sols désertiques à des teneurs en eau voisines de 1% et d'atteindre des compacités voisines de 100% de la densité AASHO modifiée.

Parmi les caractéristiques essentielles à considérer pour le choix d'un compacteur vibrant destiné à être employé sans apport d'eau, nous évoquerons la morphologie du matériel (Monocylindre, Tandem, Tracté ou automoteur, Plaques vibrantes, etc.), la masse totale du matériel, la charge statique à l'unité de longueur de la génératrice vibrante ou de surface de la plaque vibrante, le rapport masse suspendue à masse vibrante, la fréquence, le moment d'excentricité des balourds, l'orientation des vibrations, le contrôle lié ou indépendant de la fréquence et du moment d'excentricité, etc.

On voit que le sujet est complexe. Certaines expériences incitent à penser que le choix de plusieurs de ces caractéristiques est plus sensible au compactage à sec qu'au compactage à l'eau, et l'on peut expliquer cette sensibilité par le fait qu'à teneur en eau voisine de 0 la cohésion d'un sol est très faible et l'effet des impulsions de vibration se trouve accru.

Sans entrer ici dans le détail, les règles générales de choix des caractéristiques des compacteurs vibrants en fonction de la nature des sols et surtout de l'épaisseur des couches à compacter établies pour le compactage à l'eau restent semble-t-il applicables en prenant soin de tenir compete que les sols secs peuvent tous être assimilés aux sols sans cohésion.

En gros, on choisira un matériel de poids moyen ou lourd pour le compactage en profondeur, et au contratire un matériel d'autant plus léger que la couche superficielle est plus mince et le diamètre du passant à 80% du sol est plus faible.

En profondeur, les vibrations omnidirectionnelles à basse fréquence sont plus efficaces, en surface, les vibrations monodirectionnelles à haute fréquence sont préférables, avec si possible réglage indépendant de la fréquence et du moment des balourds.

Les plaques vibrantes lourdes représentent un autre type de matériel efficace pour le compactage superficiel des couches minces.

On cherchera à abaisser la teneur en eau naturelle du sol par scarification suivie d'une exposition au soleil. En effet la teneur en eau naturelle observée sur la plupart des sols du Sahara et du Sahel oscille entre 2 et 4%, plage la plus défavorable à l'efficacité du compactage vibrant.

Les précautions suivantes doivent être respectées pour atteindre de bonnes densités en ce qui concerne le choix du modèle de compacteur vibrant:

## COMPACTAGE EN PROFONDEUR SUPÉRIEURE À 20-25 CMS

Le compacteur vibrant sera choisi relativement lourd, avec un poids par cm de génératrice supérieur à 25 kgs, une force centrifuge élevée à basse fréquence réglable dans la plage 1200-1800 vibrations/minute.

COUCHES DE CHAUSSÉE NON LIÉES, SUR DES EPAISSEURS VARIANT DE 10 À 30 CMS

Pour assurer la stabilité superficielle de la couche de roulement d'une route en terre battue soumise à subir l'impact direct des pneumatiques, il est important d'obtenir des densités aussi élevées que possible non seulement dans les couches de fondation mais aussi dans les derniers cms de la couche de forme ou de roulement. Les expériences de laboratoire faites au Sahara confirment l'enseignement tiré de certaines expériences en vraie grandeur rapportées en 1974 par le Prof. Lambe (MIT) et par le Professeur Lars Forssblad (Dynapac) et plus récemment par Fritz Fisher (Clark): Le compactage des cms supérieurs à densité élevée est possible, à l'eau comme à sec, à la condition d'employer un compacteur léger présentant les caractéristiques adéquates.

Trois types de matériel vibrant sont adaptés à ce travail:

a) Le <u>rouleau vibrant léger</u> (moins de 25 kgs par cm de génératrice) <u>à vibrations orientées verticalement</u> et à haute fréquence (2000-2400 cycles/minute) avec si possible contrôle indépendant de la fréquence et de la force centrifuge.
b) Le <u>rouleau vibrant léger à vibrations omnidirectionnelles</u>, avec <u>contrôle indépendant de la fréquence et de la Force centrifuge</u>, également à haute fréquence (2000-2400 c/m).
c) La <u>plaque vibrante lourde</u> à haute fréquence de vibrations (2500-4000 c/m).

L'exécution de planches expérimentales aidera à sélectionner le type de compactage qui, pour un sol déterminé, conduit aux densités superficielles les plus fortes pour un même nombre de passes.

9. ABAISSEMENT DES TENEURS EN EAU AVANT COMPACTAGE DES SOLS

La réussite du compactage vibrant d'un sol est en partie liée, dans la plage des basses teneurs en eau, à la possibilité de s'approcher le plus près possible de la teneur en eau nulle.

Les hautes températures diurnes et la faible degré hygrométrique de l'air des régions désertiques continentales sont des facteurs favorables: Dans les régions à climat tempéré, il n'existe que quelques jours par an et quelques heures pendant ces jours au cours des quelles les conditions atmosphériques rendent possible l'abaissement des teneurs en eau naturelle d'un sol exposé à l'air à des valeurs inférieures à 2%: Dans ces régions, l'intérêt de poursuivre des essais de compactage sans eau est réduit.

En régions désertiques à climat comparable à celui du Sahara, il en va différemment. Les expériences effectuées en mars 1976 à Tamanrasset, qui n'est pas l'exemple le plus favorable (Moyennes max de températures: 26°C en mars), on permis sans difficulté et par simple exposition au soleil et à l'air libre des sols foisonnés d'abaisser leur teneur en eau de 2,5 à 1,2% en moyenne. Pour ces teneurs en eau le compactage vibrant connait une bonne efficacité. Les densités maximales obtenues sont à teneur en eau nulle mais il se révéle d'autant plus difficile d'approcher cette valeur à l'air libre que le sol contient davantage de fines.

## 10. CHAUSSÉE EN TERRE BATTUE SOUS CLIMAT DÉSERTIQUE

Les meilleurs auteurs s'accordent généralement à admettre qu'à partir du moment où l'on a pu rassembler les ressources nécessaires à l'amélioration d'une piste en terre en milieu désertique, la dégradation très rapide sous trafic d'une surface de roulement en terre battue et le coût prohibitif de l'entretien du bon état de cette surface justifient l'exécution d'un revêtement léger au niveau minimal de l'aménagement.

Après avoir parcouru et observé l'état de surface de plusieurs milliers de kms de routes et pistes en zones Sahéliennes et Sahariennes, nous avancerons l'opinion plus nuancée que voici.

La médiocre qualité de l'état de surface des pistes en terre battue et leur défaut d'entretien sont une caractéristique à peu près générale observée sur les itinéraires que nous avons parcouru dans le Sud Algérien, à l'Est du Mali, au Nord Ouest du Niger, au Nord Cameroun et dans le Nord du Nigeria.

On trouve cependant sur ces mêmes itinéraires certaines sections qui présentent un bon état de surface, après avoir subi parfois pendant plusieurs années sans entretien un trafic faible. Il a été procédé à l'identification des sols de plusieurs de ces sections dont l'état de surface est encore satisfaisant, sur la route Gao Bourem Kidal au Mali d'une part (400 kms) et sur la piste Assamaka Arlit au Niger d'autre part (250 kms). Dans les différents cas examinés, les sols se classent, soit parmi les argiles sableuses et plastiques (Banco, 1 p = 10 à 15), soit parmi les graves argileuses latéritiques (lp = 10 à 14), soit parmi les différents sables géologiquement différents (latérite, quartz, arènes granitiques, grès) mais qui ont entre eux de commun, à l'exception du banco, un pourcentage d'éléments supérieurs à 5 mm inférieur ou égal à 10, un lp variant de 0 à 10, un pourcentage d'éléments passant au tamis de 80 u compris entre 12 et 20%, une courbe de distribution granulométrique régulière et presque rectiligne (familles GA, GL, SA, Sm, SL).

Dans les régions où existent des sols de ce type, soit sur le tracé soit à faible distance, il semble justifié d'envisager un niveau minimal d'aménagement pour itinéraire à faible trafic, avec route en terre battue, car il existe de bonnes chances pour que son comportement reste longtemps satisfaisant si le trafic reste faible même avec un entretien réduit ou nul.

## 11. CHOIX ET MODALITÉS D'EMPLOI DES SOLS DÉSERTIQUES EN COUCHES DE ROULEMENT NON REVÊTUE

Le choix des sols susceptibles d'avoir un comportement acceptable en couche de roulement de route en terre sous faible trafic pourra être effectué à l'intérieur des familles ci-après: Gb, GA, GL, Gm, Sb, SL, Sm, et SA en ramenant si possible à environ 10% au maximum la proportion d'éléments supérieurs à 8 m/m pour retarder la formation de la tôle ondulée (après écrêtage de ces éléments, on tombe en général dans la famille des sables).

L'écrêtage à 8 mm des graves et des sables produira des gravillons que l'on pourra classer sans grande dépense supplémentaire au moment de l'opération de criblage. Si leur résistance Los Angeles est suffisante, ces gravillons peuvent être utilisables ultérieurement comme agrégats pour béton ou comme granulats de tapis superficiel en prenant certaines précautions pour éviter leur pollution par les vents de sables pendant leur stockage (Bâche en plastique).

Une autre utilization intéressante du refus à l'anneau de 8 mm consiste à les reincorporer après concassage au sol naturel, car l'angle de frottement élevé des éléments concassés contribue à améliorer la stabilité du mélange.

Les critères de sélection des sols seront essentiellement les densités après compactage, les valeurs de l'indice CBR et des déflexions mésurées sous la roue de camion type.

Les faibles variations des basses teneurs en eau de ces sols leur confère une aptitude à être employés aussi sous revêtement souple si leur comportement est satisfaisant en route en terre.

## 12. TENEURS EN EAU ET SOLS PLASTIQUES SOUS CLIMAT DÉSERTIQUE

En déhors des vallées d'oued où sont souvent aménagés des passages submersibles, les teneurs en eau trouvées dans les corps de chaussée de pistes Sahariennes oscillent dans la plage 2 - 4% exceptionnellement 5%: La rareté et la durée limitée des pluies ainsi que l'évaporation très rapide s'opposent à l'imbibition en profondeur des sols des plateformes routières.

Il s'ensuit qu'il est possible d'employer en chaussée non revêtue et aussi sous chaussée revêtue des sols argileux contenant une proportion élevée de fines inférieures à 80 u, en particulier les argiles sableuses du type "banco" dont les indices de plasticité varient de 8 à 15 en moyenne.

Les limons argileux plastiques, en déhors des passages submersibles, peuvent constituer l'un des sols dont la tenue sera excellente en couche de roulement de route en terre battue sous climat désertique. Pour le choix des sols argileux convenables, on se référera aux essais de laboratoire, (dénsités, CBR), en prenant soin d'effectuer les essais de stabilité à des teneurs en eau inférieures à la teneur OPM et à la limite de retrait qui ne seront généralement jamais atteintes en service.

Le compactage sans eau des sols argileux reste un problème ardu. Dans le cas où l'on ne peut pas envisager l'emploi d'eau de compactage en quantités suffisantes, (w OPM) on pourra chercher à ramener l'apparence de l'argile à celle d'un sable ou d'une grave en la brisant par passages répétés d'une charrue à disques tirée par un tracteur à chennilles. S'il existe à proximité des sables éoliens à courbe granulométrique "abrupte", donc parvres en fines, une autre solution peut aussi consister à employer les sables corrigés par adjonction de 10 à 20% d'argile. Dans ce cas, le mélange à sec de l'argile au sable exige un engin puissant du type pulvimixer ou rotomalaxer. L'opération sera conduite aux heures chaude de la journée afin de pouvoir mieux abaisser la teneur en eau du mélange de quelques points, facteur susceptible de favoriser l'exécution du compactage vibrant sans eau.

On trouvera en annexe un shéma montrant l'influence de la teneur en fines d'un sol sur sa teneur en eau en milieu désertique. (Tamanrasset mars 1976)

## 13. RÉTARDEMENT DE L'APPARITION DE LA TÔLE ONDULÉE

On peut envisager les deux types de traitement ci-après:

(a) Collage superficiel des sols - Sur les routes en terre l'apparition du phénomène "tôle ondulée", est lié à la disparition des fines qui, sous le choc du passage des véhicules, sont mobilisées et s'envolent. La solution pourra consister dans le répandage soit de cut back 0/1 suivi d'un sablage pour absorber l'excédent de bitume non absorbé par le sol, soit de lignosulfite en solution aqueuse avec un domage de l'ordre de 1 kg de lignosulfite au m2 environ.

Les Ingénieurs de l'US Corps de Vicksburg ont effectué des recherches et essais avec des produits de collage destinés à contrôler l'effet poussière, des véhicules militaires circulant en rase campagne. Certains de ces produits, appliqués sur des sols dont par ailleurs le comportement sous circulation d'un petit nombre d'essieux reste satisfaisant, permettent de reculer efficacement le seuil de trafic à partir duquel il devient nécessaire de procéder à l'exécution d'un revêtement bitumineux.

(b) Contrôle de la granulométrie des sols entrant dans la composition des couches de chaussée
Le contrôle granulométrique de la couche de roulement d'une piste en terre battue sur une épaisseur minimale de l'ordre de 10 cms, peut retarder la formation de la tôle ondulée si l'on traite comme suit les extrémités de la courbe:

i. Eléments fins - Il est bon de fixer le pourcentage d'éléments inférieurs à 80 u dans une fourchette 10-20% environ: En dessous, on risque une faiblesse de stabilité par manque de cohésion, au-dessus on risque de se heurter à des difficultés excessives de compactage à basse teneur en eau.
ii. Gros éléments - On a déjà mentionné l'intérêt de limiter le pourcentage d'éléments supérieurs à 5 mm responsables de l'apparition accélérée de la tôle ondulée (PELTIER). Pour faciliter l'opération de criblage, on pourra se fixer l'anneau de 8 mm, ce qui conduira généralement à un pourcentage d'éléments supérieurs à 5 mm de moins de 10%.
iii. Concassage des éléments supérieurs à 8 mm - On a déjà évoqué le concassage de la fraction supérieure à 8 mm générateur d'éléments plus fins à angle de frottement élevé dont le mélange au sol naturel peut améliorer de façon efficace la stabilité du sol et retarder la formation de tôle ondulée (rélèvement de l'angle de frottement interne ajouté à l'amélioration de la distribution granulométrique).

## 14. STRUCTURES DE CHAUSSÉE ROUTIÈRE (TRAFICS SUPÉRIEURS À 30 V/J DONT 50% P.L.)

(a) Portance des sols désertiques Sahariens - A l'exception des sables dunaires, et des zones argileuses en vallées inondables, la portance des sols meubles Sahariens à l'état brut est presque partout très bonne: Géotechniquement, la plupart de ces sols peuvent être rangés dans l'une ou l'autre des catégories Gm, Gl, GA, Sb, Sm, SL ou SA. Dans la grosse majorité des cas, ils appartiennent à la famille des sables plutôt qu'à celle des graves. Leur portance naturelle in situ est illustrée par une plage de résistance CBR in situ qui va de 15 à 80 et par des déflexions après compactage à la densité 100% Proctor Modifié inférieures à 10/10 de mm à de rares exceptions près. Il est clair qu'après compactage superficiel soigné, la résistance CBR de ces sols augmente, les déflexions diminuent et présentent des valeurs qui eu égard aux critères classiques, conduisent à la tentation de supprimer la couche de base.

(b) Le choix d'une couche de base et le calcul de son épaisseur - Dans leur communication à la conférence routière Africaine d'Addis Abéba en octobre 1969, les Ingénieurs Algériens ont rapporté qu'en se référant simultanément aux deux modes de calculs, CBR et déflexions, et en retenant le résultat le plus pessimiste des deux, les épaisseurs suivantes de couche de base pour des trafics de l'ordre d'une centaine de P.L. par jour, avaient été retenues:

| | |
|---|---|
| - sols caillouteux: | 10 à 15 cms |
| - sols sableux à granulométrie étalée | 15 à 20 cms |
| - sols sableux à courbe granulométrique abrupte (dunes) | 25 à 30 cms |
| - sols très plastiques soumis à imbibition | 40 cms |

A cette époque, les terrassements avaient été compactés à sec mais le matériel utilisé à rouleau pneus de 50 T, ne permettait pas d'obtenir des densités dépassant l'ordonnée à l'origine de la courbe Proctor de sorte que le compactage de la couche de forme employée en fondations et de la couche de base était effectué à l'eau suivant la technologie classique.

Il s'agissait de routes construites pour supporter un trafic pouvant atteindre 300 véhicules/jour dont la moitié de poids lourds, et l'expérience a montré l'excellent comportement général de ces structures.

Il est actuellement possible avec un matériel de compactage judicieusement choisi, d'obtenir économiquement des compacités de terrassement et de la couche de fondation bien supérieures à celles rapportées à cette conférence. Les performances minimales CBR et déflexions sont nettement meilleures, et permettent d'envisager l'économie de la couche de base. En déhors des passages en zones submersibles où la règle Algérienne des 40 cms semble justifiée, on pourrait donc, soit supprimer la couche de base, soit réduire son épaisseur aux 10 cms, sur lesquels le contrôle granulométrique reste souhaitable afin d'améliorer l'homogéneité de l'aménagement et retarder la formation de la tôle ondulée. On pourra exiger une densité voisine de 100% OPM dans la couche de forme, et cette densité pourra être obtenue par vibration à basse teneur en eau.

Les critères de valeurs maximales de déflexions recommandées par HVEEM (soit environ 1 mm pour un tapis superficiel bicouche 10/16 - 5/8 et 8/10 mm pour un enrobé dense de 4 cms d'épaisseur) peuvent ainsi être satisfaites sur de nombreuses sections de route en zone aride.

(c) Revêtement superficiel - Si l'on décide d'effectuer un revêtement hydrocarboné, la solution économique qui semble la plus adéquate consiste à réaliser un tapis superficiel bicouche structure satisfaisante pour de faibles trafics, disons trafics ingérieurs à 200 véh/jour avec une proportion de 50% de P.L.

## 15. PROFIL EN TRAVERS: CHAUSSÉES À UNE VOIE OU À DEUX VOIES

Le premier niveau d'amélioration suivant la plateforme en terre battue, consiste à réaliser une chaussée revêtue à voie unique de 4 mètres de largeur sur les itinéraires pau fréquentés. Cette disposition présente l'inconvénient de rendre difficile l'entretien des accotements non revêtus en bon état, même si les sols ont une composition favorable à leur utilisation en "route en terre". Lors des manoeuvres de croisement ou de dépassement, le passage des pneumatiques à vive allure du revêtement bitumé sur les accotements en terre battue exerce un effet destructeur qui s'accélère avec les dégradations.

Par ailleurs les chaussées à voie unqiue et à deux sens de circulation sont caractérisés par un taux d'accident élevé.

L'imprégnation au cut back 0/1 des accotements semble constituer un palliatif acceptable seulement si le trafic reste très faible.

Finalement, les conditions désertiques conduisent à abaisser fortement le niveau de trafic justifiant la chaussée revêtue à deux voies alors que les ressources financières importantes requises par ce type d'aménagement restent très difficiles à mobiliser car les justifications économiques font défaut.

Il pourra apparaître préférable pour un itinéraire à faible trafic, de choisir la chaussée à deux voies, mais en réduisant l'investissement initial qui serait alors limité à une simple imprégnation plus soignée surdeux voies, soit au lignosulfite soit au cut back, suivie dans le cas du cut back d'un sablage 2/4 mm pour absorber l'excédent de bitume.

Certes, la résistance d'une telle surface reste très inférieure à celle d'un tapis superficiel bicouche, mais en revanche la répétition des charges est réduite considérablement car la canalisation du trafic n'est pas aussi concentrée que sur la chaussée à voie unique. Si l'on fait le bilan de comparaison des deux solutions, pour un faible niveau de trafic, l'avantage peut revenir à la surface imprégnée à deux voies plutôt qu'au tapis biouche sur une voie à la condition impérative de disposer en permanence d'une équipe mobile et légère d'entretien venant réparer les désordres localisés au fur et à mesure de leur apparition.

## 16. LE CHOIX DU NIVEAU MINIMAL D'AMÉNAGEMENT ROUTIER, FACTEUR DÉCISIF DE L'AMÉLIORATION DES TRANSPORTS TERRESTRES EN MILIEU DÉSERTIQUE

### Objectifs

Nous considérons ici le cas des itinéraires à très faible projections de trafic: Inférieures à 20 véhicules/jou-, dont 50% de Poids Lourds, avec des distances de parcours considérables, de l'ordre de plusieurs centaines de kms au minimum.

Après avoir passé en revue les perspectives de développement économique et social de ces régions enclavées, les justifications que l'on arrive à rassembler pour obtenir le concours de moyens financiers nécessaires à l'amélioration de leurs dessertes routières restent limitées et conduisent à rechercher le niveau minimal d'aménagement qui entrainera un premier investissement faible.

L'objectif reste l'amélioration durable du tracé, des profils en long et en travers de l'itinéraire, avec un état de surface satisfaisant offrant un confort et une sécurité de parcours comparables à ceux que l'on trouve sur des itinéraires plus fréquentés.

On veut que cette amélioration de l'état de surface soit maintenu sans nécessiter un travail d'entretien excessif dont le seuil est rapidement dépassé compte tenu des structures souvent précaires et des ressources modestes des services responsables de l'administration de ces itinéraires.

### Définition Du Niveau Minimal Par Emploi Des Sols Naturels

(a) On a vu qu'il existe largement répandus en pays désertiques des sols présentant à leur état naturel une bonne portance et qui la conservent en toutes saisons car en déhors des passages en vallées submersibles que nous traiterons à part, leurs teneurs en eau naturelles restent faibles, toujours inférieures à leur limite de retrait.

Le parcours de nombreuses pistes en terre battue au Sud du Sahara a permis d'observer l'alternance de sections fortement déformées, de sections de tôle ondulée, de sections envahies de sables éoliens qui rendent le passage difficile, ralentissent l'allure et fatiguent véhicules et passagers. Mais on trouve aussi sur les mêmes itinéraires de longues sections dont l'état de surface est resté de qualité bonne ou acceptable, bien meilleure que celui de certaines routes revêtues en mauvais état d'entretien. Les routes Gao Kidal, au Mali, et Assamaka Arlit au Niger, correspondent à la description que nous venons d'évoquer, alors qu'elles ne présentent pas trace de travaux d'entretien important depuis de nombreuses années, et qu'elles supportent un trafic faible: inférieur à 10 véhicules/jour en moyenne.

On a observé que les sections à bon état de surface de ces deux itinéraires étaient caractérisées soit par des sols sableux ou graveleux contenant de 10 à 25% de fines argileuses, soit par des argiles sablonneuses du type Banco, la plupart de ces sols présentant des lp compris entre 8 et 15, soit par des sables de quartz à lp non mesurable.

Ces exemples montrent que pour des trafics de l'ordre de 10 véhicules/jour il est raisonnable d'imaginer un niveau minimal d'aménagement de route en terre dont la couche de roulement serait constituée des sols du même type, si de tels sols existent le long du tracé ou à proximité. Ce type d'aménagement sommaire exige cependant la mise en place d'une équipe mobile d'entretien dotée du matériel adéquat pour surveiller l'itinéraire et réparer les passages défectueux. A notre avis cette équipe doit être capable de suivre le maintien en état de 3 à 400 kms/an compte tenu du faible trafic.

(b) Le niveau d'amélioration immédiatement supérieur peut être constitué par le répandage d'une imprégnation superficielle, soit au cut back sablé, soit au lignosulfite, soit au moyen de l'un des produits expérimentés par l'US Corp of Engineers à Vicksburg pour empêcher la formation de nuages de poussières au passage des convois militaires circulant sur des sols à l'état brut. Ce traitement léger en réduisant considérablement l'effet poussières retarde l'apparition de la tôle ondulée et élève le seuil de trafic à partir duquel il devient nécessaire de réaliser un revêtement hydrocarboné plus épais. En dessous de ce seuil, on aura amélioré le confort et la sécurité du trafic pour une dépense de premier investissement réduite. Ce second niveau d'aménagement implique toutefois de façon encore plus impérative la mise en place d'une équipe mobile dotée du matériel et des matériaux nécessaires pour assurer l'entretien courant. La composition de cette cellule serait plus légère que dans le cas de la route en terre battue.

(c) On a tenté en divers pays (Nigéria, Rhodésie, Botswana, Algérie, etc.) de réaliser au niveau suivant un revêtement bicouche sur une voie unique réduite à 4,00 m de largeur. Cette disposition offre l'inconvénient de réduire sensiblement la sécurité du parcours, de créer une forte concentration de la répétition des charges sur deux bandes étroites de chaussée, et conduit à une dégradation accélérée des accotements empruntés par les véhicules qui se croisent ou lors des manoeuvres de dépassement. Aussi dans la plupart de ces pays a-t-on décidé d'abaisser fortement le seuil de trafic à partir duquel on réalise un revêtement sur deux voies.

(d) En définitive, à partir des deux niveaux "minimaux" que nous avons défini plus haut et nous référant à une étude de prix unitaires menée dans le cadre d'un projet routier récent du Sud Saharien, nous pouvons présenter comme suit l'évantail approximatif des niveaux d'aménagements possibles d'une route économique en pays désertique.

| | Coût Construction | % coût entretien anneul |
|---|---|---|
| 1. Niveau minimal: Route en terre compactée à 2 voies 6,5 m | 51 | 3 |
| 2. Route en terre compactée et imprégnée à 2 voies 6,0 m | 58 | 1 |
| 3. Corps de chaussée en terre à granulométrie contrôlée (2 voies) | 73 | 2 |
| 4. Tapis bicouche I voie 4 m sur route type 2 | 71 | 1 |
| 5. Tapis bicouche 2 voies 6 m sur route type 2 | 78 | 1,1 |
| 6. Imprégnation sur route en terre 3 à 2 voies 6,00 m | 80 | 1 |
| 7. Bicouche I voie 4 m sur route en terre 3-6 | 93 | 1 |
| 8. Bicouche 2 voies 6 m sur route en terre 3-6 | 100 | 1,1 |
| 9. Enrobés une voie 4m sur route en terre 3-6 | 103 | 0,3 |
| 10. Enrobés 2 voies 6m sur route en terre 3-6 | 122 | 0,4 |

(e) La réduction significative du coût de transport au niveau d'aménagement 2, peut rendre cette solution attrayante malgré la sujétion de mise en place d'une équipe légère d'entretien. Sous cette condition le niveau d'aménagement 2 devrait se révéler satisfaisant pour supporter de l'ordre de 2 à 5 x $10^4$ répétitions de charge, soit un trafic moyen de l'ordre de 20 véhicules/jour pendant plusieurs années. Pour des trafics supérieurs, avec un investissement de l'ordre de 20 à 25% du coût d'investissement initial, il devient techniquement nécessaire et économiquement possible de réaliser un revêtement en tapis superficiel bicouche (Niveau d'Aménagement 5).

## 17. TRAVAIL DE NUIT

En plein jour, la consommation d'eau par évaporation sur un chantier routier de compactage à l'eau en pays désertique est très élevée, car généralement la température diurne est forte et le degré hygrométrique de l'atmosphère très bas. Cette perte d'eau peut atteindre un même ordre de grandeur que l'eau strictement nécessaire au sol pour être dans la plage w OPM pendant la durée du compactage.

Par ailleurs, les vents de sables sont plus fréquents le jour que la nuit.

Dans ces conditions, il pourra être avantageux d'organiser un travail de fin de journée, de nuit et de début de journée pour les opérations d'arrosage, compactage à l'eau, imprégnation au bitume ou à la solution aqueuse de lignosulfite, ou exécution d'un tapis supérficiel bicouche. Dans ce dernier cas, il est en outre prudent d'exécuter la seconde couche immédiatement après la première couche pour éviter la pollution des interstices entre gravillons par le sable éolien.

## 18. PASSAGE DES VALLÉES D'OUED ET DE ZONES INONDABLES

(a) Vallées d'oued et thalweg à pente longitudinale non nulle
Les vallées d'oued sont en général franchies en radier submersible, en prenant soin de traiter la composition et l'épaisseur de la couche de base pour résister aux submersions temporaires.

Les pluies sont généralement rares en régions désertiques, mais l'absence totale d'une couverture végétale entraine des coefficients de ruissellement souvent élevés. Les lits d'oued sont alors le siège d'écoulements violents à régime torrentiel.

L'expérience Algérienne a montré qu'il cessait d'être économiquement justifié de construire des ouvrages à chaussée hors d'eau en zones désertiques dès que le bassin versant présente une surface égale ou supérieur à 5 km2 environ.

Au-delà, la solution qui a donné toute satisfaction au Sahara consiste à réaliser un passage submersible, avec un profil en long en léger remblai (10 à 20 cms), et un profil en travers à pente unique de l'ordre de 2%. La couche de base est généralement d'épaisseur accrue et insensibilisée à l'eau. Le revêtement hydrocarboné et en surlargeur est préférable avec lignes de gabions amont et aval pour se protéger des affouillements.

(b) Zones basses inondables à vitesses d'écoulement très faibles ou nulles
En fonction des cas, deux solutions sont possibles: Soit un passage submersible simplifié pratiquement sans protection spéciale mais avec balisage vertical latéral si le passage présente une certaine longueur, soit un passage en remblai hors d'eau si la hauteur d'eau n'est pas trop forte et si l'importance du trafic le justifie.

## 19. CONCLUSION

Le survol sommaire que nous venons d'effectuer montre que l'amélioration de la desserte routière des régions désertiques est possible à dépenses de temps, d'eau et d'energie réduites par combinaison des facteurs suivants:

(a) Etude préalable de tracé et interprétation photogéologique soignée de la couverture aérienne existante, éventuellement complétée par une couverture au 1/20 000.
(b) Etude des ressources en eau souterraines et en matériaux de construction type le long de l'itinéraire.
(c) Implantation directe et relevé topographique du profil en long de l'axe choisi.
(d) Etude au laboratoire de la réponse des sols type au compactage vibrant et exécution de planches expérimentales de compactage profond et superficiel, avec relevé des densités au nucléodensitomètre et des déflexions à la poutre Benkelemann.

(e) Ouverture du chantier capable de mettre en oeuvre indifféremment les procédés de compactage sans eau ou à l'eau, suppression ou détermination de l'épaisseur de la couche de base ou de roulement par dépouillement des résultats d'essais granulométriques, résistance CBR in situ, limite d'atterberg et mesures de déflexion.

(f) Emploi maximal des matériaux locaux améliorés ou non par criblage et mélanges. Acceptation des matériaux argileux en tenant compte de la teneur en eau maximale des sols en service en couches de chaussée.

(g) Choix de profils en travers économiques à deux voies, et traitement antipoussière de la couche de forme par simple collage superficiel du sol soit au cut back sablé, soit au lignosulfite, soit par emploi d'un nouveau produit de collage (US Corps of Engineers, Vicksburg) retardant l'apparition de la tôle ondulée tout en améliorant les conditions de transport.

(h) Consommation d'eau réduite par rapport aux chantiers routiers traditionnels, participation accrue des Techniciens locaux à la mise en oeuvre de la technologie retenue et organisation d'une cellule d'entretien légère pourvue de moyens d'intervention rapide.

# SOME ELEMENTS OF THE DEVELOPMENT AND MANAGEMENT OF DESERTS

## Recommendations of the Sacramento Conference

The Conference presented a selection of experiences and technologies from both developed and developing countries. To utilize the information presented and developed in the discussion, it will be necessary for the participants and their associates and their governments to select the strategies and techniques that will best meet the needs of their people.

This report has used the term "desert" in the broad sense of the Conference. It is recognized that both true deserts, with no vegetation whatever, and arid lands are included. Most of the recommendations apply to arid lands, some apply to both.

## DESERTIFICATION

Many millions of people are presently being endangered by desertification processes in a region of about one million square kilometers bordering on desert areas. The factors leading to this desertification are fairly well known but not always well understood. These would include:

1. The rapid growth of population which has resulted in increased flocks, overgrazing and depletion of woody vegetation for fuel.
2. Incorrect land and water management and destructive techniques of cultivation and irrigation.
3. Climatic variation and long-range climatic change.
4. Social, political and cultural factors.

Millions of hectares of formerly productive soils are lost annually, some of this loss being irreversible. Obviously steps must be taken to contain and even to reverse these destructive processes. The technologies presented at this conference are partly well known and even if there is still a great deal to be learned about them they could theoretically be put into practice today.

Populations have been growing at a rapid rate so that even where development projects have been successfully implemented, per capita income has not substantially increased and at best, these projects have provided only a partial answer to the problems of desertification.

It seems clear that development programs will have to be accompanied by an intensive program of education and by far-reaching social and political reforms. Changes in custom and tradition where necessary have to be generated from within. The developed nations can help by:

a. Providing capital and technical assistance for improvement of educational facilities and services.
b. Providing credits, subsidies and other incentives for measures leading to better land management and to reduction of population and livestock pressures.
c. Establishing produce and marketing institutions.
d. Aiding in the establishment and operation of local and regional research centers.
e. Aiding in the development of local extention services to assist in the transfer of information developed through research to the farmers and other users.

## DESERT DEVELOPMENT AND STRATEGIES

A number of the papers read at the Conference placed heavy emphasis on techniques and technology involving management of some basic components such as water, mineral, energy sources (including petroleum, solar, and geothermal) and the primary and secondary productivity of natural and agriculturally modified ecosystems.

Regional and national planners charged with strategy selection are often faced with simultaneously conflicting situations, especially when, as usual, the financial, labor, material, or other resource is limited. In such situations strategy decisions are often made with the hope of achieving maximal short term returns in order that other strategies may be activated later. However, in most cases two or more strategies are being carried out simultaneously. Even if local governmental agencies are emphasizing only a single strategy (e.g., a cash crop produced for export), local peoples are generally involved in food crop production, animal husbandry and harvesting of native fruits, berries and wildlife species as well as, perhaps, limited mining or home industry.

A number of different strategies are evident in the papers presented but most appear to be centered around the maximal acquisition and most efficient utilization of that basic resource in short supply -- WATER!

Further, most appear to involve the use of water for some agricultural purpose -- either field crops or domestic animals. Specific projects included in the reports that could be emphasized as strategies include the following:

1. Development of coastal zones and regions, utilizing the extra food resources and cheap transportation supplied by the marine interface.
2. Maximize storage of excess waters during wet seasons with inexpensive compartmented reservoirs and similar devices for use in dry seasons.

3. Where transportation to markets and technology exists, grow and export high value crops. In some cases this can include the use of controlled environment chambers (greenhouses).
4. By irrigation (from underground, imported surface waters or local impoundment in wet season) develop, where appropriate, local agriculture including field crops and animal husbandry. Although a desert area does not need to produce its own food, it must produce something resulting in an export surplus, so food can be purchased.
5. Utilize tourism (together with natural areas, game reserves and game farms) as high labor intensive source of foreign exchange.
6. Where possible, utilize extractive industries (mining).
7. In all possible situations (agriculture, mining) accomplish processing with local processing plants and labor, if water requirements are not excessive. The added value and reduced shipping charges will result in added value to local populations.
8. Creation of desert reserves in all areas is recommended to preserve a portion of the natural flora, fauna and land forms. Such reserves are valuable for study, research, and enjoyment and to preserve the indigenous germ plasm. If this is not done many species may become extinct.

## WATER RESOURCES AND MANAGEMENT

Deserts and arid lands are by definition deficient in water supplies. Any strategy for management or development, be it only the improvement of life for the existing population, the extension of agriculture and grazing, or mining, petroleum development, industrial development or the encouragement of tourism, must consider the most effective use of available water.

It is likely that a combination of strategies will be employed, depending on the resources, needs, and political attitudes of the country. There sould be a thorough survey of the water supplies available including surface supplies, ground water, geothermal water, rainfall and the possibility of imports. The type and extent of development must depend on the results of such a survey. It may be necessary to adjust quality standards and use patterns to the quality of water actually available. Maximum use must be made of water conservation measures and the reclamation and reuse of waste waters.

Long range planning is essential. For example, in some areas there are large bodies of fossil groundwater. Such water should be used benficially, even though there is no recharge, but there must be a decision as to what rate of use will be most beneficial to the people.

Some techniques to maximize the effective use of available water include improved irrigation techniques, evaporation control, salinity control and drainage, and water harvesting to maximize use of rainfall. Institutional and political solutions must be found, as well as technological answers. The economics of development must be realistic. The goal of long-term water planning should be a safe and reliable supply of water designed to shield the local economy from drought and depletion of available water.

## AGRICULTURAL DEVELOPMENT

### Cropping, Ranching, and Tree Plantations

Since some desert areas are being invaded by the plow, it seems important that this situation is given a close second look in many developing countries. This calls for a proper resources survey including soil, water and other investigations.

Lands which can sustain limited cultivation must be so identified and other lands should be considered for retirement from cultivation and diversion for other alternative land uses, e.g., ranching, tree plantation, widlife development, recreation, etc.

Grazing and harvesting of trees and other vegetation would have to be limited or the process of degradation would not be reversed.

Adequate means should be used to minimize losses to crops and native vegetation caused by pests.

Low water requirement and drought resistant crops and crop varieties including forage crops deserve special attention where cropping and grazing are considered desirable. Other practices, e.g., soil and water conservation practices, water harvesting, mulch farming, suitable conservation rotations, mixed cropping, etc., may be considered.

Scientists and extention workers should evaluate indigenous and traditional agricultural/farming practices in the region, for easy adoption and quick results. These methods should simultaneously be evaluated in the light of modern knowledge so that further improvement may be possible.

### Mixed Farming

Recognizing the place of plant species of grazing value, shrubs and trees and by-products of some field crops as may be grown in the area, animal production in the area would deserve special attention in addition to cropping. Such an approach would provide for year-round work of a productive type to the rural people in many developing countries.

Where data on relative economics of cropping vs. animal husbandry vs. mixed farming is limited, the need to generate it is indicated. Wildlife studies need special consideration. Grazing behavior of mixed herds of domestic animals and also mixed herds having a suitable mix of wildlife with domestic animals calls for systematic studies. Wildlife may be an excellent source of meat and hides, but will not provide milk or wool. Upgrading of animals and appropriate health care have to be provided for.

### Grazing and Livestock Management

In many countries, over-grazing of low rainfall desert areas is among the major factors responsible for desertification.

1. Studies on carrying capacity of such areas should be made where such information is lacking so that regulation of grazing may be considered against this background.

2. Unrestricted grazing be discouraged as much as rapidly possible. This could be enforced easily on lands held by the Government. For private lands, suitable education and, if necessary, legislative action may be needed.
3. Instead of numbers of low quality animals, every effort has to be made to build the population of quality animals adapted to the area by reducing unproductive or less productive animals. Cross-breeding to a limited extent and introduction of new breeds/species of animals may be considered in this context.
4. Both immediate and long-range programs for installing watering points have to be formulated and implemented so as to reduce excessive grazing pressures on lands around the existing or newly developed watering points.
5. A suitable program for upgrading of the degraded areas should be developed to improve vegetative cover, reduce the incidence of erosion, and to improve the economy of the areas. Revegetation of degraded areas with natural vegetation of economic value must be considered as a first step. Further improvement through reseeding and management of better types of grasses and other vegetation, both locally available and introduced species/types should be scientifically evaluated.
6. Development and improvement of marketing systems would require close attention in areas where such arrangements are lacking.

## Land Policy

Lands in many desert countries tend to be used beyond their capabilities, due to population pressures which leads to land deterioration. Land allotment policy in such areas has to be developed in keeping with the land capability. Lands which cannot sustain cultivation must be adequately safe-guarded by prescribing appropriate land use and needed practices.

## The Wildlife Resources of Deserts

One of the most abused resources of the desert areas of the world is the indigenous animal resource. Its potential has, to date, been greatly underrated.

Sizable numbers of human desert dwellers have traditionally, and still do, depend on the wildlife resource as their major and often, only, source of protein. In Botswana, for example, an estimated 45% of the human population depends upon wild animals for protein.

Many desert dwellers produce various items derived from the indigenous animals. Often this may be the only source of cash income. This economic resource requires minimal infrastructural, no physical changes to the region, little or no capital investment, and necessitates no changes in life styles.

We highly recommend that the various wildlife species be investigated further wherever sizable populations currently or formerly existed. This investigation should be set up as a large-scale, long-term, on-the-ground practical trial with careful accounting of all variables.

Not until the results of these trials are available will we know whether, with low management resources and energy input, a higher or equal level of productivity (as compared to domestic stock) can be obtained.

In many cases even with lower productivity, optimal land use may prove wildlife raising as the only long-term sustained yield protein producing system possible. In many such places the domestic animals now present should be removed.

## INDUSTRY

Industrial development can be an important factor in the transformation of an underdeveloped area into a modern society. The desert offers several advantages which could be utilized for industrial development such as: wide spaces, a dry climate and, in many cases, easy disposal of waste effluents without ecological deterioration.

The industrial potential of agriculture products could rapidly be tapped. The by-products of regular agricultural crops such as cotton seed and bulbs, molasses, bran and others could be the basis for the manufacture of many products. Desert grown industrial crops such as jojoba, guayule, acacias, etc., could be the basis for many sophisticated chemical products.

Governments can aid in establishing industry in desert areas by:

a. Granting special incentives to investors who establish their manufacturing facilities in these areas.
b. Training workers and technicians.
c. Establishing the necessary infrastructure.
d. Extending aid for marketing and export.
e. Inventorying all available resources.

Countries planning mining of minerals including oil and gas should realize that these are depletable resources whereas desert industrial plants are renewable resources. Thus, careful planning is required to avoid ghost towns and economic disruptions.

The conference recommends that UNITAR and the United Nations Industrial Development Organization (UNIDO), in cooperation with FAO organize a conference to discuss:

Arid and semi-arid zone plants as a source of industrial raw material and its processing possibilities.

Minerals including petroleum and their utilization in deserts.

Types of industries and industrial operations which most efficiently utilize water resources.

Different forms of incentives to encourage the development of the above industries.

## COASTAL RESOURCES

In addition to other natural resources in the hinterland of the desert coastal area, maritime oriented activities including off-shore resource exploration would seem to be decisive to the strategy for the development of the coastal desert and/or arid regions of the world.

Traditional human activities oriented towards exploitation of living marine resources, in the coastal desert and/or arid areas would be enhanced if successful mariculture were applied, modern fishing methods and fleets were introduced and rational fish processing and marketing schemes were developed.

Traditional maritime industries such as ship-building, drydock and ship repair and maintenance facilities should be regarded as compatible development schemes for coastal desert and/or arid region.

In the context of the emerging environmental constraints and other limitations to further industrial growth of the coastal areas in developed regions of the world, on one hand, and the need for continued and sustained economic growth for the benefit of mankind on the other, it would seem reasonable that institutional as well as economic policy guidelines for long term development of coastal desert and/or arid regions should be compatible with environmental consideration and consistent with yet to be defined policies for a new economic growth.

## TOURISM

Tourism as a means of exploiting and developing the potential of arid and semi-arid areas should be carefully considered. Such developments are embryonic in most desert areas, except in some coastal areas.

The following aspects of the desert have tourism potential:

a. Wildlife (viewing, photo safaris, big game hunting).
b. The unique environment of the desert/observation, detailed examination (study tours) of certain aspects, e.g., wild foods/scenery.
c. Health spas (tourists going to the dry region for health reasons).

The development of mineral spas or general spas is being viewed as a new potential strategy of fostering tourism in deserts.

A careful evaluation of the impact on desert residents, and the relative benefit to them (e.g., employment, investment opportunity) must be made before undertaking tourism development, to assure that there will be substantial benefit to the local people. A substantial infrastructure is needed for tourist development including transportation, food services, health services, and water supply.

## ENERGY AND MINERALS

### Energy

In desert countries energy is as vital a supply as in non-desert areas. The use of firewood for cooking still is wide-spread in desert areas and is continuing to destroy the remaining trees and bushes, and unless a country can organize on a regular basis the planting of trees or bushes for firewood supply, countries should study carefully the possibilities of replacing the use of firewood with other energy sources.

The introduction of electricity in desert areas in addition to its energy application is vital for communications and many other uses.

Wind power units should be considered in those areas where wind conditions are favorable. Unfortunately precise local wind condition surveys have not been carried out in most areas, and until such surveys are available, the application of wind power cannot be judged.

Solar energy at the present stage of development is useful for water warming, but for electricity generation the costs are too high at present. Geothermal energy is known to exist in many desert areas, and when geothermal energy can be developed it could also be used among many other applications for electricity generation.

If a country possesses sedimentary areas it should, as a matter of course, be a primary task for the country to explore its underground potential, and in such areas there is a possibility of finding fuel of many types from conventional oil, oil shale, tar sands, and gas to coal, lignite, and peat. It is equally obvious that the discovery in a desert country of large deposits of oil and gas or coal cannot only contribute to the local energy supply, but could also provide an important source of income for the country, in addition to local employment. In general, in the energy policy of a desert country the destruction of bushes and remaining forests through use as firewood should gradually be eliminated and any available source of energy discussed above should be introduced and electricity should be made available as soon as possible to the whole population.

### MINERALS

It is likely that all desert areas contain mineral resources of one type or the other. There are many mineral resources which are specific to desert areas because they are formed in the deserts, such as saline deposits, including a variety of salts such as potash, as well as other minerals, such as lithium. Depending on the geology of a desert area one can expect to find materials from gold to silver, lead, zinc, and many others, including bauxite, the raw material for aluminum. Unfortunately, mineral exploration requires high skill and risk capital; moreover, many of the desert areas have limited geological information, frequently no detailed geological mapping, and modern geochemical and geophysical surveys covering only very limited areas. It is recommended, therefore, that desert areas consider mining exploration as one of their primary tasks, including training of their own geologists and geochemists and geophysicists so that the government at a fairly early stage will be able to evaluate its own underground resources, and on the basis of this, can make policy decisions on which resources to develop first.

Modern mineral development in desert areas need not to be water intensive, and the present situation where mineral deposits are left undeveloped because of water shortages could in many cases be overcome, by application of dry processing methods. In addition, the reuse and reprocessing of mining water can sharply reduce the total quantity of fresh water which may be required. Furthermore, most mining processes allow the use of low quality water, and so on; and thus, if the mine is within proximity of agricultural areas or urban centers, the mining development, except for the need of the personnel, could be covered by waste water.

The same policy should be applied to the oil industry in desert areas which still use in some countries large quantities of fresh water. Thus, the development of mineral resources in desert areas need not be a burden on the limited water resources of a desert country, but could make a considerable contribution to the development of such a country through the income effect, the employment effect, and supply of energy, and other material also needed in the country.

## HEALTH

### Introduction

This topic falls into two categories:

1. Maintaining the health, or improving the health, of the resident and immigrant desert population; and
2. Health care as a focus for tourists potential. The inherent dry nature of deserts, it should be noted, means health problems are probably generally of a less chronic nature than in more humid areas.

### Recommendations

a. Examination of the traditional desert-peoples diet could reveal foods and means of keeping a desert population healthy;
b. Strategies for bringing health-care to resident desert populations should be further investigated and different country experiences collated. Mobile health services (ranging from flying doctors to barefoot doctor type) is particularly stressed.
c. Tackling nutritional needs and desert-specific health problems and diseases is critical to other developments.

## INFRASTRUCTURE

In some desert areas there may be large supplies of ground water or other resources that have not been developed because there is no infrastructure. To make development possible there must be transportation systems (roads, railroads, airports) and communications (telephone or radio). There must be energy resources, a marketing structure and community services. The latter includes health services, education and vocational training, and government functions.

The infrastructure should be self-supporting eventually, but it is probable that outside capital will be needed to provide the initial services so development can begin.

## PARTICIPANT LIST

- A -

Dr. Mansur M. Aba-Husayn
Assistant Deputy Minister for Research and Development
Ministry of Agriculture and Water
Riyadh, Saudi Arabia

Adly Abdel-Meguid
Senior Industrial Development Officer
United Nations Industrial Development Organization
United Nations, Room 2766
New York, New York 10017

Dr. Ahmed Gamal Abdel Samie
Vice President
Academy of Scientific Research and Technology
101 Kasr El Eini Street
Cairo, Egypt

Adolfo Aguilar
Center for Third World Economic and Social Studies
Cor. Profirio Diaz 80
Mexico 20, D.F.

Rhousmane Ahmoud
Chef de Service Planification Sectorielle, Ministere du Plan
B. P. 862
Niamey, Niger

Shahid Akbar
Chief (Corporate Planning)
Oil and Gas Development Corporation
Shafi Chambers, Club Road
Karachi-4, Pakistan

Kola Aladejana
Acting Director of Forestry
Federal Department of Forestry
P.M.B. 5011
Ibadan, Nigeria

Omar A. Mughram Al-Ghamdi
Mechanical Engineer, Saline Water Conversion Corporation
P.O. Box 4931
Jeddah, Saudi Arabia

Dr. Abdullah Al-Jahmi
Director General of Plant Production
Ministry of Agriculture
Sidi Misri
Tripoli, Libya

Laika Khidir Al-Jibury
Scientific Research Foundation
Jadiriya
Baghdad, Iraq

Ahmad A. Al-Mansour
P.O. Box 2
Alkhobar, Saudi Arabia

Nasser Mohamed Amer
Director, Irrigation Department
Ministry of Agriculture
P. O. Box No. 1161, Khormksar
Aden, P.D.R. Yemen

Abdullah Abdir-Razik Arar
Regional Land and Water Development Officer
Food and Agriculture Organization
P. O. Box 2223
Cairo, Egypt

Mazhar Aslam
Manager, Groundwater Development
Resources Development Corporation
Dawood Centre
Karachi, Pakistan

Mohammad Osman Atif
Officer in Charge of Personnel Training, Forestry Department
Ministry of Agriculture
Kabul, Afghanistan

El Hag Makki Awouda
Head, Management and Services Division, GUM Arabic, Sudan Forest Service
c/o Forestry Department
Khartoum, Sudan

Mesfin Aytenffisu
Regional Manager, Harar Regional Office
Ethiopian Water Resources Authority
P. O. Box 2
Harar, Ethiopia

Eduardo Azuara Salas
Center for Third World Economic and Social Studies
Cor. Profirio Diaz 80
Mexico 20, D.F.

- B -

Dr. Omar Ba
General Director
Sahalien Institute, CILSS
c/o I.E.R.
Bamako, Mali

Dr. Amanclyte M. Babayev
Desert Institute
744012 Ashkhabad
U.S.S.R.

E. Bakutina
c/o State Committee for Science & Technology
11 Gorkey Street
Moscow
U.S.S.R.

Dr. Joseph Bamea
Co-Director of Conference
United Nations Institute of Training and Research
801 United Nations Plaza
New York, New York 10017

Roshan B. Bhappu
Vice President and General Manager, Mountain States Research and Development Co.
P.O. Box 17960
Tucson, Arizona 85731

Michael Bloome
United Nations Institute of Training and Research
801 United Nations Plaza
New York, New York 10017

Sergio E. Bonilla
Vice President, INIA
Santiago, Chile

Boualem Bousseloub
Resources Planner, California
Department of Water Resources
P.O. Box 388
Sacramento, California 95802

Jakov Bradanovic
Centre for National Resources, Energy and Transport
United Nations
P.O. Box 20 G.C.P.O.
New York, New York 10017

Dr. Michael Bradley
Associate Professor
Department of Hydrology and Water Resources
University of Arizona
Tucson, Arizona 85721

Dr. Donat B. Brice
Staff Chemical Engineer
California Department of Water Resources
P. O. Box 388
Sacramento, California 95802

John E. Bryson
Chairman, California State Water Resources Control Board
1416 Ninth Street
Sacramento, California 95814

William E. Bullard, Jr.
4931 N.E. Roselawn
Portland, Oregon 97218

Joseph I. Burns
President, World Affairs Council of Sacramento
600 Forum Building
Sacramento, California 95814

William E. Bye
Environmental Engineer
U.S. Environmental Protection Agency
401 "M" Street, S.W.
Washington, D.C. 20460

- C -

Professor Chang Sen-dou
Department of Geography
University of Hawaii
Honolulu, Hawaii 96822

Bertrand H. Chatel
Chief, Technology Applications Section, Office of Science and Technology
United Nations
New York, New York 10017

Dr. C. Brent Cluff
Water Resources Research
University of Arizona
Tucson, Arizona 85721

Dr. E. Lendell Cockrum
Head, Department of Ecology and Evolutionary Biology
University of Arizona
Tucson, Arizona 85721

A. Gene Collins
Project Leader, Petroleum Production Research
U.S. Energy Research and Development Administration
P.O. Box 1398
Bartlesville, Oklahoma 94003

Keith R. Cooley
United States Water Conservation Laboratory
4331 East Broadway
Phoenix, Arizona 85040

Carlos Correa Sanfuentes
Regional Secretariat for Planning and Coordination
Prat 350 La Serena
P.O. Box 387
La Serena, Chile

Kit Cullen
California Office of Economic Opportunity
555 Capitol Mall
Sacramento, California 95814

- D -

Sami R. Daghistani
Institute for Research on Natural Resources
Jederiah
Baghdad, Iraq

Dr. B. C. Dando
Epidemiologist
World Health Organization
Box 54
Gaborone, Botswana

Ali Darag Ali
Ministry of Agriculture and Natural Resources
P.O. Box 199
Khartoum, Sudan

Christian DeClerq
Economist, U.N. Economic Commission for Western Asia
P. B. 4656
Beirut, Lebanon

Dr. K. James DeCook
Associate Hydrologist, Water Resources Research Center
University of Arizona
Tucson, Arizona 85721

Claire T. Dedrick
Secretary for Resources
California Resources Agency
1416 Ninth Street
Sacramento, California 95814

Georges Cohen deLara
Chief, Department of Water Quality, Sogreah, Inc.
6 rue de Lorraine
Grenoble, France

Philippe de Seynes
Director, Programme of Future Studies
United Nations Institute of Training and Research
801 United Nations Plaza
New York, New York 10017

Aliou M. Diallo
Programme Management Office
Sahelian Unit, Office of Technical Co-operation
United Nations
New York, New York 10017

Efren Diaz
Pergolas 192
Xochimilco, Mexico D.F.

John R. Donnell
Chief, Oil Shale Section
Chemical Research Branch
U.S. Geological Survey
P. O. Box 25046; Stop 939
Denver, Colorado 80225

\- E -

Lic. Rodolfo Echeverria Zuno
Center for Third World Economic and Social Studies
Cor. Profirio Diaz 80
Mexico 20, D.F.

Stahrl W. Edmunds
Dean, Graduate School of Administration, University of California Riverside
Riverside, California 92521

Dr. William C. Ellet
Director, Marketing Research
Northrop Corporation
1800 Century Park East
Los Angeles, California 90067

Azouz Ennifar
Director, Tunisian National Tourist Office
630 Fifth Avenue
New York, New York 10020

Dr. Mohamed Aly Ezzat
Undersecretary for Desert Irrigation
Ministry for Irrigation
70 Gumhuriya Street
Cairo, Egypt

\- F -

Michael W. Fall
Fish and Wildlife Service
U.S. Department of the Interior
Denver Wildlife Research Center
Lakewood, Colorado 80255

Egidio Feliu Acevedo
SERPLAC
Edificio Gobierno Regional
Iquique, Chile

Dr. Elias Fereres-Castiel
Area Soil and Water Specialist
LAWR: Water Science and Engineering
University of California, Davis
Davis, California 95616

Robert F. Fingado
Chief of Implementation, Delta Branch, Central District
California Department of Water Resources
P. O. Box 160088
Sacramento, California 95816

Walter W. Fisher
Metallurgist
Arizona Bureau of Mines
University of Arizona
Tucson, Arizona 85721

Carlton E. Forbes
Chief Engineer, California Department of Transportation
1120 N Street
Sacramento, California 95814

\- G -

Kay Gary
University of California Davis
Davis, California 95616

James J. Geraghty
President, Geraghty and Miller, Inc.
P. O. Box 17174
Tampa, Florida 33682

Professor McGuire Gibson
Oriental Institute
University of Chicago
Chicago, Illinois 60637

Herbert W. Greydanus
Chief, Division of Planning
California Department of Water Resources
P. O. Box 388
Sacramento, California 95802

Dr. Joseph G. Gringof
Director, Middle Asian Regional Hydrometeorological Research Institute
Uzbekistan, Tashkent
U.S.S.R.

- H -

Dr. Robert M. Hagan
Professor of Water Science
University of California Davis
Davis, California 95616

Dr. Hassan H. Hajrah
Head of Biology Department
Faculty of Science
King Abdulaziz University
P. O. Box 1540
Jeddah, Saudi Arabia

Walter G. Halset
United Nations International Labour Organization
P. O. Box 1966
Lusaka, Zambia

Dr. Adnan Hardan
Agricultural Consultant
Ministry of Planning
P. O. Box 2018
Baghdad, Iraq

Edward L. Hastey
State Director
U.S. Bureau of Land Management
2800 Cottage Way
Sacramento, California 95825

Thomas J. Henderson
President, Atmospherics, Inc.
5652 E. Dayton
Fresno, California 93727

Lee F. Hermsmier
Agricultural Engineer
Imperial Valley Conservation Research Center
4151 Highway 86
Brawley, California 92227

Lonnie Hiebert
Saline Processors
Box 21
Claremont, California 91711

Diane Hilden
University of California Davis
Davis, California 95616

Dr. Colin W. Holloway
International Union for Conservation of Nature and Natural Resources (IUCN)
1110 Morges
Switzerland

Dr. David Hopcraft
Director, Wildlife Ranching and Research
Athi River, Box 47272
Nairobi, Kenya

H. E. Horton
Assistant Regional Director
U. S. Bureau of Reclamation
2800 Cottage Way
Sacramento, California 95825

Abdoulgader Houdairi
Hydrogeologist
P.O. Box 4411
Tripoli, Libya

Dr. William L. Hughes
Director, Engineering Energy Laboratory
Oklahoma State University
Stillwater, Oklahoma 74074

Dr. John L. Hult
President, Application Concepts and Technology Association
P. O. Box 1731
Santa Monica, California 90406

- I -

El Mahdi H. Ismail
Director, Agricultural and Industrial Section
Ministry of Planning
P. O. Box 600
Tripoli, Libya

Dr. Shawkat Abdel-kader Ismail
Advisory to H.R.H. Governor
Saline Water Conversion Corporation
P. O. Box 4931
Jeddah, Kingdom of Saudi Arabia

- J -

Dr. Merle H. Jensen
Environmental Research Laboratory
University of Arizona
Tucson International Airport
Tucson, Arizona 85706

E. Mostafa Jermoumi
Chef de Service de la Mise en Valeur Agricole de Marrakech
Ministère de l'Agriculture et de la Réforme Agraire
Direction Provinciale de l'Agriculture
B. P. 21
Marrakech, Morocco

Darshan Johal
Centre for Housing, Building and Planning
United Nations
New York, New York 10017

Dr. Jack D. Johnson
Director, Arid Lands Studies
University of Arizona
845 N. Park Avenue
Tucson, Arizona 85719

Professor Arthur W. Jokela
Department of Landscape Architecture
California State Polytechnic University
3801 West Temple Avenue
Pomona, California 91768

Professor Paul H. Jones
Department of Geology
Louisiana State University
Baton Rouge, Louisiana 70803

- K -

Dr. Michel Keita
Institute de Recherche en Sciences Humaines
B. P. 318
Niamey, Niger

Roy J. Kelly
Planning Economist
Metropolitan Water District of Southern California
P. O. Box 54153
Los Angeles, California 90054

Ngare Kessely
Conseiller
Ministère des Affaires Etrangères
N'djamena, Tchad

M. Said Kettanah
Scientific Research Foundation
Jadiriya
Baghdad, Iraq

Dr. Nikolai G. Kharin
Biologist, Desert Institute
Academy of Sciences of Turkmen S.S.R.
GSP 24, 744012 Ashkhabad
U.S.S.R.

Ali Kholdbarin
Expert, Soil Conservation Bureau
Forest and Range Organization
Ministry of Agriculture and Natural Resources
Farahg Onobi-Mansour
Farahnal 25
Tehran, Iran

Robert E. King
Consultant
580 Shore Acres Drive
Mamaroneck, New York 10543

Ira E. Klein
Consulting Geologist
1521 El Nido Way
Sacramento, California 95825

- L -

Frank L. Lambrecht, D.T.M.L.H.
International Health Program, Health Sciences
University of Arizona
Tucson, Arizona 85724

Everard Mervyn Lofting
Engineering-Economics Associates
1700 Solano Avenue
Berkeley, California 94707

Hussain A. K. Lolo
Undersecretary, Ministry of Agriculture
Sanaa, Yemen Arab Republic

Janece Long
Administrative Assistant
California Secretary of State
925 "L" Street, Suite 605
Sacramento, California 95814

Manuel Lopez, Jr.
Regional Director, U.S. Bureau of Reclamation
P. O. Box 427
Boulder City, Nevada 89005

\- M -

F. G. "Phil" Macias
Pacific Southwest Planning Officer, U.S. Department of the Interior
450 Golden Gate Avenue
San Francisco, California 94102

Dr. Joel Maltos Romo
Secretaria de Agricultura Zonas Aridas
Apartado Postal 426
Saltillo, Coahuila, Mexico

Billy E. Martin
Regional Director, U. S. Bureau of Reclamation
2800 Cottage Way
Sacramento, California 95825

John W. Masier
Chief, Planning Investigations Branch, San Joaquin District
California Department of Water Resources
P. O. Box 5710
Fresno, California 93755

Dr. W. Gerald Matlock
International Agriculture
University of Arizona
Tucson, Arizona 95721

Melvin Mattson
General Engineer
Membrane Processes Division
Office of Water Research and Technology, U.S. Department of the Interior
Washington, D.C. 20240

Richard L. Maullin
Chairman, California Energy Resources Conservation and Development Commission
1111 Howe Avenue
Sacramento, California 95825

Dr. Fernando Medellin-Leal
Director, Instituto de Investigacion de Zonas Deserticas
Universidad Autonoma de San Luis Potosi
Apartado Postal 458
San Luis Potosi, S.L.P., Mexico

Ing. Salvador Mejia-Acevedo
Center for Third World Economic and Social Studies
Cor. Profirio Diaz 80
Mexico 20, D.F.

Professor Aden B. Meinel
Optical Sciences Center
University of Arizona
Tucson, Arizona 85721

Dominick Mendola
Vice President and Director of Operations
Solar Aqua Systems, Inc.
P. O. Box 88
Encinitas, California 92024

Bashkar P. Menon
Center for Economic and Social Information
United Nations
New York, New York 10017

S. W. O. Menzel
Chief Executive
Reed Irrigation Systems
8400 K Magnolia
Santee, California 92071

Jewell L. Meyer
Area Soil and Water Specialist
University of California
Cooperative Extension
433 County Center, III Court
Modesto, California 95350

Robert W. Miller
Chief, Interstate Planning
California Department of Water Resources
P. O. Box 388
Sacramento, California 95802

Mohamed Khattar Mohamed
General Director for Planning
Secretariat of Municipalities
P. O. Box 2739
Tripoli, Libya

Lucien Monition
Scientific Advisor, Bureau de recherches geologiques et minieres
B. P. 6009
45018 Orleans Cedex, France

Virginia S. Mueller
Vice President, World Affairs Council of Sacramento
Stanford Brothers Warehouse
106 L Street
Sacramento, California 95814

- N -

Stephen J. Nicola
Senior Fishery Biologist
California Department of Fish and Game
1416 Ninth Street
Sacramento, California 95814

F. Niknam
Director General, Soil Conservation Bureau, Forestry and Range Organization
Ministry of Agriculture and Natural Resources
Tehran, Iran

Rose M. Nonini
Special Assistant
California Department of Water Resources
P. O. Box 388
Sacramento, California 95802

- O -

Mohamed J. Omar
Planning Director, Settlement Development Programme
P. O. Box 1407
Mogadishu, Somalia

Dr. James E. Osborn
Chairman, Department of Agricultural Economics
Texas Tech University
P.O. Box 4169
Lubbock, Texas 79409

Dr. B. F. Osorio-Tafall
Director General
Center for Third World Economic and Social Studies
Cor. Profirio Diaz 80
Mexico 20, D.F.

Abdullah Ouderdan
Forest Director
Forestry Department
Kabul, Afghanistan

John Dalmas Owvor Onyango
Deputy Chief Conservator of Forests, Forest Department
P.O. Box 30513
Nairobi, Kenya

- P -

Patricia Paylore
Office of Arid Lands Studies
University of Arizona
845 N. Park
Tucson, Arizona 85719

Professor Zvi Pelah
Head, R. and D. Authority
Ben-Gurion University of the Negev
P. O. Box 1025
Beer-Sheva, Israel

Dr. Ishwar Prakash
Principal Animal Ecologist
Central Arid Zone Research Institute
Jodhpur, India

Dr. Robert M. Pratt
Special Assistant
California Department of Food and Agriculture
1220 N. Street
Sacramento, California 95814

Edgar P. Price
Special Assistant to the Regional Director
U.S. Bureau of Reclamation
2800 Cottage Way
Sacramento, California 95825

- Q -

Dr. William J. Quirk
Deputy Chief Scientist for the Global Atmospheric Research Project
NASA Goddard Institute for Space Studies
2880 Broadway
New York, New York 10027

- R -

Joseph Ramamonjisoa
Directeur du Project de Development de l'Androy
Ministere du Development Rural et de la Reforme Agraire
Anosy
Tananarive, Madagascar

Robin R. Reynolds
Co-Director of Conference and
Deputy Director, California Department of Water Resources
P. O. Box 388
Sacramento, California 95802

Robin R. Reynolds III
California State Water Resources Control Board
1416 Ninth Street
Sacramento, California 95814

Ronald B. Robie
Director, California Department of Water Resources
P. O. Box 388
Sacramento, California 95802

Frank E. Robinson
Water Scientist, Imperial Valley Field Station
University of California
1004 E. Holton Road
El Centro, California 92243

Bruce C. Rogers
Special Assistant
California Secretary of State
925 L Street, Suite 605
Sacramento, California 95814

Dr. Virgilio Roig
Director, Instituto de Investigaciones de Zonas Aridas
Universidad Nacional de Cuyo
Gregorio Torres 2085
Godoycruz
Mendoza, Argentina

Richard E. Rominger
Director, California Department of Food and Agriculture
1220 N Street
Sacramento, California 95814

Dr. Adam Rose
Assistant Professor,
Department of Economics
University of California Riverside
Riverside, California 92521

Professor Boris G. Rozanov
Assistant to the Secretary-General, United Nations Conference on Desertification
U. N. Environment Programme
P. O. Box 30552
Nairobi, Kenya

- S -

Antonio Sainz Unzueta
Jefe Dpto. de Suelos
Ministerio de Asuntos Campesinos y Agropecuarios
Carlos Blanco No 841
Casilla de Correo No 5780
La Paz, Bolivia

Nampaa N. Sanogho
Ingenieur Principal, Direction Nationale des Eaux et Forets
B. P. 275
Bamako, Mali

Dr. Albert Sasson
Division of Ecological Sciences
U.N.E.S.C.O.
Paris 75007, France

Margaret W. Savage
Thermo. Mist Company
9528 Lemoran Avenue
Downey, California 90240

Glenn B. Sawyer
Supvr. Land and Water Use Analyst,
California Department of Water Resources
P. O. Box 388
Sacramento, California 95802

Daniel P. Schadle
Supervisor, Lands, Development and Maintenance
Arizona Game and Fish Department
2222 W. Greenway Road
Phoenix, Arizona 85023

Dr. Joel Schechter
Director, Applied Research Institute
Ben-Gurion University
P. O. Box 1025
Beer-Sheva, Israel

Frank J. Schober, Jr.
Commanding General, California Military Department
P. O. Box 214405
Sacramento, California 95821

Abdul Gasim Seif El Din
Senior Research Officer
GUM Research Division
P. O. Box 302
El Obeid, Sudan

Gonsalo Sevilla Rodriguez
Instituto de Reforma Agraria y Colonizacion
Carrio 1040, Casilla 146-A
Quito, Ecuador

Richard P. Sheldon
Chief Geologist
U. S. Geological Survey
Reston, Virginia 22092

Dr. Marilyn L. Shelton
Assistant Professor of Geography
University of California Davis
Davis, California 95616

Clark W. Shraeder
Sierra Club
Box 621
Broderick, California 95605

Dr. Ousmane Silla
C.I.L.S.S.
B. P. 7049
Ouagadougou, Upper Volta

Dr. Ranbir Singh
Director of Agriculture
Government of Rajasthan
Jaipur, India

Dr. Subrata Sinha
Director, Environmental Geology (W.R.)
Geological Survey of India
C-33, Lajpat Marg
Jaipur 302001, India

George I. Smith
Geologist, U.S. Geological Survey
345 Middlefield Road
Meno Park, California 94025

Professor Norman S. Smith
Arizona Cooperative Wildlife Research Unit
219 Biological Science Building
University of Arizona
Tucson, Arizona 85721

Dr. J. Herbert Synder
Director, Water Resources Center
University of California Davis
Davis, California 95616

Alberto A. Sojit
Project Director
United Nations Development Program/ Inter-American Development Bank UNDP
P. B. O. 197D
Santiago, Chile

William Spaulding, Jr.
Geothermal Advisor, U.S. Fish and Wildlife Service
U.S. Department of the Interior
1730 K Street
Washington, D.C. 20240

Professor Brian Spooner
Department of Anthropology
University of Pennsylvania
Philadelphia, Pennsylvania 19174

Terry G. Spragg
A.C.T.A.
221 Third Place
Manhattan Beach, California 90266

Dr. Preston Stegenga
Director, International Center
California State University, Sacramento
6000 J Street
Sacramento, California 95819

Wesley E. Steiner
Executive Director
Arizona Water Commission
222 North Central Avenue, Suite 800
Phoenix, Arizona 85004

C. L. Stetson
Chief, San Joaquin District
California Department of Water Resources
P. O. Box 5710
Fresno, California 93755

Professor H. H. Stoevener
Department of Agriculture and Resource Economics
Oregon State University
Corvallis, Oregon 97331

- T -

Michael Takla
Project Engineer, Bechtel Inc.
P. O. Box 3965
San Francisco, California 94119

Harold E. Thomas
1339 Portola Road
Woodside, California 94062

A. Toure
Directeur de la Protection et Amélioration Efficace
Espèce Agro.-Pastorale
B. P. 170
Nouackchott, Mauritania,
West Africa

P.A. Towner
Chief Counsel, California Department of Water Resources
P. O. Box 388
Sacramento, California 95802

Dr. Thaddeus C. Trzyna
President, Center for California Public Affairs and
Chairman, Sierra Club International Committee
P. O. Box 30
Claremont, California 91711

George L. Turcott
Associate Director
Bureau of Land Management
U.S. Department of the Interior
Washington, D.C. 20240

- U -

Zafar Uddin
Deputy Director of Research
Agricultural Research Council
F-7/2 Islamabad, Pakistan

- V -

Frank P. Vandemaele
Senior Technical Adviser
United Nations Development Programme
Room DC-2062
New York, New York 10017

- W -

Kenneth Wahl
Water Resources Division
U. S. Geological Survey
855 Oak Grove Avenue
Menlo Park, California 94025

William E. Warner
Consultant
2090 Eighth Avenue
Sacramento, California 95818

Robert O. Watkins
District Director, California Department of Transportation
P. O. Box 847, 500 South Main
Bishop, California 93574

Gary D. Weatherford
Deputy Secretary for Resources
California Resources Agency
1416 Ninth Street
Sacramento, California 95814

Richard A. Weaver
Wildlife Manager, California Department of Fish and Game
1416 Ninth Street
Sacramento, California 95814

Ernest M. Weber
Engineering Geologist
Colorado River Board of California
107 South Broadway, Room 8103
Los Angeles, California 90012

James L. Welsh
Chief, Statewide Planning Branch,
California Department of Water Resources
P. O. Box 388
Sacramento, California 95802

Ruth Welsh
California Department of Water Resources
P. O. Box 388
Sacramento, California 95802

Elizabeth Wily
Ministry of Local Government and Lands
P. Bag 006
Gaborone, Botswana

Dr. Vicente M. Witt
Chief, Environmental Program Support
World Health Organization
Pan American Health Organization
525 Twenty-Third Street, N.W.
Washington, D.C. 20037

- X Y Z -

Dr. James A. Young
Range Scientist
Agricultural Research Service
U.S. Department of Agriculture
Reno, Nevada 89512

INDEX

Volume I: Energy and Minerals includes page 1-254.
Volume II: Agriculture includes pages 255-481.
Volume III: Water includes page 482-946.
Volume IV: Desert Management includes page 947-1396.